Reading the Human Body

Studies on the Texts of the Desert of Judah

Edited by

Florentino García Martínez

Associate editors

Peter W. Flint
Eibert J.C. Tigchelaar

VOLUME 67

This book is printed on acid-free paper.

ISSN 0169-9962
ISBN 978 90 04 15717 0

PRINTED IN THE NETHERLANDS

Reading the Human Body

Physiognomics and Astrology in the Dead Sea Scrolls and Hellenistic-Early Roman Period Judaism

by

Mladen Popović

BRILL

LEIDEN • BOSTON
2007

This book is dedicated to my parents

Nedeljko Popović and Marija Popović-Maravić

and in loving memory of my cousin

Jelenko Popović (17.09.1972 – 03.09.1994)

CONTENTS

ACKNOWLEDGMENTS

This study was submitted in the summer of 2006 as a doctoral dissertation at the University of Groningen. One of the great pleasures when you have finished writing a book like this is to thank the many people who helped in different ways during the years of research. Without the assistance of various people this book would not have turned out as it has. Needless to say, however, I remain solely responsible for the content of this work and any mistakes there may be.

I owe a tremendous debt to my supervisors Eibert Tigchelaar and Florentino García Martínez for their confidence, their help, suggestions, challenges, criticism, their time for and attention to all my questions, and for the many stimulating discussions. Eibert and Florentino, each in their own way, represent great examples for young scholars and I am fortunate and feel very proud to have had them as my teachers. I hope our friendship lasts *ad multos annos*!

The Faculty of Theology and Religious Studies in Groningen has been a wonderful place to work and I thank everyone for their support, especially my fellow PhD students and the staff at the faculty library and administration.

At Yale University I wish to thank John Collins, Eckart Frahm, Dale Martin, Matt Neujahr, and Michael Peppard for their interest in my work during our stay at the Divinity School in the spring of 2004. I thank John and Adela Collins for their hospitality and kindness towards us while we were at Yale.

My research has benefited from presentations on various occasions, but one especially has been very important. I wish to thank the Nordic Network in Qumran Studies, its board Torleif Elgvin and Hanne von Weissenberg, and all the participants during the Jerusalem symposium in September 2005 for the opportunity to present a paper, for the wonderful atmosphere, and a very special meeting. When I stayed at the École Biblique, Émile Puech was very kind to devote some of his time to discussing 4Q534.

I am most grateful to Professor Wolfgang Hübner for sharing his great knowledge of Hellenistic astrology with me, providing valuable suggestions, and for his willingness to read an early draft of Chapter Three. Wytse Keulen I thank for discussing the Greco-Roman physiognomic material during my presentation at the Ancient World Seminar in Groningen and afterwards, and for reading Chapter Two. I wish to thank Herman Vanstiphout (Stip) for reading the Babylonian physiognomic material in Chapter Two and discussing it with me. I thank Ton Hilhorst for improving my translation of Hephaestion and Jan Bremmer for reading the entire manuscript. Many thanks are due to Julia Harvey for quickly checking and improving the English.

I wish to thank Francis Schmidt and Jonathan Ben-Dov for discussing various issues of ancient Jewish astronomy and astrology and Wout van Bekkum and Ronit Nikolsky for discussing the medieval Jewish physiognomic texts.

Finally, but above all, I thank my friends, family, and Annemieke ter Brugge.

Mladen Popović
Groningen/Leuven, August/October 2006

ABBREVIATIONS

AAWGPHK	Abhandlungen der Akademie der Wissenschaften in Göttingen, Philologisch-Historische Klasse
ABAWPHK	Abhandlungen der Bayerischen Akademie der Wissenschaften, Philosophisch-Historische Klasse
ABD	*Anchor Bible Dictionary*. Edited by D. N. Freedman. 6 vols. New York, 1992
AfO	*Archiv für Orientforschung*
AfO.B	Archiv für Orientforschung Beiheft
AGP	Arbeiten zur Geschichte des Pietismus
AGJU	Arbeiten zur Geschichte des antiken Judentums und des Urchristentums
AHAWPHK	Abhandlungen der Heidelberger Akademie der Wissenschaften, Philosophisch-historische Klasse
AIUON	*Annali dell'Istituto Universitario Orientale di Napoli*
AJEC	Ancient Judaism and Early Christianity
AJP	*American Journal of Philology*
AMD	Ancient Magic and Divination
AnBib	Analecta Biblica
ANSupp	*Ancient Narrative, Supplementum*
ANRW	Aufstieg und Niedergang der römischen Welt: Geschichte und Kultur Roms im Spiegel der neueren Forschung. *Edited by H. Temporini and W. Haase. Berlin, 1972–*
AOAT	Alter Orient und Altes Testament
AOS	American Oriental Series
AS	Assyriological Studies
AWLMAGSK	Akademie der Wissenschaften und der Literatur Mainz, Geistes- und Sozialwissenschaftlichen Klasse
BA	Babylonische Archive
BABesch	*Bulletin Antieke Beschaving*
BAC	Bochumer Altertumswissenschaftliches Colloquium
BaF	Baghdader Forschungen
BASORSup	Bulletin of the American Schools of Oriental Research Supplement Series
BAWPHK.S	Bayerische Akademie der Wissenschaften, Philosophisch-Historische Klasse, Sitzungsberichte
BBB	Bonner Biblische Beiträge
BEBM	Biblioteca de estudios bíblicos minor
BETL	Bibliotheca ephemeridum theologicarum lovaniensium
BibOr	Biblica et orientalia
BibSem	The Biblical Seminar
BJA	Berliner Juristische Abhandlungen
BJSUCSD	Biblical and Judaic Studies from the University of California, San Diego

BKAT	Biblischer Kommentar Altes Testament
BKP	Beiträge zur klassischen Philologie
BSOAS	*Bulletin of the School of Oriental and African Studies*
Budé	Collection des universités de France, publiée sous le patronage de l'Association Guillaume Budé
BWANT	Beiträge zur Wissenschaft vom Alten und Neuen Testament
BZNW	Beihefte zur Zeitschrift für die neutestamentliche Wissenschaft
CAD	Assyrian Dictionary of the Oriental Institute of the University of Chicago
CahRB	Cahiers de la Revue Biblique
CBET	Contributions to Biblical Exegesis and Theology
CBQ	*Catholic Biblical Quarterly*
CBQMS	Catholic Biblical Quarterly Monograph Series
CCAG	Catalogus codicum astrologorum Graecorum
CCTC	Cambridge Classical Texts and Commentaries
CDOG	Colloquien der Deutschen Orient-Gesellschaft
CEA.EtB	Collection d'études anciennes: études bibliques
CEJL	Commentaries on Early Jewish Literature
CF	Colleción Fundamentos
CH	Collectanea Hellenistica
CJAS	Christianity and Judaism in Antiquity Series
CM	Cuneiform Monographs
CNIP	Carsten Niebuhr Institute Publications
ConBOT	Coniectanea Biblica: Old Testament Series
CP	*Classical Philology*
CPJ	Corpus Papyrorum Judaicarum. Edited by V. Tcherikover. 3 vols. Cambridge, 1957–1964
CRAI	Comptes rendus de l'Académie des inscriptions et belles-lettres
CREJ	Collection de la Revue des Études Juives
CRINT	Compendia Rerum Iudaicarum ad Novum Testamentum
CTN	Cuneiform Texts from Nimrud
DAWBSSA	Deutsche Akademie der Wissenschaften zu Berlin, Schriften der Sektion für Altertumswissenschaft
DJD	Discoveries in the Judaean Desert
DMAHA	Dutch Monographs on Ancient History and Archaeology
DNP	*Der Neue Pauly: Enzyklopädie der Antike*
DSD	*Dead Sea Discoveries*
DSSR	The Dead Sea Scrolls Reader
DV	*Dieu Vivant*
EH	Europäische Hochschulschriften
EPRO	Études préliminaires aux religions orientales dans l'Empire romain
ER	*Encyclopedia of Relgion, Second Edition.* Edited by L. Jones and M. Eliade. 15 vols. Detroit, 2005

FAS	Freiburger Altorientalische Studien
FAT	Forschungen zum Alten Testament
FJB	*Frankfurter Judaistische Beiträge*
FJS	Frankfurter Judaistische Studien
FO	*Folia Orientalia*
GMS	Grazer Morgenländischer Studien
GR	*Greece and Rome*
GRBS	*Greek, Roman, and Byzantine Studies*
HBS	Herders Biblische Studien
HdA	Handbuch der Altertumswissenschaft
HDAC	Histoire des doctrines de l'antiquité classique
HdA	Handbuch der Altertumswissenschaft
HdO	Handbuch der Orientalistik
HOS	Harvard Oriental Series
HS	Scholars Press Homage Series
HSAO	Heidelberger Studien zum Alten Orient
HSCP	*Harvard Studies in Classical Philology*
HSS	Harvard Semitic Studies
HThKAT	Herders Theologischer Kommentar zum Alten Testament
HWP	Historisches Wörterbuch der Philosophie. Edited by J. Ritter, K. Gründer and G. Gabriel. 12 vols. Darmstadt, 1971-2004
HTR	*Harvard Theological Review*
IEJ	*Israel Exploration Journal*
IOS	*Israel Oriental Studies*
IPTS	Islamic Philosophy, Theology, and Science (Texts and Studies)
JANESCU	*Journal of the Ancient Near Eastern Society of Columbia University*
JAOS	*Journal of the American Oriental Society*
Jastrow	Dictionary of the Targumim, Talmud Babli, Yerushalmi and Midrashic Literature. New York, 1996 (1903)
JCS	*Journal of Cuneiform Studies*
JdC	*Jahrbuch der Charakterologie*
JJS	*Journal of Jewish Studies*
JNES	*Journal of Near Eastern Studies*
JPSP	*Journal of Personality and Social Psychology*
JSHRZ-NF	Jüdische Schriften aus hellenistisch-römischer Zeit – Neue Folge
JSJ	*Journal for the Study of Judaism*
JSJSup	Supplements to the Journal for the Study of Judaism
JSNTSup	Journal for the Study of the New Testament Supplement Series
JSOT	*Journal for the Study of the Old Testament*
JSOTSup	Journal for the Study of the Old Testament Supplement Series
JSP	*Journal for the Study of the Pseudepigrapha*
JSPSup	Journal for the Study of the Pseudepigrapha Supplement Series

JSS	*Journal of Semitic Studies*
LCL	Loeb Classical Library
LSc	Letteratura della Siria cristiana
LSJ	Liddell, H. G., R. Scott, H. S. Jones, A Greek-English Lexicon. 9th ed. with revised supplement. Oxford, 1996
LSTS	Library of Second Temple Studies
KAW	Kulturgeschichte der antiken Welt
MAPS	Memoirs of the American Philosophical Society Held at Philadelphia for Promoting Useful Knowledge
MBFJTFL	Mitteilungen und Beiträge der Forschungsstelle Judentum der Theologischen Fakultät Leipzig
MBPF	Münchener Beiträge zur Papyrusforschung und antiken Rechtsgeschichte
MdB	*Le Monde de la Bible*
MMed	The Medieval Mediterranean Peoples, Economies and Cultures, 400-1500
MMS	Münstersche Mittelalter-Schriften
MS	Melammu Symposia
MSL	Materials for the Sumerian Lexicon
MVAG	Mitteilungen der Vorderasiatisch-Aegyptischen Gesellschaft
NABU	*Nouvelles Assyriologiques Brèves et Utilitaires*
NHMS	Nag Hammadi and Manichaean Studies
NTOA	Novum Testamentum et Orbis Antiquus
NTOA.SA	Novum Testamentum et Orbis Antiquus, Series Archaeologica
NTS	*New Testament Studies*
ÖAWPH.SB	Österreichische Akademie der Wissenschaften, Philosophisch-Historische Klasse, Sitzungsberichte
OCD	*Oxford Classical Dictionary*. Edited by S. Hornblower and A. Spawforth. 3d ed. Oxford, 1996
Oeuvres	Les oeuvres de Philon d'Alexandrie
OLA	Orientalia Lovaniensia Analecta
OPSNKF	Occasional Publications of the Samual Noah Kramer Fund
Or	*Orientalia*
OrA	Orbis Antiquus
OrAnt	*Oriens Antiquus*
OTS	Oudtestamentische Studiën
OTF	Oriental Translation Fund
OTP	*Old Testament Pseudepigrapha*. Edited by J. H. Charlesworth. 2 vols. New York, 1983
PACS	Philo of Alexandria Commentary Series
PAPS	*Proceedings of the American Philosophical Society Held at Philadelphia for Promoting Useful Knowledge*
PEQ	*Palestine Exploration Quarterly*
PG	Patrologia Graeca. Edited by J.-P. Migne. 162 vols. Paris, 1857–1886

PGM	*Papyri graecae magicae: Die griechischen Zauberpapyri*. Edited by K. Preisendanz. Berlin, 1928
PH	Papyrusinstitut Heidelberg
PhA	Philosophia Antiqua
Phil	*Philologus*
RA	*Revue d'Assyriologie et d'Archéologie Orientale*
RAC	*Reallexikon für Antike und Christentum*. Edited by T. Klauser et al. Stuttgart, 1950–
RC	*Ricerca Papirologica*
RB	*Revue Biblique*
RE	*Paulys Realencyclopädie der classischen Altertumswissenschaft*. Edited by G. Wissowa et al. 49 vols. Munich, 1894-1997
REL	*Revue des Études Latines*
RelSoc	Religion and Society
ResOr	Res Orientales
RevAnt	Revealing Antiquity
RevQ	*Revue de Qumrân*
RHPR	*Revue d'histoire et de philosophie religieuses*
RHR	*Revue de l'histoire des religions*
RI	Recherches Intertestamentaires
RlA	*Reallexikon der Assyriologie und Vorderasiatischen Archäologie*
RVV	Religionsgeschichtliche Versuche und Vorarbeiten
RWRL	Rombach Wissenschaft – Reihe Litterae
SAA	State Archives of Assyria
SAAA	Studies on the Apocryphal Acts of the Apostles
SAAS	State Archives of Assyria Studies
SA.B	Sudhoffs Archiv Beihefte
SBLDS	Society of Biblical Literature Dissertation Series
SBLEJL	Society of Biblical Literature Early Judaism and Its Literature
SBLSP	*Society of Biblical Literature Seminar Papers*
SBLSymS	Society of Biblical Literature Symposium Series
SBO	Studia Biblica et Orientalia
SBT	Studies in Biblical Theology
SBW	Studien der Bibliothek Warburg
ScrHier	Scripta Hierosolymitana
SDSSRL	Studies in the Dead Sea Scrolls and Related Literature
SGLG	Studia Graeca et Latina Gothoburgensia
SHC	Studies in Hellenistic Civilization
SHMPS	Studies in the History of Mathematics and Physical Sciences
SHR	Studies in the History of Religions (Supplements to *Numen*)
SOR	Serie Orientale Roma
SSN	Studia Semitica Neerlandica

STAR	Studies in Theology and Religion
STDJ	Studies on the Texts of the Desert of Judah
STS	Semitic Texts and Studies
STT	Semitic Texts with Translations
SUNT	Studien zur Umwelt des Neuen Testaments
SVBOGLP	Studia Varia Bruxellensia ad Orbem Graeco-Latinum Pertinentia
SVTP	Studia in Veteris Testamenti Pseudepigrapha
SWC	Sammlung Wissenschaftlicher Commentare
TAPA	*Transactions of the American Philological Association*
TAPS	Transactions of the American Philosophical Society Held at Philadelphia for Promoting Useful Knowledge
Teubner	Bibliotheca scriptorum graecorum et romanorum teubneriana
Theophaneia	Theophaneia: Beiträge zur Religions- und Kirchengeschichte des Altertums
TICP	Travaux de l'Institut Catholique de Paris
TRE	*Theologische Realenzyklopädie*. Edited by G. Krause and G. Müller. Berlin, 1977–
TUGAL	Texte und Untersuchungen zur Geschichte der altchristlichen Literatur
TVOa	Testi del Vicino Oriente antico
TWAT	*Theologisches Wörterbuch zum Alten Testament*. Edited by G. J. Botterweck and H. Ringgren. Stuttgart, 1970–
TSAJ	Texte und Studien zum Antiken Judentum
TWNT	*Theologisches Wörterbuch zum Neuen Testament*. Edited by G. Kittel and G. Friedrich. Stuttgart, 1932–1979
UNHAII	Uitgaven van het Nederlands Historisch-Archaeologisch Instituut te Istanbul
UNT	Untersuchungen zum Neuen Testament
UUÅ	Uppsala Universitets Årsskrift
UVB	Vorläufige Berichte über die ... Ausgrabungen in Uruk-Warka. Berlin, 1930–
VisRel	Visible Religion
VT	*Vetus Testamentum*
VTSup	Vetus Testamentum Supplement
WBC	Word Biblical Commentary
WIST	Warburg Institute Surveys and Texts
WMANT	Wissenschaftliche Monographien zum Alten und Neuen Testament
WS	*Wiener Studien, Zeitschrift für klassische Philologie*
WUNT	Wissenschaftliche Untersuchungen zum Neuen Testament
YJS	Yale Judaica Series
YSHSM	Yale Studies in the History of Science and Medicine
ZA	*Zeitschrift für Assyriologie*
ZAH	*Zeitschrift für Althebraistik*

ZAW	*Zeitschrift für die alttestamentliche Wissenschaft*
ZThK	*Zeitschrift für Theologie und Kirche*

INTRODUCTION

Your face, my thane, is as a book, where men may read strange matters.
William Shakespeare, *Macbeth* 1.5.62–63

If, at some time, from a triplicity there are two zodiacal signs above the earth to which we assume the horoscope applies, then we also pay attention to the shape of the man, which one of the zodiacal signs he resembles more [...].
Hephaestion of Thebes, *Apotelesmatica* 2.2.27

This study deals with two manuscripts from the Dead Sea Scrolls whose contents are physiognomic and physiognomic-astrological. These manuscripts contain material that was unknown to have existed in this form in Hellenistic-Early Roman period Judaism (ca. third century BCE–first century CE) before the discovery of the Dead Sea Scrolls near Qumran.[1] The two manuscripts are fragmentary, the style of both texts is succinct, and some of the terminology is enigmatic. Despite these difficulties, the manuscripts contain enough to understand some of the arcane things these texts reveal to their intended reader.

The two texts considered here are the Hebrew manuscript 4Q186 (hereafter named *4QZodiacal Physiognomy*) and the Aramaic manuscript 4Q561 (hereafter named *4QPhysiognomy ar*).[2] The intention of this study is the reconstruction and understanding of what remains of these texts; what sort of texts they represent, what their sense is, and which functions they may have had and in what contexts. The approach is comparative, understanding these two texts in relation to other physiognomic and astrological writings, mainly from Babylonian and Greco-Roman traditions. The ancient Jewish manuscripts from Qumran share certain features with texts from these other traditions, but, not surprisingly, they also exhibit some peculiar and distinct features of their own.

[1] The Dead Sea Scrolls are a collection of over 900 reconstructed manuscripts. These were found between 1947–1956 in eleven caves in the immediate vicinity of the ancient settlement of Qumran near the northwestern shore of the Dead Sea. Most scholars attribute the texts to a religious Jewish sect that occupied the Qumran settlement during the first century BCE–first century CE. Besides some typical sectarian writings, many texts from this collection were not composed by the sect itself. The most obvious examples are the books from the Hebrew Bible. The sect shared these writings with other Jews. The collection is, therefore, not only relevant for understanding an ancient Jewish sect, it also sheds new light on different aspects of Hellenistic-Early Roman period Judaism in general. See e.g. F. Mébarki and É. Puech (eds.), *Les Manuscrits de la mer Morte* (Rodez: Rouergue, 2002); F. García Martínez and E. Tigchelaar (eds.), *Fragmenten uit de woestijn: De Dode-Zeerollen opnieuw bekeken* (Zoetermeer: Meinema, 2003); J.C. VanderKam and P.W. Flint, *The Meaning of the Dead Sea Scrolls: Their Significance for Understanding the Bible, Judaism, Jesus, and Christianity* (London: T&T Clark, 2005).

[2] The number 4 stands for Cave 4; Q stands for Qumran. The fragments found from the cave were collected, joined, numbered, and photographed at The Palestine Archaeological Museum (PAM). In addition to this formal notation, the reconstructed texts were given modern names. For an explanation of the names used for 4Q186 and 4Q561, see Chapter One.

PHYSIOGNOMICS: THINGS SIGNIFIED BY THE HUMAN BODY

In Shakespeare's tragedy *Macbeth*, Duncan, ignorant of his impending doom, says "there's no art to find the mind's construction in the face." Lady Macbeth, however, proves him wrong and accurately tells her husband, when he is brooding, that his face "is as a book, where men may read strange matters."[3]

Judging by appearances is an everyday practice used in social communication and interaction between human beings. People observe each other's appearances and expressions in order to guess one another's thoughts and feelings. People may also try to go beyond these temporary states of mind; many form an opinion of what kind of person someone else is on the basis of how such an individual looks, dresses, behaves, talks, and moves. These outward features are taken as indicative signs of an individual's personality and character.[4] Although preconceptions undoubtedly play a role, this is hardly applying in a conscious way a fixed set of rules for judging the physical traits of someone else as indications of his or her personality. It is precisely this conscious reflection on the body as signifier and what is signified by it that characterizes the art of physiognomics in a more formal sense.

The textual evidence for physiognomic interests extends from the early second millennium BCE until modern times, and includes divination, literature, medicine, art, and philosophy. On the one hand, there is a level of continuity in the transmission of physiognomic lore and there are many similarities between physiognomic writings from different cultures and ages. On the other hand, the actual expression, application, and justification of physiognomic knowledge varies, due to different cultural, political, philosophical, religious, and social settings.[5]

[3] William Shakespeare, *Macbeth* 1.4.11–12, 1.5.62–63. Lady Macbeth's ability to read the face of her husband is known by the term metoposcopy. This is considered a branch of physiognomics, which flourished greatly during the modern period. Cf. A.G. Clarke, "Metoposcopy: An Art to Find the Mind's Construction in the Forehead," in *Astrology, Science, and Society: Historical Essays* (ed. P. Curry; Woodbridge, Suffolk: Boydell, 1987), 171-95; K.J.H. Bergland, "Reading Character in the Face: Lavater, Socrates, and Physiognomy," *Word and Image* 9 (1993): 252-69.

[4] See e.g. R. Foerster, *Die Physiognomik der Griechen* (Kiel: Schmidt & Klaunig, 1884), 6; R. Hassin and Y. Trope, "Facing Faces: Studies on the Cognitive Aspects of Physiognomy," *JPSP* 78 (2000): 837-52.

[5] Cf. S. Vogt, *Aristoteles: Opuscula VI: Physiognomonica* (Aristoteles, Werke in deutscher Übersetzung 18.6; Berlin: Akademie Verlag, 1999), 39-41. See also M. Blankenburg, "Physiognomik, Physiognomie," HWP 7 (1989), 955-63; R. Campe and M. Schneider (eds.), *Geschichten der Physiognomik: Text, Bild, Wissen* (RWRL 36; Freiburg im Bresgau: Rombach, 1996); L. Rodler, *Il corpo specchio dell'anima: Teoria e storia della fisiognomica* (Milan: Bruno Mondadori, 2000).

For example, at the modern end of the spectrum there is the Swiss theologian Johann Kaspar Lavater (1741–1801), who caused a popular revival of physiognomics in the eighteenth and nineteenth centuries through the publication of his *Physiognomische Fragmente zur Beförderung der Menschenkenntniß und Menschenliebe* in four volumes (1775–78). He defined physiognomics as "die Fertigkeit durch das Aeußerliche eines Menschen sein Inneres zu erkennen."[6] In the eighteenth and nineteenth centuries physiognomics went hand in hand with newly developing disciplines such as phrenology and craniometry. The most reliable basis for physiognomic study, according to Lavater, was the use of silhouettes, because "the decisive clues to character are found in the outline of the skull, not in the transitory play of facial features, with their countless shadings and blurry details, their constant interplay of dissimulation and disclosure."[7]

The scientific context of physiognomics in Greco-Roman antiquity was different. Skulls were not measured; the outline of the skull was not deemed as significant as in Lavater's days.[8] Instead, Greco-Roman physiognomics should be understood against another (physiological) background. The second-century CE physician Galen stated that "the faculties of the soul depend on the mixtures of the body."[9] Over the course of several centuries, physicians and philosophers (such as Hippocrates, Aristotle, and Galen) developed a theory of the human body according to which four fluids constituted the form and shape of the human body. Each of these fluids was ascribed two of four elementary qualities.[10] The appearance of the human body was the result of the specific mixture (κρᾶσις, *temperamentum*) of these fluids and their qualities. At the same time, this mixture also affected people's characters and souls. Thus, the second-century CE Skeptic Sextus

[6] J.C. Lavater, *Physiognomische Fragmente zur Beförderung der Menschenkenntniß und Menschenliebe: Erster Versuch* (Leipzig und Winterthur: M.G. Weidmanns Erben und Reich, und Heinrich Steiner und Compagnie, 1775), 13. For general studies on Lavater, see e.g. E. Shookman (ed.), *The Faces of Physiognomy: Interdisciplinary Approaches to Johann Caspar Lavater* (Columbia, South Carolina: Camden House, 1993); K. Pestalozzi and H. Weigelt (eds.), *Das Antlitz Gottes im Antlitz des Menschen: Zugänge zu Johann Kaspar Lavater* (AGP 31; Göttingen: Vandenhoeck & Ruprecht, 1994).

[7] C. Siegrist, "'Letters of the Divine Alphabet' – Lavater's Concept of Physiognomy," in *The Faces of Physiognomy*, 25-39, at 34.

[8] The physician Soranus of Ephesus (first century–second century CE) gives advice to the midwife on how to shape the skull of the newborn into good proportions so that it may become neither too lengthy nor pointed. See Soranus, *Gynecology* 2.33. Cf. D.B. Martin, *The Corinthian Body* (New Haven: Yale University Press, 1995), 26-27.

[9] Galen, *Quod animi mores corporis temperamenta sequantur* 7 = K. IV 767. Translation from P.N. Singer, *Galen: Selected Works* (Oxford: Oxford University Press, 1997), 150.

[10] Blood was hot and moist, phlegm cold and moist, yellow bile warm and dry, and black bile cold and dry.

Empiricus could say that "the general form of the soul is the body, as is shown by the science of physiognomics."[11]

These two examples demonstrate that the basic notion of physiognomics according to which the human soul and character can be known by observing the appearance of the body has remained the same over time. However, they also reveal the different views of the human body that constitute the context in which physiognomic knowledge is expressed.

The term physiognomics derives from the Greek word φυσιογνωμονία, which means judging a man's character from his bodily features.[12] This, however, only partly covers what people may believe to be signified by the human body. According to Babylonian tradition, the shape and appearance of the body reveal signs communicated by the gods about people's fates that have been ordained by these same gods. Consequently, there can be differences between physiognomic traditions as to what things may be signified by the body.

Another example is a remark by the late antique astrologer Hephaestion of Thebes, which provides evidence for the existence of the notion that astrological matters could be learned through physiognomic inquiry. If one cannot ascertain which of two zodiacal signs at a given time above the horizon represents someone's horoscope sign (= the ascendant), it is possible, says Hephaestion, to determine it by looking at the shape of the person's body and see which of the two zodiacal signs he resembles more. This observation by Hephaestion is important for this study. It sheds new light on *4QZodiacal Physiognomy*, one of the most intriguing texts from Qumran, which combines physiognomic and astrological learning.

Physiognomic Learning and Physiognomic Consciousness in the Dead Sea Scrolls and Second Temple Period Judaism

A number of texts from the Dead Sea Scrolls have in common that they give descriptions and characterizations of the human body. These descriptions do not display a mere interest in human anatomy or in the body per se. Rather, they impart knowledge of how to read strange matters from the human body; the interest is in what the body can reveal about people. The

[11] Sextus Empiricus, *Outlines of Pyrrhonism* 1.85.

[12] Its earliest occurrences are from the fourth century BCE with Demosthenes and Aristotle. The occurrence of the term in the Hippocratic corpus is probably due to medieval copyists who inserted the two captions in *Epidemics* 2.5 (φυσιογνωμονίη), and *Epidemics* 2.6 (φυσιογνωμονικόν). See Vogt, *Physiognomonica*, 37 n. 3.

Although etymologically "physiognomonics" would be more correct, "physiognomics" is used in this study because it is the more familiar term in English. Cf. T.S. Barton, *Power and Knowledge: Astrology, Physiognomics, and Medicine under the Roman Empire* (Ann Arbor, Michigan: University of Michigan Press, 1994), 95 n. 1.

human body is a signifier, full of signs that, taken together, show the competent reader certain things (the signified) concerning individual types of people whose bodies are scrutinized and described.

There are at least four texts from the Dead Sea Scrolls that demonstrate a physiognomic interest, but the texts differ from each other regarding the ways in which this interest is expressed. In addition to the two texts *4QZodiacal Physiognomy* and *4QPhysiognomy ar*, there are two literary texts: parts of a physical description of a figure called the "elect of God" in *4QBirth of Noah*^{a} *ar* (4Q534) and of Sarai in the *Genesis Apocryphon* (1QapGen ar 20:2–8).[13] Apart from these texts from the Dead Sea Scrolls, there are other writings from the Second Temple period that have passages showing physiognomic interest, such as, for example, the *Wisdom of Ben Sira*, *1 Enoch*, and the *Testaments of the Twelve Patriarchs*.[14]

In contrast with *4QZodiacal Physiognomy* and *4QPhysiognomy ar*, these other texts have what is perhaps best described by the words "physiognomic consciousness." This terminology was introduced by Elizabeth Evans to characterize those genres of Greco-Roman writings, such as epic, history and biography, drama, and satire, that do not deal with physiognomics on a theoretical or technical level, but make a more general use of physiognomic notions, and to distinguish these literary forms from the theoretical and formal Greco-Roman physiognomic treatises and catalogues.[15]

[13] Furthermore, two more texts, *4QBarkhi Nafshi*$^{a,\ c}$ (4Q434 and 4Q436) and *4QWiles of the Wicked Woman* (4Q184), have some passages with a possibly physiognomic meaning.

[14] For New Testament and early Christian texts, see e.g. A.J. Malherbe, "A Physical Description of Paul," in *Paul and the Popular Philosophers* (A.J. Malherbe; Minneapolis: Fortress, 1989), 170-75 (originally published in *HTR* 79 [1986]: 170-75); J. Massyngbaerde Ford, "The Physical Features of the Antichrist," *JSP* (1996): 23-41; J. Bollók, "The Description of Paul in the Acta Pauli," in *The Apocryphal Acts of Paul and Thecla* (ed. J.N. Bremmer; SAAA 2; Kampen: Kok Pharos, 1996), 1-15; J.N. Bremmer, "Magic, Martyrdom and Women's Liberation in the Acts of Paul and Thecla," in *Apocryphal Acts of Paul and Thecla*, 36-59, at 38-39; B.J. Malina and J.H. Neyrey, *Portraits of Paul: An Archaeology of Ancient Personality* (Louisville, Kentucky: Westminster John Knox, 1996), 100-52; G. Frank, *The Memory of the Eyes: Pilgrims to Living Saints in Christian Late Antiquity* (Berkeley: University of California Press, 2000), 134-70; M.C. Parsons, "'Short in Stature': Luke's Physical Description of Zacchaeus," *NTS* 47 (2001): 50-57; J.A. Harril, "Invective against Paul (2 Cor 10:10), the Physiognomics of the Ancient Slave Body, and the Greco-Roman Rhetoric of Manhood," in *Antiquity and Humanity: Essays on Ancient Religion and Philosophy Presented to Hans Dieter Betz on His 70th Birthday* (eds. A. Yarbro Collins and M.M. Mitchell; Tübingen: Mohr Siebeck, 2001), 191-213.

[15] E.C. Evans, *Physiognomics in the Ancient World* (TAPS 59/5; Philadelphia: American Philosophical Society, 1969), 5-6. Suetonius (ca. 70–after 126), for example, was familiar with physiognomic principles, which he used in his *Lives of the Caesars*. Descriptions of the shape and appearance of an emperor's body were meant to convey his character and moral personality. The context is the literary form of a biography, not a technical treatise on physiognomics, or a long catalogue of physiognomic characters. See e.g. E.C. Evans, "Roman Descriptions of Personal Appearance in History and Biography," *HSCP* 46 (1935): 43-84; E.C. Evans, "Literary Portraiture in Ancient Epic: A Study of the Descriptions of Physical Appearance in Classical Epic," *HSCP* 58-59 (1948): 189-217; E.C. Evans, "A Stoic Aspect of

The characterization "physiognomic consciousness" seems appropriate and useful for distinguishing between, on the one hand, *4QZodiacal Physiognomy* and *4QPhysiognomy ar*, which can be characterized as physiognomic writings on a more formal, technical level, and, on the other hand, texts that show in certain passages a more general physiognomic interest.

This study is limited to reconstructing and understanding what remains of *4QZodiacal Physiognomy* and *4QPhysiognomy ar*.[16] Moreover, of these two manuscripts, *4QZodiacal Physiognomy* will receive the most attention since the remaining fragments preserve more text and raise more questions.

4QZodiacal Physiognomy and *4QPhysiognomy ar*: Physiognomic-Astrological and Physiognomic Lists

The extant text of the Hebrew manuscript *4QZodiacal Physiognomy* lists several entries that are made up of various elements. Most clearly, the entries contain descriptions of the human body from head to toe that belong to different types of people. Following these descriptions, the text provides enigmatic data about a person's spirit having certain numbers in the "house of light" and the "house of darkness," as well as astrological information in terms of a section of the person's zodiacal sign at the time of his birth.

The Aramaic manuscript *4QPhysiognomy ar* is even more fragmentary than *4QZodiacal Physiognomy*. Like that text, it lists in different entries the physical descriptions of several types of people from head to toe. *4QPhysiognomy ar* also provides some evidence for predictions that presumably follow on from these descriptions. Unlike *4QZodiacal Physiognomy*, however, the remaining fragments of this text do not contain any references to zodiacal signs. This, of course, does not mean that these references could not have been there in the original, complete text. But it is significant that the remaining fragments do not provide any evidence that in this text the purpose of physiognomic inquiry was to reveal the division of people's spirits between the "house of light" and the "house of darkness."

The two texts are similar in structure and style. Both represent lists or catalogues with different entries that are structured according to descriptions of the human body. The style of these texts is terse and succinct. These two characteristics set them apart from the other texts amongst the Dead Sea Scrolls and Second Temple period writings that demonstrate physiognomic

---

Senecan Drama: Portraiture," *TAPA* 81 (1950): 169-84; J. Couissin, "Suétone physiognomiste dans les *Vies des XII Césars*," *REL* 31 (1953): 234-56; Evans, *Physiognomics*, 46-58; J.W. Drijvers, "'Hij was een uitzonderlijke schoonheid...': Keizerlijke lichamen in beeld en geschrift," in *Machtige lichamen: Het vingertje van Luns en andere politieke wapens* (eds. C. Santing, H. te Velde and M. Wilke; Amsterdam: Wereldbibliotheek, 2005), 191-207.

[16] For a brief discussion of texts showing physiognomic consciousness, see Appendix II.

interest. Representing the intriguing remains of physiognomic-astrological and physiognomic lists or catalogues, the manuscripts of *4QZodiacal Physiognomy* and *4QPhysiognomy ar* are the earliest Jewish texts of their kind known to date. Their relevance, however, extends further than that.

The text of *4QZodiacal Physiognomy* is important evidence for the history of physiognomic and astrological learning in antiquity in general because it combines physiognomic and astrological learning in the form of a list that was structured according to physiognomic criteria. Like the observation by Hephaestion, this Jewish text testifies to the existence of the belief that the human body could signify astrological matters, whereas in every other example known to us of ancient lists that combine astrology and physiognomics the signifying relationship is reversed.

The manuscript of *4QZodiacal Physiognomy* is not only remarkable for its content, but also for its manner of writing. It immediately catches the reader's attention due to the way it is written. Contrary to the regular direction of writing in Hebrew, which is right to left, the text is written from left to right. In addition, characters from different scripts have been used. In addition to the regular, so-called square script, the writer or copyist used ancient Hebrew, Greek, and cryptic letters. These two features – inverted and mixed writing – make this text exceptional. There are no other known examples of Jewish texts written entirely in reversed order and in mixed scripts.

As already said, these two lists from Qumran provide the earliest Jewish evidence for technical physiognomic learning, predating medieval physiognomic manuscripts by a thousand years.[17] Some scholars have suggested that the physiognomic learning from the Qumran sect was passed on into the hands of those Jewish circles eventually responsible for the medieval physiognomic tradition, the groups perhaps even being historically related to each other.[18] However, the terminological overlap between the texts is

[17] For medieval Jewish physiognomic and astrological-physiognomic texts, see G. Scholem, "הכרת פנים וסדרי שרטוטין (Physiognomy and Chiromancy)," in *ספר אסף: קובץ מאמרי מחקר* (eds. M.D. Cassuto, J. Klausner and J. Gutmann; Jerusalem: Mossad Harav Kook, 1953), 459-95; G. Scholem, "Ein Fragment zur Physiognomik und Chiromantik aus der spätantiken jüdischen Esoterik," in *Liber Amicorum: Studies in Honour of Professor Dr. C.J. Bleeker* (SHR 17, Leiden: E.J. Brill, 1969), 175-93; I. Gruenwald, "קטעים חדשים מספרות הכרת־פנים וסדרי־שרטוטין (Further Jewish Physiognomic and Chiromantic Fragments)," *Tarbiz* 40 (1970-1971): 301-19; P. Schäfer, *Geniza Fragmente zur Hekhalot-Literatur* (TSAJ 6; Tübingen: J.C.B. Mohr [Paul Siebeck], 1984), 135-39; P. Schäfer, "Ein neues Fragment zur Metoposkopie und Chiromantik," in *Hekhalot-Studien* (P. Schäfer; TSAJ 19; Tübingen: J.C.B. Mohr [Paul Siebeck], 1988), 84-95 (originally published in *FJB* 13 [1985]: 61-82); P.S. Alexander, "Incantations and Books of Magic," in *Emil Schürer: The History of the Jewish People in the Age of Jesus Christ (175 B.C.-A.D. 135)* (vol. 3.1; eds. G. Vermes, F. Millar and M. Goodman; Edinburgh: T&T Clark, 1986), 333-79, at 366-69.

[18] Gruenwald, "Jewish Physiognomic," 304-6; I. Gruenwald, *Apocalyptic and Merkavah Mysticism* (AGJU 14; Leiden: E.J. Brill, 1980), 218-24; L.H. Schiffman, *Reclaiming the Dead*

not, as is claimed, technical in any way, but applies to physiognomic descriptions that are too general to be conclusive for a direct connection. Instead, it might be worthwhile to consider these medieval Jewish physiognomic traditions against the background of late antique and early medieval culture in the Near East as a continuation of ancient Babylonian and Greco-Roman traditions, similar to Aramaic astrological-physiognomic traditions in the Mandean *Book of the Zodiac*.[19]

TWO NON-SECTARIAN COMPOSITIONS, ONE POSSIBLY SECTARIAN COPY

Qumran scholars distinguish between sectarian and non-sectarian writings amongst the Dead Sea Scrolls.[20] It is argued that sectarian texts should be discernable by a distinctive terminology that is linked with a particular set of ideas.[21] The criteria for this distinction as well as the precise understanding of what is meant by the term "sectarian text," however, are not entirely clear. There are four possible levels of understanding this term.[22]

Firstly, a "sectarian text" is one used by members of the community. This implies that the mere presence of a text at Qumran is enough to conclude that it was used and read there. This is the broadest definition because

---

Sea Scrolls: The History of Judaism, the Background of Christianity, the Lost Library of Qumran (Philadelphia: The Jewish Publication Society, 1994), 362-64; J.R. Davila, "4QMess ar [4Q534] and Merkavah Mysticism," *DSD* 5 (1998): 367-81; J.R. Davila, "The Dead Sea Scrolls and Merkavah Mysticism," in *The Dead Sea Scrolls in their Historical Context* (eds. T.H. Lim et al.; Edinburgh: T&T Clark, 2000), 249-64, at 253-55, 263-64; M.D. Swartz, "The Dead Sea Scrolls and Later Jewish Magic and Mysticism," *DSD* 8 (2001): 182-93, at 192-93.

19 Cf. Chapter Two nn. 23, 205.

20 See e.g. D. Dimant, "Qumran Sectarian Literature," in *Jewish Writings of the Second Temple Period: Apocrypha, Pseudepigrapha, Qumran Sectarian Writings, Philo, Josephus* (ed. M.E. Stone; CRINT 2.2; Assen: Van Gorcum and Philadelphia: Fortress Press, 1984), 483-550, at 487; C.A. Newsom, "'Sectually Explicit' Literature from Qumran," in *The Hebrew Bible and its Interpreters* (eds. W.H. Propp, B. Halpern and D.N. Freedman; BJSUCSD 1; Winona Lake, Indiana: Eisenbrauns, 1990), 167-87; E.G. Chazon, "Is *Divrei ha-me'orot* a Sectarian Prayer?," in *The Dead Sea Scrolls: Forty Years of Research* (eds. D. Dimant and U. Rappaport; STDJ 10; Leiden: E.J. Brill and Jerusalem: Magnes, 1992), 3-17; D. Dimant, "The Qumran Manuscripts: Contents and Significance," in *Time to Prepare the Way in the Wilderness: Papers on the Qumran Scrolls by Fellows of the Institute for Advanced Studies of the Hebrew University, Jerusalem, 1989-1990* (eds. D. Dimant and L.H. Schiffman; STDJ 16; Leiden: E.J. Brill, 1995), 23-58; A. Lange, *Weisheit und Prädestination: Weisheitliche Urordnung und Prädestination in den Textfunden von Qumran* (STDJ 18; Leiden: E.J. Brill, 1995), 6-20; J.M. Jokiranta, "'Sectarianism' of the Qumran 'Sect': Sociological Notes," *RevQ* 78/20 (2001): 223-39; J. Jokiranta, *Identity on a Continuum: Constructing and Expressing Sectarian Social Identity in Qumran* Serakhim *and* Persharim (Diss. Helsinki, 2005), 65-73.

21 Dimant, "Qumran Sectarian Literature," 487; Dimant, "Qumran Manuscripts," 27-29.

22 For the following, see Newsom, "'Sectually Explicit,'" 172-79; Jokiranta, "'Sectarianism,'" 236-38. It is also possible to distinguish another level between the first and second ones, namely texts that were copied at Qumran (or that display the specific characteristics of the so-called Qumran scribal practices), see below.

it encompasses all the texts found at Qumran. Secondly, a "sectarian text" is one authored by people from the sect. This definition already restricts the number of writings that can be characterized as sectarian, the biblical manuscripts being the prime example not fitting this category. It cannot, however, help to deal with the possibility that texts were indeed composed by members of the community, but cannot be identified as such because they lack the combination of a distinctive terminology with a particular set of ideas. Content and style should not be conflated with authorship. In such a case we have no way of knowing whether a text was authored within the community.[23] Thirdly, a "sectarian text" is one containing a distinctive set of ideas that concern the unique structures and history of the group and that are expressed by means of a self-consciously polemical rhetoric referring to the separation from a larger community. Fourthly, a "sectarian text" is one that shows clear evidence of a sectarian stance, which consists of a self-understanding as uniquely legitimate and of a negative tension with the sociocultural environment.

Returning to *4QZodiacal Physiognomy* and *4QPhysiognomy ar*, it seems that there is nothing particularly sectarian about these texts in any of the three latter senses. Like the majority of Aramaic texts from Qumran, *4QPhysiognomy ar* can be regarded a non-sectarian composition. It could have been read and used by the community, but was most probably not composed by it.[24] With regard to the Hebrew text *4QZodiacal Physiognomy* the matter is less clear.[25] Many scholars assume it to be a sectarian composition due to the text's alleged light and darkness dualism, which they regard as typical of the Qumran sect worldview. The words "house of light" and "house of darkness," however, are not dualistic in the context of the astrological framework of *4QZodiacal Physiognomy*.[26] There is, therefore, no evidence suggesting that the text was authored within the community.

According to the first definition of a Qumran "sectarian text" above, it is possible to understand *4QZodiacal Physiognomy* and *4QPhysiognomy ar* as

[23] Cf. Dimant, "Qumran Manuscripts," 28 n. 14.

[24] Cf. S. Segert, "Die Sprachenfrage in der Qumrāngemeinschaft," in *Qumran-Probleme: Vorträge des Leipziger Symposions über Qumran-Probleme vom 9. bis 14. Oktober 1961* (ed. H. Bardtke; DAWBSSA 42; Berlin: Akademie Verlag, 1963), 315-39; Dimant, "Qumran Manuscripts," 34-35, 55; J.A. Fitzmyer, "Aramaic," in *Encyclopedia of the Dead Sea Scrolls* (eds. L.H. Schiffman and J.C. VanderKam; Oxford: Oxford University Press, 2000), 48-51, at 50.

[25] For example, Dimant, "Qumran Sectarian Literature," 531, considers it a sectarian text. However, Dimant, "Qumran Manuscripts," 34, 51, classifies it as a text that does not use Qumran community terminology, but D. Dimant, "Dualism at Qumran: New Perspectives," in *Caves of Enlightenment: Proceedings of the American Schools of Oriental Research Dead Sea Scrolls Jubilee Symposium (1947-1997)* (ed. J.H. Charlesworth; North Richland Hills, Texas: Bibal, 1998), 55-73, at 62-64, regards it again as part of the community's literature.

[26] See Chapters Three and Four.

sectarian texts in the sense that they were found in one of the caves near Qumran. By their mere presence, they were part of the, presumably deliberately gathered, collection of manuscripts of the Qumran community, which may imply that they were used and read by (some) sectarian members.[27]

In addition to the four levels of distinction for a "sectarian text," the aspect of manuscript copying may provide further information. One should allow for the possibility that many, if not most, of the Dead Sea Scrolls manuscripts were not copied at Qumran but elsewhere in Judea or Palestine.[28] It is also possible that many manuscripts were not even copied by members of the community. If one assumes, however, that the cryptic A script was a sectarian writing system,[29] then the possible use of one Cryptic A *yod* in בבית ("in the house") in 4Q186 1 ii 7 may be evidence that *4QZodiacal Physiognomy* was copied within the Qumran community, but this is not certain.[30]

[27] Contrary to some interpretations of the Qumran settlement that treat it separately from the caves with the scrolls (see e.g. Y. Hirschfeld, *Qumran in Context: Reassessing the Archaeological Evidence* [Peabody, Massachusetts: Hendrickson, 2004]), I agree with Roland de Vaux's assessment that scroll caves and settlement belong together. See M. Popović, "De archeologie van Qumran," in *Fragmenten uit de woestijn*, 48-77. The intricate connection between caves and settlement is now corroborated by INAA of pottery at both locations. See J. Gunneweg and M. Balla, "Neutron Activation Analysis: Scroll Jars and Common Wares," in *Khirbet Qumrân et 'Aïn Feshkha II: Études d'anthropologie, de physique et de chimie, Studies of Anthropology, Physics and Chemistry* (eds. J.-B. Humbert and J. Gunneweg; NTOA.SA 3; Fribourg: Academic Press Fribourg and Göttingen: Vandenhoeck & Ruprecht, 2003), 3-53.

[28] E. Tov, *Scribal Practices and Approaches Reflected in the Texts Found in the Judean Desert* (STDJ 54; Leiden: Brill, 2004), 5.

[29] In his different articles on the Cryptic A script, Stephen Pfann has emphasized that this script was written for the internal purposes of the community and he has ascribed the use of this script to the *Maskil*. See S.J. Pfann and M. Kister, "4QcryptA Words of the Maskil to All Sons of Dawn," in *Qumran Cave 4.XV: Sapiential Texts, Part 1* (eds. T. Elgvin et al.; DJD 20; Oxford: Clarendon Press, 1997), 1-30; S. Pfann, "4Qpap cryptA Midrash Sefer Moshe," in *Qumran Cave 4.XXV: Halakhic Texts* (eds. J. Baumgarten et al.; DJD 35; Oxford: Clarendon Press, 1999), 1-24; S.J. Pfann, "Introduction," in *Qumran Cave 4.XXVI: Cryptic Texts and* Miscellanea, *Part 1* (eds. P. Alexander et al.; DJD 36; Oxford: Clarendon Press, 2000), 515-46; S.J. Pfann, "The Writings in Esoteric Script from Qumran," in *The Dead Sea Scrolls: Fifty Years After Their Discovery: Proceedings of the Jerusalem Congress, July 20-25, 1997* (eds. L.H. Schiffman, E. Tov and J.C. VanderKam; Jerusalem: Israel Exploration Society, 2000), 177-90. Cf. also Dimant, "Qumran Manuscripts," 45 n. 55; Tov, *Scribal Practices*, 50, 204.

[30] The use of Cryptic A script in a manuscript is a problematic criterion for determining whether a text is a sectarian composition or even a copy (the moment this script turns up outside Qumran, it will be difficult to argue that it was a sectarian writing system). If it was a sectarian writing system, why is it that texts that are evidently sectarian, like the *Rule of the Community* or the *Pesharim*, are not written in this script, while texts that do not have a distinct terminology and a particular set of ideas, like some calendrical texts (4Q317 or 4Q324^{d-i}) or sapiential texts (4Q298), are? Why are texts that are similar in content, like the calendrical or sapiential texts, written in both the square script and the Cryptic A script? The Cryptic A script is not a clear criterion for distinguishing between sectarian and non-sectarian writings. For example, on the one hand Devorah Dimant assumes that texts written in the cryptic script belong to the community, and she therefore includes the calendrical text 4Q317, but on

As non-sectarian compositions, this means that the origins of *4QZodiacal Physiognomy* and *4QPhysiognomy ar* lie outside the immediate context of the Qumran community. In addition to a sectarian context, one should, therefore, also allow for the possibility that these lists had a non-sectarian context. As is the case with the biblical writings and other non-sectarian texts, the community shared these texts with other Jews. *4QZodiacal Physiognomy* and *4QPhysiognomy ar* provide us with a valuable window on an intriguing aspect of Second Temple period Judaism in general. The important fact that these two technical physiognomic and physiognomic-astrological lists were found at Qumran is evidence that Jewish people in the Hellenistic-Early Roman period had knowledge of and access to physiognomic and astrological learning on more than a general level. These texts show that Jewish culture in Palestine participated in sophisticated disciplines of learning that were current in Babylonian and Greco-Roman cultures.

Outline and Main Theses of the Present Study

This inquiry is not the first to try to reconstruct and understand different aspects of *4QZodiacal Physiognomy* and *4QPhysiognomy ar*, but it is the first attempt at a comprehensive treatment of both texts in comparison with similar writings from Babylonian and Greco-Roman cultures.

John Allegro published the extant fragments of *4QZodiacal Physiognomy* more than forty years ago, but he himself did not comment extensively on the text.[31] Allegro clearly could not make much of it, something he expresses vividly in a letter to his wife Joan, written several days after deciphering the inverted and mixed writing:

> I worked this morning on my piece of the cryptic script, and after puzzling all morning decided that the script was the least cryptic thing about it. It doesn't make sense, and I think some bored Essene was amusing himself making life difficult for a later generation. [letter December 13, 1953][32]

---

the other hand she argues that none of the calendrical and chronological texts belong to the community compositions. See Dimant, "Qumran Manuscripts," 34, 45, 51.

[31] J.M. Allegro, "Le travail d'édition des fragments manuscrits de Qumrân: Communication de J.M. Allegro," *RB* 63 (1956): 62-64, at 63-64; J. Allegro, *The Dead Sea Scrolls: A Reappraisal* (2d ed.; Harmondsworth, Middlesex: Penguin, 1964), 127; J.M. Allegro, "An Astrological Cryptic Document from Qumran," *JSS* 9 (1964): 291-94; J.M. Allegro (ed. with the collaboration of A.A. Anderson), *Qumrân Cave 4.I (4Q158-4Q186)* (DJD 5; Oxford: Clarendon, 1968), 88-91.

[32] J.A. Brown, *John Marco Allegro: The Maverick of the Dead Sea Scrolls* (Grand Rapids, Michigan: Eerdmans, 2005), 30.

Although Allegro's edition was criticized,[33] many scholars accept his understanding of *4QZodiacal Physiognomy* as an astrological text dealing with the influence of the stars on the human body and spirit, which was related to the so-called *Two Spirits Treatise* in the sectarian *Rule of the Community* (1QS 3:13–4:26).

Scholars such as Roland Bergmeier, Francis Schmidt, and Matthias Albani focused on elucidating the astrological background of the text, while others such as Jacob Licht and Philip Alexander stressed its physiognomic nature.[34] These approaches have contributed many valuable insights to our understanding of *4QZodiacal Physiognomy*. However, instead of paying attention to just one approach and neglecting the other, this study will integrate them in order to do justice to both the physiognomic and astrological elements in the text and to further our understanding of its sense.

The manuscript of *4QZodiacal Physiognomy* has been known to Dead Sea Scrolls scholars for fifty years now. Allegro's rapid publication of it has contributed to the familiarity of scholars with this text, as is demonstrated by the many references to it in books and articles. The manuscript of *4QPhysiognomy ar* has also been known for nearly fifty years,[35] but until recently only part of the text had been published, and the final publication by Émile Puech in *DJD* 37 is still forthcoming.[36] Undoubtedly, this has

[33] Cf. J. Strugnell, "Notes en marge du volume V des 'Discoveries in the Judaean Desert of Jordan,'" *RevQ* 7/26 (1970): 163-276, at 274-76; M. Popović, "A Note on the Reading of שמונה and עמוד השני in 4Q186 2 i 7," *RevQ* 21/84 (2004): 635-41.

[34] J. Licht, "שוקיים סימן לבחירה (Legs as Signs of Election)," *Tarbiz* 35 (1965-1966): 18-26; R. Bergmeier, *Glaube als Gabe nach Johannes: Religions- und theologiegeschichtliche Studien zum prädestinatianischen Dualismus im vierten Evangelium* (BWANT 112; Stuttgart: W. Kohlhammer, 1980), 78-79; P.S. Alexander, "Physiognonomy [sic!], Initiation, and Rank in the Qumran Community," in *Geschichte – Tradition – Reflexion: Festschrift für Martin Hengel zum 70. Geburtstag, Band I Judentum* (eds. H. Cancik, H. Lichtenberger, and P. Schäfer; Tübingen: J.C.B. Mohr [Paul Siebeck], 1996), 385-94; F. Schmidt, "Astrologie juive ancienne: Essai d'interprétation de *4QCryptique (4Q186)*," *RevQ* 18/69 (1997): 125-41; M. Albani, "Horoscopes in the Qumran Scrolls," in *The Dead Sea Scrolls After Fifty Years: A Comprehensive Assessment* (eds. P.W. Flint and J.C. VanderKam; vol. 2; Leiden: E.J. Brill, 1999), 279-330; M. Albani, "Horoscopes," in *Encyclopedia of the Dead Sea Scrolls*, 370-73.

[35] In photograph PAM 43.598, taken June 1960, all fragments have been put together.

[36] The original editor of *4QPhysiognomy ar*, Jean Starcky, gave a translation of part of it. See J. Starcky, "Un texte messianique araméen de la grotte 4 de Qumrân," in *Mémorial du cinquantenaire 1914-1964, École des Langues Orientales Anciennes de l'Institut Catholique de Paris* (TICP 10; Paris: Bloud & Gay, 1964), 51-66, at 64-65. See Appendix I for references to other partial transcriptions and translations.

Prior to Puech's final edition, all of the fragments have recently been published. See M. Wise, "4Q561 (4QPhysiognomy/Horoscope ar)," in *Additional Genres and Unclassified Texts* (eds. D.W. Parry and E. Tov; DSSR 6; Leiden: Brill, 2005), 228-31; S. Holst and J. Høgenhaven, "Physiognomy and Eschatology: Some More Fragments of 4Q561," *JJS* 57 (2006): 26-43. Whereas Michael Wise considers all fragments in photograph PAM 43.598 part of one manuscript, Søren Holst and Jesper Høgenhaven follow, I think rightly, Starcky in seeing the three fragments in the upper half of the photograph as part of another manuscript and composition, termed *ar S* by Starcky.

impeded scholars taking full account of all the manuscript fragments, hindering the proper study of *4QPhysiognomy ar*. In the case of *4QZodiacal Physiognomy* I had the opportunity to study the fragments themselves at the Israel Antiquities Authority Dead Sea Scrolls Laboratory in Jerusalem, but as the fragments of *4QPhysiognomy ar* are not yet officially published in the *DJD* series I did not have access to these. My reading of the text is based on the available photographs,[37] as well as previous transcriptions. Nevertheless, my reading and understanding of *4QPhysiognomy ar* differ at some points from earlier publications.

In Chapter One, introducing *4QZodiacal Physiognomy* and *4QPhysiognomy ar*, I will argue that *4QZodiacal Physiognomy* represents the remains of a physiognomic-astrological list: the structure of the different elements in the entries of this list demonstrate that it is organized according to physiognomic descriptions from which some astrological data can be discerned, possibly in relation to magico-medicinal stones. Regarding *4QPhysiognomy ar* it will be argued that it represents the remains of a physiognomic list and that the text originally provided predictions for each physiognomic type, but that there is no evidence for references to zodiacal signs or other astrological notions. As for the relationship between the two manuscripts, I will argue that there is no evidence that suggests they contain the same literary composition; there is no reason to assume that the Hebrew is a translation of the Aramaic.

In Chapter Two, discussing various aspects of physiognomic writings and their backgrounds in Babylonian and Greco-Roman traditions, it will be argued from a comparative perspective that the Qumran lists cannot be directly linked to either a Babylonian or Greek *Vorlage*; despite some similarities in form, there are significant differences. As for the possibility of tracing a cultural influence from either Mesopotamia or the Greek world on the Jewish physiognomic tradition witnessed by *4QZodiacal Physiognomy* and *4QPhysiognomy ar*, I will argue that neither can be excluded as far as the physiognomic aspect is concerned. With regard to the combination of physiognomic and astrological aspects in *4QZodiacal Physiognomy*, comparison with other texts demonstrates that this text is familiar with the notion that the zodiacal signs have an effect on the appearance of the human body, but also that the relationship between the two elements is expressed differently: lists from Babylonian as well as Greco-Roman traditions that combine astrology with physiognomics are arranged according to astrological criteria, whereas *4QZodiacal Physiognomy* is organized according to physiognomic descriptions. Remarks from Greco-Roman (astrological) lit-

[37] T.H. Lim (ed. in consultation with P.S. Alexander), *The Dead Sea Scrolls Electronic Reference Library: Volume I* (Oxford: Oxford University Press and Leiden: Brill, 1997).

erature demonstrate that people believed it possible to reason the other way round: discerning from a person's physiognomy his ascendant zodiacal sign. It is this line of reasoning that informs the textual arrangement of the physiognomic and astrological aspects in *4QZodiacal Physiognomy*.

In Chapter Three I will investigate in detail the astrological framework behind *4QZodiacal Physiognomy*, reviewing previous hypotheses against the background of notions from Babylonian and Hellenistic astrology. I will support Albani's so-called ascendant interpretation as the most likely hypothesis, but also propose some modifications and elaborations. Most importantly, it will be argued that *4QZodiacal Physiognomy* can be situated within an astrological tradition that combines the notions of *dodecatemoria* and *melothesia*. When the text mentions a specific part of the zodiacal sign, saying in 4Q186 1 ii 9 that a person was born "in the foot of *Taurus*," it refers to that part of the zodiacal sign ascending above the eastern horizon (this is what "horoscope" literally meant in antiquity). According to the ascendant interpretation, the realization of the numbers in the "house of light" (area above the earth) and the "house of darkness" (area below the earth) is thus the result of the division of the zodiacal sign during its ascendancy over the eastern horizon. This division of the sign and the names given to its different parts, such as "foot of *Taurus*," is, I suggest, due to merging the astrological concepts of *dodecatemoria* and *melothesia*. The original text of *4QZodiacal Physiognomy* would have represented an elaborate catalogue that listed physiognomic typologies, which lead the intended reader to the various subdivisions between light and darkness of each of the twelve signs of the zodiac. In addition, I will argue that this astrological framework points to a Hellenistic background for *4QZodiacal Physiognomy*: the concern with the ascendant is typical of Hellenistic astrology and is lacking in Babylonian astrology.

In Chapter Four, focusing on what *4QZodiacal Physiognomy* means when it connects the numbers in the "house of light" and the "house of darkness" with the רוח ("spirit") there is said to be for the type of person described, I will argue against the interpretations of רוח as a reference to the space or room occupied by the zodiacal sign or as a reference to the human spirit. Within the latter interpretation many scholars have related *4QZodiacal Physiognomy* to the *Two Spirits Treatise*, arguing that the former text provides the arithmetic for the notion of two spirits of light and darkness fighting over human beings as attested by the latter text. However, this connection between the two texts is not feasible in light of a numerical discrepancy between the number of twelve zodiacal signs and the number of alleged divisions of the human spirit that are possible on the nine-point scale that is supposedly used in *4QZodiacal Physiognomy*. Instead, it will be argued that רוח ("spirit") refers to the zodiacal spirit, because if the numbers listed in *4QZodiacal Physiognomy* are a result of the ascendant zodiacal sign, then the "spirit" should also be related to the zodiacal sign. The notion of zodiacal spirits will be demonstrated by adducing other texts, most notably the *Testament of Solomon*. Although the division between the "house of light" and the "house of darkness" is astrologically the result of the ascendant zodiacal sign's position vis-à-vis the eastern horizon, I suggest that this was understood in *4QZodiacal Physiognomy* in terms of the zodiacal spirit being divided between light and darkness.

In Chapter Five, addressing the social and cultural background of the physiognomic and physiognomic-astrological lists from Qumran in terms of their status, function, and context, I will argue that these texts represent forms of ancient Jewish science, suggesting further that in the case of *4QZodiacal Physiognomy* a notion of cosmic sympathy may be behind the combination of different aspects of learning: physiognomics, astrology, stones of a possibly magico-medicinal nature. Taking account of the ways and possibilities for transmission of and education in physiognomic and astrological arts in Babylonian and Greco-Roman traditions, I will argue that the persons who were interested in such texts and the knowledge they contain were part of a well-educated body of people in ancient Jewish society; both priestly and secular scribes or scholars could have been responsible for the dissemination of these learned arts in Hellenistic-Early Roman period Palestine. It is also argued that familiarity with these learned forms of physiognomics and astrology came from outside Palestine; *4QPhysiognomy ar* possibly from the Babylonian cultural realm, although a Hellenistic origin cannot be excluded, and *4QZodiacal Physiognomy* most likely from the Hellenistic realm, possibly Greco-Egyptian. In relation to the scientific interests evinced by parts of *1 Enoch* and the lists of revealed things in apocalyptic texts, I will suggest, first, that the physiognomic and physiognomic-astrological lists from Qumran were possibly not framed by religious interests, acknowledging the danger of an anachronism here, and, second, that they show that during the Hellenistic-Early Roman period Jews in Palestine were interested in contemporary scientific learning, and not just in "outdated" forms of Mesopotamian astronomy as in the *Astronomical Book*. Although astrology attracted ambivalent attitudes, I will argue that *4QZodiacal Physiognomy* reflects an interest in astrological matters on the part of members of the Qumran community, as well as other Jews. The inverted and mixed writing of this manuscript signifies, I suggest, the high status that was accredited to its learning; the use of these writing techniques being a scribal means to limit accessibility to and availability of this expert knowledge to those who were suitable to understand it. Finally, realizing that these texts may simply have been read as pieces of speculative, scientific learning, I will also hypothesize on contexts and functions of practical use. The predictions in *4QPhysiognomy ar* may have been used in a divinatory practice, while the information in *4QZodiacal Physiognomy*, concerning the nature of people's zodiacal spirits as divided between light and darkness, may have served a diagnostic function both in and outside the sectarian community of Qumran to indicate the harmful nature of these spirits. In general, the learning in the latter text may have been used in a magico-medicinal context, diagnosing which zodiacal spirit was troubling a person and which measures, such as apotropaic or magico-medicinal stones, should be used against it. In a sectarian context, it may have been used, as part of the Qumran community's fight against the evil spirits of Belial and the sons of darkness, to control the admission of potential candidates for community membership by keeping people with more dangerous and harmful zodiacal spirits outside the group.

CHAPTER I

READING *4QZODIACAL PHYSIOGNOMY* (4Q186) AND *4QPHYSIOGNOMY AR* (4Q561): TEXTS, GENRE, AND STRUCTURE

INTRODUCTION

This chapter will introduce the remains of the two physiognomic catalogues from Qumran. Attention is paid to matters such as material reconstruction, arrangement of columns, cryptic writing, paleography, and date, as well as to the contents of both texts. Although transcriptions and translations are provided, the detailed notes and comments on readings can be found in Appendix I. Some elements have been singled out for discussion in this chapter. Other elements, most notably the astrological framework and the understanding of רוח ("spirit") in 4Q186, will be dealt with in subsequent chapters.

Most scholars have assumed that the Hebrew manuscript 4Q186 is an astrological text structured according to astrological criteria, but the different entries of this catalogue-text consist of a set arrangement of elements in which the astrological data are provided subsequently to the physiognomic descriptions. Is this, therefore, an appropriate characterization of 4Q186? In order to reassess this understanding of the text, this chapter will investigate the structure of the text. How are the physiognomic and astrological elements in the entries related to each other, and what light does their arrangement shed on the nature of the text? Also, is it possible to reconstruct the beginning of the entries in the catalogue?

Another issue that demands consideration is the enigmatic phrase "the second column" (העמוד השני), which scholars have pondered at great length. What does it mean here? Is it the second zodiacal sign; the second quadrant; the second phase of the moon; or the second column of a scroll.

The clearest indication for astrological concerns in 4Q186 is the reference to a person's horoscope (מולד), but scholars differ in their understanding of the word: birth-time; horoscope of birth; horoscope of conception. How is it to be understood in this context?

Finally, a much overlooked element in the entries is the mention of specific sorts of stones. Although only one occurrence is preserved, it may have been an important aspect of the entries in the original text. How does it relate to other elements in the text such as the descriptions of the human body and the zodiacal signs? Does it shed further light on the nature of the list in 4Q186? Is there any significance in the fact that the words "a granite stone" (אבן צונם) are not written in an inverted manner like the rest of the text?

The Aramaic manuscript 4Q561 primarily preserves physiognomic descriptions. It has been argued that the text did not list predictions in relation to these descriptions, but the occurrence of certain words seems to suggest otherwise. Does 4Q561 represent the remains of a physiognomic catalogue that listed predictions?

According to Jean Starcky, the first editor, 4Q561 establishes a relation between body and spirit. On one small fragment Starcky reconstructed the words "his [sp]irit" ([ר]וחלה). He suggested that this is the equivalent of רוח לו in 4Q186 and that it implies the context of a reference to the "house of light" and the "house of darkness." Moreover, Starcky assumed 4Q561 to be the Aramaic version of 4Q186. However, if [ר]וח לה is the correct reading, is it materially possible to assume it to be equivalent to רוח לו, and have the same context as 4Q186? If this is not the case, is it possible to understand the nature of the relationship between body and spirit established in 4Q561 in a different way? Also, what is the evidence for seeing both 4Q561 and 4Q186 as representing the same literary composition?

4QZodiacal Physiognomy (4Q186)

Name and Genre

In Dead Sea Scrolls research, 4Q186 has primarily been interpreted from an astrological perspective. An indication for the presence of astrological notions is found in 4Q186 1 ii 8–9: "And this is the horoscope under which he was born: in the foot of *Taurus* [...]. And this is his animal: *Taurus*." When the text was first mentioned, Allegro characterized it as "une *melothesia* planétaire apparentée d'une certaine manière à l'ouvrage de Ptolémée d'Alexandrie."[1] The genre designation "horoscope" was quickly applied,[2] and the text came to be officially known as *4QHoroscope*.[3]

[1] Allegro, "Le travail d'édition," 64; cf. also J.T. Milik, *Ten Years of Discovery in the Wilderness of Judaea* (2d ed.; SBT 26; trans. J. Strugnell; London: SCM Press, 1963), 119; Allegro, "Astrological Cryptic Document," 291: 4Q186 "deals with the influence of the stars on the human body and spirit."

[2] Cf. e.g. J. Carmignac, "Les Horoscopes de Qumrân," *RevQ* 5/18 (1965): 199-217, at 199-206; M. Delcor, "Recherches sur un horoscope en langue hébraïque provenant de Qumrân," in *Religion d'Israël et Proche Orient Ancien: Des Phéniciens aux Esséniens* (M. Delcor; Leiden: E.J. Brill, 1976), 298-319 (originally published in *RevQ* 5/20 [1966]: 521-42).

[3] Cf. E. Tov et al. (eds.), *The Texts from the Judaean Desert: Indices and an Introduction to the* Discoveries in the Judaean Desert *Series* (DJD 39; Oxford: Clarendon, 2002), 51.

The name *4QHoroscope*, however, creates the wrong impression that the text represents a horoscope or a collection of horoscopes, which it does not. Notwithstanding the presence of certain astrological notions, 4Q186 as such cannot be characterized as a horoscope text.[4] It does not contain the actual horoscopes of particular individuals. The text lacks many elements that would qualify it as belonging to the genre of horoscopes as known from Babylonian, Greek or later Jewish examples.[5] Most significantly, 4Q186 does not have any explicit reference to the zodiacal position of the sun, moon or any of the other five planets known in antiquity.

Rather, 4Q186 represents a sort of list or compendium of physiognomic and astrological content.[6] Since the text is primarily of physiognomic content, Alexander suggests renaming it "4QAstrological Physiognomy."[7] Schmidt aptly refers to 4Q186 as "un texte de physiognomonie zodiacale."[8]

[4] Cf. O. Neugebauer, *Ethiopic Astronomy and Computus* (ÖAWPH.SB 347; Vienna: Österreichischen Akademie der Wissenschaften, 1979), 21; Bergmeier, *Glaube als Gabe*, 78-79; Alexander, "Incantations and Books of Magic," 364-65; Alexander, "Physiognonomy," 385; Albani, "Horoscopes in the Qumran Scrolls," 287-88, 309.

[5] For Babylonian horoscopes see e.g. A. Sachs, "Babylonian Horoscopes," *JCS* 6 (1952): 49-75; F. Rochberg, *Babylonian Horoscopes* (TAPS 88/1; Philadelphia: American Philosophical Society, 1998); F. Rochberg, *The Heavenly Writing: Divination, Horoscopy, and Astronomy in Mesopotamian Culture* (Cambridge: Cambridge University Press, 2004), 98-163. For Greek horoscopes see e.g. O. Neugebauer and H.B. van Hoesen, *Greek Horoscopes* (MAPS 48; Philadelphia: American Philosophical Society, 1959); D. Baccani, *Oroscopi Greci: Documentazione papirologica* (RC 1; Messina: Sicania, 1992); A. Jones, *Astronomical Papyri from Oxyrhynchus (P. Oxy. 4133-4300a)* (2 vols; MAPS 233; Philadelphia: American Philosophical Society, 1999), 1:247-95, 2:372-447.

Comparison with texts from the Cairo Genizah similar to *4QZodiacal Physiognomy* is illuminating. These astrological-physiognomic texts differ in genre from the actual horoscopes found at the Cairo Genizah. For astrological-physiognomic texts from the Cairo Genizah, see Gruenwald, "Jewish Physiognomic." For Cairo Genizah horoscopes, see B.R. Goldstein and D. Pingree, "Horoscopes from the Cairo Geniza," *JNES* 36 (1977): 113-44; B.R. Goldstein and D. Pingree, "More Horoscopes from the Cairo Geniza," *PAPS* 125 (1981): 155-89. Cf. also S.D. Goitein, *A Mediterranean Society: The Jewish Communities of the Arab World as Portrayed in the Documents of the Cairo Geniza* (vol. 5; Berkeley: University of California Press, 1988), 420 n. 28; H. Ben-Shammai, "Saadia's Introduction to Daniel: Prophetic Calculation of the End of Days vs. Astrological and Magical Speculation," *Aleph* 4 (2004): 11-87, at 36-37.

[6] The list-like character of the entire composition, like that of 4Q561 (*4QPhysiognomy ar*), is comparable to another text from Qumran: 4Q560 (*4QExorcism ar*), which is an example of a part of a magical "book of recipes." Cf. D.L. Penny and M.O. Wise, "By the Power of Beelzebub: An Aramaic Incantation Formula from Qumran," *JBL* 113 (1994): 627-50, at 630, 650; P.S. Alexander, "'Wrestling Against Wickedness in High Places': Magic in the Worldview of the Qumran Community," in *The Scrolls and the Scriptures: Qumran Fifty Years After* (eds. S.E. Porter and C.A. Evans; JSPSup 26; Sheffield: Sheffield Academic Press, 1997), 318-37, at 329-30; P.S. Alexander, "Contextualizing the Demonology of the Testament of Solomon," in *Demons: The Demonology of Israelite-Jewish and Early Christian Literature in Context of their Environment* (eds. A. Lange, H. Lichtenberger and K.F.D. Römheld; Tübingen: Mohr Siebeck, 2003), 613-35, at 614.

[7] Alexander, "Physiognonomy," 385.

[8] Schmidt, "Astrologie juive ancienne," cf. also the English translation "Ancient Jewish Astrology: An Attempt to Interpret 4QCryptic (4Q186)," in *Biblical Perspectives: Early Use*

Schmidt's more limited characterization fits the content better, because the only certain astrological element in the extant text is the reference to a sign of the zodiac in 4Q186 1 ii 9. Therefore, I suggest the name *4QZodiacal Physiognomy* for 4Q186.

Material Reconstruction

4QZodiacal Physiognomy consists of ten fragments.[9] Allegro joined these to make two larger fragments and a third small fragment which he suggested might belong above 4Q186 1 iii.[10] He described the leather of the manuscript as "a soft reddish-brown skin,"[11] but did not discuss the paleography or the date of *4QZodiacal Physiognomy*.

On inspecting the museum plate that shows *4QZodiacal Physiognomy* (IAA #109), it is evident that the fragment is now more damaged than in the photograph in *DJD* 5 (Plate XXXI).[12] Gaps have appeared in places where before there were none, and gaps that were there already on the *DJD* 5 photograph have become larger.

The leather is a rather dark brown. Of the fragments, only 4Q186 4 has something like a whitish layer of "dust-like" substance on its surface.[13] Allegro did not comment on this in his short description of the leather of *4QZodiacal Physiognomy*, perhaps because this is not very unusual. Other manuscripts have a "dust-like" white substance also. It is possibly a clue that 4Q186 4 does not belong together with 4Q186 1 or 2 on the same sheet or on the same part of the sheet, because no such whitish layer is found on the leather of those fragments.

The placement of the fragments on the museum plate confirms my arguments against Allegro's material reconstruction of the second fragment,[14] except perhaps for the join between 4Q186 4 and 4Q186 5.[15]

The two parts of 4Q186 2 cut in half by Allegro have been put together again, clearly after the photograph for *DJD* 5 was taken.[16] Due to the fact

---

and Interpretation of the Bible in Light of the Dead Sea Scrolls: Proceedings of the First International Symposium of the Orion Center for the Study of the Dead Sea Scrolls and Associated Literature, 12-14 May, 1996 (eds. M.E. Stone and E.G. Chazon; STDJ 28; Leiden: E.J. Brill, 1998), 189-205.

[9] See PAM 40.615; 41.314; 41.804; 41.892; 42.616; 43.344; 43.438.

[10] Allegro, DJD 5.91 and Plate XXXI.

[11] Allegro, "Astrological Cryptic Document," 291.

[12] September 22, 2005 at the Dead Sea Scrolls Laboratory of the Israel Antiquities Authority, Jerusalem. I wish to thank the curators Tamar Rabbi-Salhov and Lena Liebman for their kind assistance during my visit.

[13] See Figure 1 for the numbering of the fragments.

[14] Popović, "A Note."

[15] See Plate I. Courtesy of the Israel Antiquities Authority.

[16] It seems unlikely that Allegro is responsible for this because it contradicts his reconstruction. The current curators at the IAA have informed me that the scrolls laboratory recently placed the fragments thus on IAA #109, but the reason why is not clear.

that the two halves have been rejoined, 4Q186 4 1 is no longer in line with 4Q186 2 i 7.[17] 4Q186 4 1 has shifted down slightly, and thus does not continue in 4Q186 2 i 7 nor provide evidence for the numerical sequence reconstructed by Allegro ("eight and o[ne"). But the placement by the IAA is also highly improbable because it does not take into account the spacing between the lines in either this fragment or 4Q186 2 i 7–9.[18] 4Q186 4 cannot, therefore, be placed directly below 4Q186 2 i as it has been by Allegro and, more recently, by the IAA.

Finally, the current placement on IAA #109 seems to show a clear join between 4Q186 4 and 4Q186 5, something I have previously argued against. However, if the whitish layer is a clue it argues against such a join because 4Q186 5 does not have this. Moreover, it is difficult to assess whether the two fragments really fit. After inspection with the microscope it seems that someone at the IAA has put them together in such a way that part of fragment 5 lies under fragment 4, especially in the upper part, whereas in the lower part the edges of both fragments curl up and stand back to back to each other.

The current material reconstruction has resulted in an extant text of *4QZodiacal Physiognomy* consisting of two main fragments, 4Q186 1 and 2, and four smaller, separate fragments.

Columns and Measurements

Notwithstanding the inverted writing and the reversed order for reading the lines, Allegro presents the columns as if they should read from right to left, as has every scholar since then. However, there is no clear indication as to the order in which the columns have to be read, from right to left or vice versa.

4Q186 1 contains the remains of four columns. Of these four columns three have a bottom margin measuring 1.5 cm, but the top margin of all four columns is lacking. This makes it difficult to establish the direction in which a bottom line continued in the next column: whether the sentence resumed at the top of the column to its left or to its right.

[17] Below יושבות in 4Q186 2 i 6 the leather of the manuscript extends slightly further, revealing the blank between the lines. There is clearly not enough space to place 4Q186 4 and 4Q186 5 between this piece of leather and the leather of 4Q186 2 i 7–9, nor join the latter to 4Q186 2 i 7–9, as I have argued previously. It is obvious that precisely because of this the IAA curators must have placed 4Q186 4 *below* the leather extension and not next to it to the right as Allegro did in *DJD* 5.

[18] The placement of 4Q186 4 below 4Q186 2 i 6 in the way it is now on IAA #109 results in too much space between l.6 and l.1 of 4Q186 4 (ה]עْמוד השניש[ן). Moreover, 4Q186 4 1 has been joined to 4Q186 2 i 8. This leaves 4Q186 2 i 7 isolated. The discontinuation of the lines is also clear in the case of 4Q186 4 2 (מולדו) and 4Q186 2 i 9, where the traces of the upper parts of two letters stand significantly lower than 4Q186 4 2.

4Q186 2 preserves the top margin of one column (measuring 2.0 cm), and also contains the meager remnants of a second column at the lower left part of the fragment. Furthermore, 4Q186 2 preserves the edge of a sheet. Not only is part of the margin on the right preserved, the stitches by means of which another sheet was attached are unmistakably visible also. Unfortunately, this does not enable the order in which the columns were organized to be reconstructed.

The blank space in 1.6 of the first column at the right of 4Q186 1 is not indicative for establishing the reading order of the columns. The next line (1.7) is certainly the beginning of a new entry for a typological description (ואיש אשר יהיה קֹ[, "And someone [whose] ... will be"). It is possible that this entry represents the beginning of the actual composition, which means that the columns should be read from right to left. One should also allow for the possibility that different types of texts were copied on the same sheet of leather.[19] The scant remains of letters in the two lines above the blank in 4Q186 1 i 6, however, do not provide any clues as to the content of these lines. Consequently, it cannot be determined whether these lines represent the end of another type of text on the same sheet. It is equally plausible that the blank in 1.6 functions as a marker of an inner division in this zodiacal physiognomic text, setting off the new entry of a typological description from a preceding one.

Schmidt is the only scholar who has commented on the order of the columns, but his argument is not convincing. Based on his reconstruction of the astrological framework, he assumes the text moves from the person with the most parts in the "house of light" to the one with the most parts in the "house of darkness." This would mean that the columns have to be read from right to left and that 4Q186 2 precedes 4Q186 1.[20]

Schmidt's conjecture would mean that the manuscript would have had an extraordinary column height in relation to its width.[21] This is because of

[19] 4Q318 (*4QZodiology and Brontology ar*) contains, in the words of the editors, two types of texts: a selenodromion and a brontologion. Both texts occupy the same line in the manuscript, but they are visibly separated by means of a blank in 4Q318 viii 6. The beginning of that line includes the last word of the selenodromion and after the blank the brontologion begins. It is evident that they represent two different types of text, see J.C. Greenfield et al., "4QZodiology and Brontology ar," in *Qumran Cave 4.XXVI: Cryptic Texts and* Miscellanea, *Part 1* (eds. P. Alexander et al.; DJD 36; Oxford: Clarendon, 2000), 259-74. For the possible relationship between both text types, see M.O. Wise, "Thunder in Gemini: An Aramaic Brontologion (4Q318) from Qumran," in *Thunder in Gemini And Other Essays on the History, Language and Literature of Second Temple Palestine* (M.O. Wise; JSPSup 15; Sheffield: Sheffield Academic Press, 1994), 13-50, at 35-39; M. Albani, "Der Zodiakos in 4Q318 und die Henoch-Astronomie," *MBFJTFL* 7 (1993): 3-42, at 13-19.

Regarding the possible occurrence of different, possibly related, compositions written in the same scroll, see Tov, *Scribal Practices*, 39.

[20] Schmidt, "Astrologie juive ancienne," 137 n. 23.

[21] See Tov, *Scribal Practices*, 82-104.

the number of accounts that have to be assumed to stand between those preserved in 4Q186 1 ii and 4Q186 1 iii in conjunction with an estimated minimum amount of lines of which an account consists.

If one were to follow Schmidt's hypothesis, four other accounts (5:4, 4:5, 3:6, and 2:7) would stand in between that of the person described at the end of 4Q186 1 ii as having six parts in the "house of light" and three parts in the "house of darkness" and that of the individual portrayed at the end of 4Q186 1 iii as having one part in the "house of light" and eight parts in the "house of darkness."[22]

As no complete account has been preserved, it is impossible to establish a set number of lines for an entire account. But it seems reasonable to assume that an average account had no less than seven or eight lines. I suggest three examples to be indicative, if not for the complete text, at least for the remaining fragments. Assuming 4Q186 1 ii 9 to be the end of the account, it at least began in 4Q186 1 ii 4, but probably before that. 4Q186 2 i 1 is most probably not the beginning of an account that continues at least until 4Q186 2 i 7. Finally, the account in 4Q186 1 iii at least includes l.5 and probably continued in another column following the end of l.9.

A minimum number of 7-8 lines per account results in 28-32 lines for the four accounts in between the one at the end of 4Q186 1 ii and 4Q186 1 iii, if one were to accept Schmidt's argument. Taken together with the six lines of the final account in 4Q186 1 iii, this means a minimum column height of 34-38 lines for *4QZodiacal Physiognomy* would have to be assumed.

An average leather height can be suggested on the basis of the vertical space covered by a certain number of lines in combination with the measurements of the top and bottom margins of the manuscript. The height of 5 lines in 4Q186 1 ii and 4Q186 2 i measures ca. 3.5 cm (including the space between the lines at ca. 0.5 cm). Taken together with a hypothetical minimum column height of 34-38 lines, this means that a column measured ca. 24-27 cm in height. Adding to this a top margin of 2.0 cm and a bottom margin of 1.5 cm,[23] the resulting leather height would have been ca. 27.5-30.5 cm. This would place *4QZodiacal Physiognomy* in Emanuel Tov's categories of leather scrolls with a large or very large writing block.[24] This seems too large in relation to the width of ca. 8-9 cm of the columns in *4QZodiacal Physiognomy*. Because of these considerations it seems unlikely that four other accounts stood between those in 4Q186 1 ii and 4Q186 1 iii, as Schmidt proposed.

22 4Q186 1 iii 8–9 actually mentions first the "house of darkness" and secondly the "house of light." It is unclear whether this reversal is significant, and, if so, in what way.

23 Cf. also Tov, *Scribal Practices*, 101.

24 See Tov, *Scribal Practices*, 87-89.

Due to the material condition of the fragments of *4QZodiacal Physiognomy*, it is not possible to determine the sequence in which the columns have to be read with any degree of certainty. One argument for reading the columns from left to right is that this would be in accordance with the overall inverted nature of the text, as is evident from the way the lines and words are written. But to present the columns of *4QZodiacal Physiognomy* in an inverted manner would confuse my presentation unnecessarily with previous scholarship, and, more importantly, such an inverted sequence does not significantly enhance our understanding of the text. For these reasons, I have retained Allegro's ordering of the columns.

A restoration of the entire amount of text cannot be determined on material grounds alone. On the basis of content, however, an educated guess as to the original length of the text of *4QZodiacal Physiognomy* can be proposed.

Assuming that the physiognomic entries in *4QZodiacal Physiognomy* are intrinsically related to the number of subdivisions of each zodiacal sign, the result is a huge catalogue. *4QZodiacal Physiognomy* would have been an elaborate catalogue that listed physiognomic typologies leading the reader to the various subdivisions between light and darkness of each of the twelve signs of the zodiac. These divisions were determined by the ascendancy of particular sections of the signs, understood as a part of each sign's body. Depending on which section of the zodiacal sign was identified rising (in the case of 4Q186 1 ii 9 "the foot of *Taurus*") a division was established of parts already above the earth, in the "house of light," and parts still below the earth, "the house of darkness." Assuming, for the sake of argument, that each zodiacal sign was divided into nine parts (corresponding to the nine-point scale of *4QZodiacal Physiognomy*), this results in a total of 96 physiognomic entries.[25]

If one assumes three entries to a column, such a column would consist of 21-24 lines. The column height would be between 15-17 cm and the leather height ca. 18.5-20.5 cm.[26] This would mean that 32 columns would be needed for 96 physiognomic accounts. Bearing in mind a column width of ca. 8-9 cm together with a margin of ca. 0.5-1 cm, *4QZodiacal Physiognomy* could have been a scroll of ca. 2.7-3.2 meters.

[25] Albani, "Horoscopes in the Qumran Scrolls," 312, suggests that the entire text of *4QZodiacal Physiognomy* could have contained 108 physiognomic types (9×12). But if one assumes that one part of a zodiacal sign must at least be visible above the eastern horizon for it to radiate its influence one would rather expect the number of 96 kinds of physiognomic types (8×12). It makes no sense to draw a horoscope based on a sign, or part of it, if none of its nine parts has yet ascended. See Chapter Three for a more detailed discussion.

[26] According to Tov's classification a scroll with a medium-sized writing block, cf. Tov, *Scribal Practices*, 86-88.

If one limits the number of accounts per column to two, the height measurements are lowered and the scroll lengthened. For two accounts the number of lines per column would be between 14 and 16, the column height ca. 10-11 cm, and the leather height ca. 13.5-14.5 cm.[27] The number of columns needed for 96 accounts would then be 48! Allowing for a column width of ca. 8-9 cm together with a margin of ca. 0.5-1 cm, *4QZodiacal Physiognomy* would then have been a scroll of ca. 4.1-4.8 meters. This seems very long. Maybe an average of between two and three entries is better. Another possibility is that not all accounts were individually listed, but this would diminish the usefulness of the text. Be that as it may, these measurements give a good impression of the material consequences of the ascendant interpretation.

Writing, Paleography, and Date

The copying of the manuscript has been executed in a fine hand. The writing gives the impression that trouble was taken with it, not strange considering the reversed direction of writing.

The way that *4QZodiacal Physiognomy* is written is unique among the Dead Sea Scrolls. The lines as well as the words have to be read in reverse order from left to right. Another scribal peculiarity is that the scribe wrote the medial instead of the final form of the letters *kap*, *mem*, and *nun* at the end of a word.[28] The reason for this is not entirely clear. One might suppose that although the text is basically read from left to right, the scribe actually wrote in the usual way from right to left. But this seems unlikely since, first of all, the flush left is straight which suggests that the scribe started writing there. Secondly, there are no ligatures in the manuscript where this may have been expected had the direction of writing been right to left.[29] Finally, had the direction of writing been right to left, the scribe could still have written final forms at the beginning of words read from left to right, which is not the case.[30]

Only two words have to be read in the regular order from right to left. In 4Q186 1 ii 2 the scribe wrote the words אבן צונם ("a granite stone") from

[27] According to Tov's classification a scroll with a small or medium-sized writing block, cf. Tov, *Scribal Practices*, 84-88.

[28] 4Q186 1 i 9: מעורבימ;1 ii 6: מנ; 1 ii 8: החושכ; 1 iv 7: תוכ; 2 i 1: בינ; 2 i 3: ארוכ.

[29] 4Q186 1 i 9: מעורבים; 1 ii 5: ארוכות; 1 iii 7: עבות; 1 iii 9: ש]מ֯ונה; 1 iv 7: תוכ; 2 i 1: בין; 2 i 3: סרכמה and ארוך.

[30] 4Q186 1 i 9: מעורבים; 1 ii 6: מן; 1 iii 6: מיראות; 1 iii 9: מבית; 2 i 2: ממע֯נט; 2 i 4: ממולד֯ו֯; 4 2: מולדו.

right to left.[31] Also, in this case the copyist did make use of the final forms of the characters.[32]

In addition to the inverted writing, the manuscript exhibits another distinct scribal feature, namely the use of mixed scripts.[33] Characters from other scripts were used alongside the usual square script: paleo-Hebrew, Greek, and cryptic letters. The paleo-Hebrew letters in the extant text are *gimel*, *he*, *waw*, *ḥet*, *yod*, *lamed*, *mem*, *nun*, *samek*, *ṣade*, *reš*, *šin*, and *taw*.[34] The use of paleo-Hebrew *taw* is salient. It occurs frequently and only in final position. Out of the 34 words ending in *taw*, 11 are written entirely in square characters;[35] 18 are written in square characters except for the final *taw*, which is in paleo-Hebrew;[36] and 4 are written entirely in non-square characters.[37] Two Greek letters were used, i.e. *alpha* and *beta*.[38] Only one Cryptic A letter seems to appear in the remaining fragments, namely *yod*,[39] but it is also possible that it represents a simplified paleo-Hebrew *yod*. Finally, there is one letter that cannot be identified. In 4Q186 1 iii 4 a letter occurs, the third one from left, which many scholars have read as a Greek *beta*, but this is extremely unlikely. Its exact identification remains unknown.[40] Although difficult to detect a "cryptographic" system, it is interesting to note that when non-square characters are used the entire word is written in non-square characters.[41]

[31] יפי in 4Q186 3 3 was perhaps also written from right to left, because of the ligature.

[32] Carmignac, "Les Horoscopes," 199 n. 1, argues that this occurrence of the regular order of writing and the final form of characters suggests that it was the scribe of *4QZodiacal Physiognomy* who initiated the inverted writing and that this was not yet the way that the manuscript he copied was written. But this is far from clear (see below the comments on 4Q186 1 ii 2).

[33] There is only one other manuscript from Qumran that perhaps has mixed scripts in the running text, namely *4QcryptC Unidentified Religious Text* (4Q363a). The text uses the paleo-Hebrew script (note also the dots that function as word dividers) and characters unknown to us that have been called Cryptic C script. Only a photograph has been published of 4Q363a, see M. Bernstein et al., *Qumran Cave 4.XXVIII: Miscellanea, Part 2* (DJD 28; Oxford: Clarendon, 2001), Plate XLIII. However, whether the scribe who wrote them also regarded these unknown characters as cryptic is not clear, nor whether he regarded them to be two distinct scripts.

[34] 4Q186 1 i 8: רחבים and סגלגלים; 1 i 9: רׄושׄ[ו]; 1 ii 4: נׄצרׄ[ו]ת; 1 ii 7: האור; 1 ii 8: החׄושך; 1 iii 4: וא·ה; 1 iii 8: בבית; 2 i 2: ממעׄ[נ]ט.

[35] 4Q186 1 ii 6: דקות; 1 iii 7: עבות, ומלאות, and לאחת; 1 iii 8: ואצבעות and עבות; 1 iii 9: ואחת; 2 i 2: ובת; 2 i 5: וכפות; 3 1: מבית̊; 6 2: זות.

[36] 4Q186 1 ii 5: ארוכות, ודקות, and ואצבעות; 1 ii 6: וארוכות; 1 iii 6: מיראות, רומות, and ואצבעות; 1 iii 8: וקצרׄות; 1 iii 9: מבית; 2 i 1: שחורות and מׄנׄמׄריות; 2 i 3: דקות and ויושבות; 2 i 4: אצבעות and דקות; 2 i 5: וארׄו[כ]ות and חלקות; 2 i 6: יושבות.

[37] 4Q186 1 ii 4: נׄצרׄ[ו]ת; 1 ii 7: בבית (twice); 1 iii 8: בבית.

[38] 4Q186 1 i 8: רחבים; 1 ii 7: בבית (twice); 1 ii 7: האור; 1 iii 4: וא·ה; 1 iii 8: בבית. The one remaining instance of *taw* in final position, 4Q186 1 ii 4: (ו)אר[ו]כׄות, cannot be classified.

[39] 4Q186 1 ii 7: בבית ("in the house"). Cf. Pfann, "Introduction," 528.

[40] See the comments on 4Q186 1 iii 4 in Appendix I.

[41] Exceptions, however, are 4Q186 1 iii 7: בבית̊; 1 iii 8: החׄושך; 2 i 2: ממעׄ[נ]ט; 4 2: מולדו.

The square script of *4QZodiacal Physiognomy* can be characterized, according to the typology of Frank Moore Cross, as a Herodian "Round" semiformal hand, showing both earlier and more developed forms.[42] *ʾAlep* has a right arm thickened at the top, which develops into a *serif*,[43] and the left leg is bent to the right. *Bet* still has its tick at the right, upper shoulder in some cases,[44] but is losing it in others.[45] *Gimel* has a right down stroke that is gently curved and bent to the right at the top, without *serif*, and to the left at the bottom. The left leg is connected low on the right down stroke. *Dalet* has an "s"-shaped right leg characteristic of the semiformal style.[46] *He* has a crossbar or roof that is thick and heavily shaded. Also, at the top of the right down stroke above the roof a small projection to the right appears. *Waw* and *yod* are not always easily distinguished. Sometimes *yod* is longer than *waw*. *Zayin* has both forms that appear in the early Herodian style: a simple stroke thickened or slightly bent to the right at the top,[47] and a doubly curved down stroke.[48] *Ḥet* has a right leg curved inward, and a crossbar set rather low in some cases. *Ṭet* is broad and squat, tending more to the developed Herodian formal script. *Kap* has a down stroke that curves outward to the right, and the base is sometimes rather broad.[49] *Lamed* has a large, sometimes rounded hook. *Mem* conforms to the late Hasmonean style according to which the left oblique is penned last. *Nun* appears with a down stroke bent to the right and thickened at the top, but without *serif*.[50] *Samek* is fully closed. *ʿAyin* has a right down stroke that is sometime thickened.[51] *Pe* has a sharp head. *Ṣade* appears with a left arm that characteristically curves inward to the right at the top,[52] and the right arm is bent up and thickened at the tip.[53] *Qop* has two forms of the down stroke: straight,[54] and "s"-shaped, resembling that of *dalet*.[55] *Reš* has some variation in the width of the head. *Šin* has a left down stroke that

[42] See F.M. Cross, Jr., "The Development of the Jewish Script," in *The Bible and the Ancient Near East: Essays in Honor of William Foxwell Albright* (ed. G.E. Wright; Garden City, New York: Doubleday, 1961), 133-202, at 173-81.

[43] 4Q186 1 iii 6: לאבר.

[44] 4Q186 1 i 9: מעורבים; 1 ii 9: ברגל; 1 iii 6: לאבר; 2 i 6: .

[45] 4Q186 1 iii 7: עבות; 3 2: בכתפ֯ני.

[46] 4Q186 1 ii 5: ודקות; 1 ii 8: המולד.

[47] 4Q186 2 i 1: וזקנו; 6 2: זות.

[48] 4Q186 1 ii 8: וזה.

[49] Perhaps implying a final form? Cf. 4Q186 1 ii 8: החושכ; 1 iv 7: תוכ; 2 i 3: ארוכ. The base is broad when the letter is in final position. This shows that the scribe did have the tendency to give some letters different forms when they are in word-final or line-final position.

[50] 4Q186 1 ii 2: צונם; 2 i 2: ענוה.

[51] 4Q186 1 i 9: מעורבים; 1 iii 8: עבות; 2 i 3: על.

[52] 4Q186 2 i 4: קצר.

[53] 4Q186 1 iii 8: ואצבעות.

[54] 4Q186 2 i 5: חלקות.

[55] 4Q186 1 ii 6: דקות; 2 i 2: קולו.

continues below the right arm, and the middle arm is gently curved. *Taw* was not drawn in a continuous stroke, and is not yet squat and broad.

On paleographic grounds a date for the manuscript of *4QZodiacal Physiognomy* between ca. 30 BCE–20 CE, according to the Cross dating, seems probable.[56]

Contents

4QZodiacal Physiognomy lists different accounts that, as far as can be observed from the extant fragments, consist of three set elements at least.

First, the accounts contain the physiognomic descriptions of "ideal types" of individuals. These descriptions of the human body are structured according to the *a capite ad calcem* principle, i.e. they run from head to toe.

Second, they register a division of numbers with regard to the person's רוח ("spirit") in the "house of light" and the "house of darkness." It is possible that this division is made on a nine-point scale, but this is neither certain nor necessary.[57]

Third, the different sections provide certain zodiacal information with regard to the moment of birth of the aforementioned person: his horoscope (מולד).

In addition, the text lists predictions concerning the described type's future state or gives an indication of people's characters. But, due to the fragmentary state of the manuscript, this can only be verified for one type of description (4Q186 1 ii 9: עני יהיה: "he will be poor," or ענו יהיה, "he will be humble").

Although one should allow for the possibility that the words העמוד השני ("the second column") in the phrase והואה מן העמודהשני ("and he is from the second column")[58] are a set element,[59] it is questionable whether the phrase represents a fixed part of the text. Only in 4Q186 1 ii 6 is its position clearly set between the physiognomic description and the part concerning the "house of light" and the "house of darkness." This is certainly not the case in 4Q186 1 iii and 4Q186 2 i where the words ורוח לו ("and there is a spirit for him"), introducing the part concerning the "house of light" and the "house of darkness," follow immediately upon the physiognomic sections. This means that the phrase העמוד השני ("the second column"), whatever its

[56] Matthias Delcor mentions in passing that Joseph Milik and Jean Starcky suggested to him a date in the second half of the first century BCE for the manuscript. See Delcor, "Recherches sur un horoscope," 319 n. 36.

[57] See Chapter Three for a detailed discussion of the ascendant interpretation.

[58] 4Q186 4 1 only has ה[עֹ]מוד השני, but it is likely that the same phrase is implied as in 4Q186 1 ii 6. See also 4Q186 6 2: העמוד ה[שֹׁ]ני.

[59] Cf. Schmidt, "Astrologie juive ancienne," 126; Albani, "Horoscopes in the Qumran Scrolls," 284.

exact sense may be, is perhaps optional in the physiognomic accounts in *4QZodiacal Physiognomy*.

Finally, it is possible that the different accounts in *4QZodiacal Physiognomy* listed certain stones in relation to a person's physiognomic traits and zodiacal sign. Both Babylonian and Greco-Roman astrology were familiar with various connections between the zodiacal signs and stones. Perhaps the stones mentioned in *4QZodiacal Physiognomy* were deemed to have some sort of magical powers. It is clear that a stone (אבן צונם) is listed in 4Q186 1 ii 2, but its exact sense is uncertain due to the fragmentary context.

Transcription and Translation

In the following transcription the presentation of the text has already been adapted to the regular manner of reading from right to left. For notes and comments on readings see Appendix I.

4Q186 1 i

4 הׄ·[
5 הׄ[
6 *vacat*]
7 ואיש אשר יהיה קׄ[
8 רחבים<ו>סגלגלים]
9 מעורבים ולוא שאר רׄוׄשׄ[ו

4. …[
5. …[
6. *blank* [
7. And someone (whose) … will be[
8. broad <and> rounded [
9. mixed, but the rest of [his] head is not [

4Q186 1 ii

1]]ג טמא
2]]אבן צונם
3] [איש עוׄ]ר
4] (ו)אר]וׄכׄות ו/יהׄ[]·· נׄצרׄ[ו]ת
5 ושוקיו ארוכות ודקות ואצבעות רגליו
6 דקות וארוכות והואה מן העמוד השני
7 רוח לו בבית האור שש ושלוש בביתׄ
8 החׄושך וזה הואה המולד אש⌉הואה ילוד עליו
9 ברגל השור ענו יהיה וזה בהמתו שור

1. [ ]… unclean
2. [ ] a granite stone
3. [ ] a bli[nd (?)] man

4. (and) lo]ng, …[…]… sec[re]t parts (?)
5. and his thighs are long and slender, and his toes are
6. slender and long. And he is from the second column.
7. There is a spirit for him in the house of light (of) six (parts), and three (parts) in the house of
8. darkness. And this is the horoscope under which he was born:
9. in the foot of *Taurus*. He will be humble, and this is his animal: *Taurus*.

4Q186 1 iii

4 וא·ה[ ]
5 וראושו [ ]··[ ]
6 מיראות [ ] ושניו רומות לאבר ואצבעות
7 ודיו <ע>בות ושוקיו עבות ומלאות ש̇ער לאחת
8 ואצבעות רגליו עבות וקצר̇ות ורוח לו בבית
9 [החושך ש]מ̇ונה ואחת מבית האור ו··ש

4. and… [ ]
5. and his head [ ]…[ ]
6. terrifying [ ] and his teeth are protruding. And the fingers of
7. his hands are <th>ick, and his thighs are thick and each one is hairy.
8. His toes are thick and short. And there is a spirit for him in the house of
9. [darkness (of) ei]ght (parts), and one (part) from the house of light. And…

4Q186 1 iv

6 [ ]שמה אלה
7 [ ]יהיה תוך
8 [ ]ל

6. [ ]there. These
7. [ ](he/it) is inside
8. [ ]…

4Q186 2 i

[ ]על[
1 סרכ̇מ̊[ה וע]̊יניו בין שחורות ובנ[ין]מ̊נ̇מריות וזקנו
2 ממע̊[ט]והואה תרגל ובת קולו ענוה ושניו
3 דקות ויושבות על סרכמה והואה לוא ארוך
4 ולוא קצר והואה̊[ ]ממולד̊ו̊[ ]אצבעות ידיו דקות
5 וא̊ר̊ו[כ]ות ושוקיו חלקות וכפות רגליו
6 [ואצבעות רג]ל[יו] יושבות על סרכמה ורוח ל[ו]
7 [ ]ס̊/ע̊/ש̇ה̇ וא[ ]

8 [ ]ד הו̇[אה]
9 [ ]··[ ]

[…]
1. [are] well ordered. [And] his [ey]es are between black and speckled (?). And his beard
2. is sp[arse]and it is wavy (?). And the sound of his voice is kind. And his teeth
3. are fine and well ordered. And he is neither tall
4. nor short, and that[]because of his horoscope.[]His fingers are slender
5. and lo[n]g, and his thighs are smooth. And the soles of his feet
6. [and the toes of his]f[eet] are well ordered. And there is a spirit for [him]
7. [ ]… and…[ ]
8. [ ]… h[e]
9. [ ]…[ ]

4Q186 2 ii

4 [ ]·
5 [ ]הוא
6 [מ[עו̇רב
7 [ ]שלוג/גולש

4. [ ]…
5. [ ]he/it
6. [m]ixed
7. [ ]fair/flowing

4Q186 3

1 מבית̊[האור/החושך וזה (הואה) המולד אשר הואה ילוד עליו]
2 בכתפ̊[י
3 []·[] יפי[

1. from the house[of light/darkness. And this is the horoscope under which he was born:]
2. in the shoulder[s of…
3. []…[] beautiful[

4Q186 4

1 והואה מן ה[ע̊מוד השני ש[
2] מולדו [
3 וזה ה[ואה בה<ב>(מ)ת̊[ו

1. and he is from the] second column …[
2.] his horoscope [
3. and th]at is [his] animal[

4Q186 5

1]מון[
2]יׄ לוׄ[
3]יׄ [

4Q186 6

1]··[
2 והואה מן העמוד ה]שׁׄני זות[
3]·גׄ/נׄהׄ[]לׄ[

1.]…[
2. and he is from the] second [column]. This[
3.]…[

THE STRUCTURE OF THE TEXT AND THE PRIMACY OF PHYSIOGNOMICS

The contents of *4QZodiacal Physiognomy* appear to be ordered according to a specific sequence of elements, which supports the inference that the text as a whole is structured according to physiognomic descriptions.

Many scholars regard *4QZodiacal Physiognomy* as an astrological text that deals with the influence of the stars on the human body and spirit. Albani explicitly states that it is "a list or compilation of options for astrological interpretations systematically arranged according to certain astrological criteria," and as such "an auxiliary astrological resource for creating horoscopic prognostications."[60] In another recent explanation Schmidt argues that *4QZodiacal Physiognomy* provides predictions regarding the physiognomic and spiritual character of categories of individuals on the basis of the date of conception.[61] However, like Albani, he does not explain how the text is structured to facilitate access to this kind of information.

Given the text's structure and the sequence of its different elements, the understanding of *4QZodiacal Physiognomy* as a primarily astrological auxiliary text arranged according to certain astrological criteria cannot be maintained.

It is evident that, first, the descriptions of the physical characteristics are given, next, the division of light and darkness with regard to the person's רוח ("spirit") is considered, and only then does the text mention the individual's horoscope (מולד) and zodiacal sign. This is at least the case in 4Q186 1 ii, where ll.5–6 describe the bodily features, ll.7–8 list the numbers in the "house of light" and the "house of darkness", and ll.8–9, finally, provide astrological information. 4Q186 1 iii and 2 i demonstrate that the

60 Albani, "Horoscopes in the Qumran Scrolls," 309.
61 Schmidt, "Astrologie juive ancienne."

physiognomic descriptions precede the references to the "house of light" and the "house of darkness." It is therefore likely that here, too, the astrological element followed subsequently in the original text.

If the astrological interpretation of עמוד ("column") as a reference to a zodiacal sign is accepted, then 4Q186 1 ii provides astrological information prior to the listed numbers in the "house of light" and the "house of darkness," but still after the physiognomic descriptions. The position of עמוד in *4QZodiacal Physiognomy*, however, is variable. In 4Q186 1 iii and 2 i it does not occur between the descriptions of the human body and the statement concerning the person's רוח ("spirit").[62] It may have stood more towards the end of those accounts. One should perhaps allow for the possibility that the reference to עמוד can also occur at the beginning of each account, thereby securing the idea that *4QZodiacal Physiognomy* is ordered according to astrological criteria. However, the phrase ואיש אשר יהיה קׄן ("And someone (whose) … will be[") in 4Q186 1 i 7 provides the best indication of the way sections begin in *4QZodiacal Physiognomy*.[63] Variation is possible, but it is unlikely that (ו)הואה מן העמוד השני ("[And] he is from the second column"), or any other numbered "column" up to twelve if it is a reference to zodiacal signs, represents the opening of a typological account. Furthermore, it seems superfluous to have astrological data mentioned twice in one account; once at the beginning and again at the end.

The sequence of both elements, first physiognomics and then astrology, has consequences for our understanding of the relationship between physiognomics and astrology in *4QZodiacal Physiognomy* and of the sense of this text in general.

The text's structure suggests that physiognomic learning is of primary importance. *4QZodiacal Physiognomy* is a text that is structured according to certain physiognomic criteria. The different accounts in the catalogue begin with descriptions of the human body. They describe the appearance and shape of the different parts of the human body from head to toe, following the *a capite ad calcem* principle. These descriptions of the human body lead the reader to further information regarding the division of the person's רוח ("spirit") between the "house of light" and the "house of darkness," his horoscope, and zodiacal sign.

4QZodiacal Physiognomy is not an example of the divinatory use of astrology to discern the human body and spirit, but rather the other way around. The text is an example of the divinatory use of physiognomics to determine people's horoscope and the nature of the zodiacal spirit of their horoscope. *4QZodiacal Physiognomy* acknowledges, on the one hand, the

[62] See below, the section on "the second column" (העמוד השני).
[63] See below for the reconstruction of the beginning of the physiognomic entries.

influence of the zodiac on bodily form and appearance,[64] but, on the other hand, the text's structure reveals its direction of interest. This observation is crucial for grasping the sense of *4QZodiacal Physiognomy*. The descriptions of the human body point the reader to certain zodiacal information concerning the type of person described. Thus, that the individual in 4Q186 1 ii is born "in the foot of *Taurus*" (ברגל השור) can be learned from the appearance of his body, that being a consequence of the astral influences at the time of his birth. *4QZodiacal Physiognomy* assumes that the physical traits of people are known and that on the basis of these one can discern a person's ruling zodiacal sign.[65] In this regard it is a truly physiognomic text, because the reader gains astrological knowledge about people on the basis of physiognomic criteria. It should, therefore, be characterized as a physiognomic-astrological catalogue.

RECONSTRUCTING THE BEGINNING OF THE PHYSIOGNOMIC ENTRIES

The extant text of *4QZodiacal Physiognomy* does not preserve an entire account from beginning to end. This impedes any modern understanding of the text. Fortunately, the first column of 4Q186 1 provides in l.7 a clue regarding the form according to which the entries began. Together with the following lines that preserve descriptive words for body parts, this suggests that in the original text of *4QZodiacal Physiognomy* the accounts began with physiognomic descriptions. Apart from the explicit mention of the head in 4Q186 1 i 9, there are no further concrete references to the body. It is, however, probable that the descriptive words in ll.8–9 describe certain aspects of a type's face and head. 4Q186 1 i thus nicely follows the *a capite ad calcem* principle (from head to toe) by beginning the account with a description of certain parts of the face and head.

4Q186 1 i 7: Start of an Entry

4Q186 1 i 7, below the previous blank line, certainly represents the entry for another typological description. The remaining letter *qop*, before the manuscript breaks off, hardly reveals the nature of the data presented in this opening line. I suggest two possibilities.

First, the account began with a certain part of the body. As the bodily descriptions seem to be structured *a capite ad calcem*, the account in 4Q186 1 i 7 might have begun with the head or crown of the head (ואיש אשר יהיה

[64] Cf. 4Q186 2 i 4: והואה֯[]ממולדו֯ ("and that[]because of his horoscope").

[65] Cf. also L. Ness, *Astrology and Judaism in Late Antiquity* (Diss. Oxford, Ohio, 1990), http://www.smoe.org/arcana/diss4.html; Alexander, "Physiognonomy," 386; M.O. Wise, "A Horoscope Written in Code (4Q186)," in *The Dead Sea Scrolls: A New Translation* (eds. M.O Wise, M.G. Abegg, Jr. and E.M. Cook; New York: HarperCollins, 2005), 275-78, at 276.

קֹ]נדקדו, "And someone whose head…"), with a qualification of the described person as being bald (קֹ]נרח), or with a description of the tone of his voice (קֹ]נולו). But it is also possible that the description began with a general bodily characterization about the individual's height (קֹ]נומתו), or a more direct qualification, such as that the person is small (קֹ]נטן), short (קֹ]נצר), or has a dark complexion (קֹ]נדר).

A second possibility is that the account opened not with a physical description, but with a character trait, such as that a jealous (ואיש אשר יהיה קֹ]נא) or angry (קֹ]נצף) person was the subject of the typological account. The first treatise of the pseudo-Aristotelian text *Physiognomonica* has a catalogue that lists twenty-two physiognomic accounts according to character. But, in connection with the astrological element and the ascendant interpretation, it is unlikely that *4QZodiacal Physiognomy* listed as many as 96 character types.[66] In Greek physiognomics, for example, one can observe the elaboration of physiognomic distinctions between the pseudo-Aristotelian *Physiognomonica* and Polemo's physiognomic treatise, but the number of set character types does not increase concomitantly.[67]

Regarding the first possibility there is interesting comparative material in a later medieval Jewish text, *The Book of the Reading of the Hands by an Indian Sage*.[68] This physiognomic text lists various physical characteristics. It is primarily concerned with divining the lines on the palm of the hand (chiromancy or palmistry), but the text also devotes attention to other parts of the body. The form in which it introduces some of its physiognomic data is similar to the phrase in 4Q186 1 i 7. An almost exact stylistic parallel for the phrase ואישאשריהיהקֹ] is the following sentence: ודע והבן כל איש אשר יהיה בכף רגלו כתואר יין אדום או כתואר שלג ("Know and understand that everyone whose foot sole has an appearance like red wine or an appearance like snow"). Other parts of the human body are introduced in a similar form; for example, the head and forehead (וכל אשר יהיה ראשו ומצחו יבשים, "And everyone whose head and forehead will be dry"), or the eyes (וכל אשר יהיה יפה עינים, "And everyone who will have beautiful eyes").[69] These descriptions are followed by predictions, such as that one will fare well and be loved.

It is possible to understand the sentence ואיש אשר יהיה קֹ] in 4Q186 1 i 7 as a *protasis* in the sense of "and if someone (whose) … will be, then…," expressing the conditional clause similar to that used in Babylonian omen lists. Support for such an understanding of the beginning of an entry in

[66] This does not mean, of course, that statements concerning people's character do not occur in *4QZodiacal Physiognomy*. See the comments on 4Q186 1 ii for two possible characterizations.

[67] See Chapter Two.

[68] See Scholem, "Physiognomy," 488-92.

[69] Scholem, "Physiognomy," 491.21, 8, 13.

4QZodiacal Physiognomy comes from material in the Qumran penal codes, where the construction (ו)אישאשר or (ו)האיש אשר is used to introduce the protasis.[70] If the entries in *4QZodiacal Physiognomy* opened with protases such as in 4Q186 1 i 7, this does not necessarily mean that the text was structured in the same manner as the Babylonian physiognomic omen compendium. It is likely that this phrase introduced a string of physiognomic descriptions such as one finds in 4Q186 1 ii, iii, and 2 i,[71] not just of one body part. Furthermore, the question is where the apodosis would have started, with the element following the physiognomic part, such as the reference to the "second column" (4Q186 1 ii), or with the spirit in the "house of light" and the "house of darkness" (4Q186 1 iii; 2 i)?[72]

4Q186 1 i 8–9: Describing Face and Head

It is plausible that 4Q186 1 i 7 starts with a description of the human body. There is additional evidence in the expressions רחבים ("broad"), סגלגלים ("rounded"), and מעורבים ("mixed") used in the following lines of 4Q186 1 i. These expressions are used in later Jewish (physiognomic) texts to describe various parts of the human body, making it likely that 4Q186 1 i 7 concerns an opening description of the human body that was continued in 4Q186 1 i 8–9.

The root רחב ("to be broad") can be used to describe different parts of the body. In *b. Šabb.* 30b–31a two men decide to wager on who can make Hillel angry. One of them goes up to Hillel three times and asks him three questions. His third question for Hillel is why the feet of Africans are wide (מפני מה רגליהם של אפרקיים רחבות). In *b. Ned.* 66b a man is willing to accept his wife only if she has not shown anything of her body that is beautiful to R. Ishmael. The answer to the question by R. Ishmael whether the wife's feet are beautiful is that they are as broad as those of a duck (שמא רגליה נאות? רחבות כשל אווזא). In the medieval text *The Physiognomy of R. Ishmael*, רחב refers to the eyebrows (גבות עיניו רחבות), the tip of the nose (ראש חוטמו [ ] ורחב), the nose holes (נקובי חוטמו רחבים), the ears (אזניו רחבות), and the shape of the face (צורת פניו [ ]לארחבה).[73] *The Book of the Reading of the Hands by an Indian Sage* describes someone's shoulders as broad (כתיפיו רחבות).[74]

[70] See 1QS 6:12.24; 7:4.13.15.17.18.22; 4Q270 i 12. Cf. also CD 9:9; 1QM 7:5–6.

[71] Cf. also Scholem, "Physiognomy," 491.8–9.

[72] See the section on the textual format of physiognomic descriptions in Chapter Two.

[73] Scholem, "Physiognomy," 480.10, 486.77–78, 81, 91–92; Scholem, "Fragment," 182 §3, 186 §§34, 36, 42.

[74] Scholem, "Physiognomy," 491.10.

The word סגלגל from the root סגל ("to be round") can be used as a description of the head.[75] In *b. Šabb.* 30b–31a, the first question posed to Hillel is why the heads of the Babylonians are round (מפני מה ראשיהן של בבליים סגלגלות). In *b. Ned.* 66b R. Ishmael asks whether the wife's head is beautiful, but the answer is that it is round (שמא ראשה נאה? אמרו לו סגלגל). But סגלגל ("round") can also refer more specifically to the eye. A physiognomic text from the Cairo Genizah portrays a person's eye as זגלגל ("round").[76]

Given these other occurrences of רחב ("broad") and סגלגל ("round"), I suggest that 4Q186 1 i 8 describes an individual's face, or perhaps even more specifically his eyes, as broad and round.[77]

The final line of 4Q186 1 i describes part of the head with the word מעורבים. There are two possible interpretations. First, מעורבים is a *pu'al* participle derived from the rootערב, "to be sweet, pleasing."[78] Second, it is a *pu'al* participle from ערב, "to mix, confuse."[79] In both possibilities it can qualify a bodily feature.[80] I follow the second interpretation. Interestingly, in *4QPhysiognomy ar* the passive participle מערבין occurs too, and it is also followed by the negative adverb ולא. In this Aramaic text the sense "mixed,

75 See also M.R. Lehmann, "New Light on Astrology in Qumran and the Talmud," *RevQ* 8/32 (1975): 599-602, at 601-2.

76 T.-S. K 21.88 1/ב 22, see Gruenwald, "Jewish Physiognomic," 314.

77 Cf. also Delcor, "Recherches sur un horoscope," 299. Unfortunately, no body part is mentioned in the extant text in 4Q561 5 3 where סגלגל also occurs, but it is possible that it refers to the eye.

78 Allegro, DJD 5.90; Delcor, "Recherches sur un horoscope," 299, 309; F. García Martínez and E.J.C. Tigchelaar, *The Dead Sea Scrolls Study Edition: Volume 1 (1Q1-4Q273)* (Leiden: E.J. Brill and Grand Rapids, Michigan: Eerdmans, 2000), 381; B. Ejrnæs, S. Holst and M. Müller (eds.), *Dødehavsskrifterne og de antikke kilder om essæerne* (2d ed.; Copenhagen: ANIS, 2003), 496.

79 R. Gordis, "A Document in Code from Qumran – Some Observations," *JSS* 11 (1966): 37-39, at 39; J. Maier, *Die Qumran-Essener: Die Texte vom Toten Meer: Band II: Die Texte der Höhle 4* (vol. 2; Munich: Ernst Reinhardt, 1995), 135; Wise, "Horoscope Written in Code," 277.

80 This is not the case with Carmignac's interpretation. He understands מעורבים as the dual of the noun עֶרֶב, "evening", preceded by the preposition מן ("depuis (la) soirée"), but he does not explain its meaning in this context. Perhaps Carmignac had some sort of astrological data in mind, but this remains unclear. Carmignac, "Les Horoscopes," 201, 205 (cf. in Appendix I his suggestion to reconstruct י(ו)ר]ח שאר at the end of this line, "since (the) evening, but not the rest of the moo[n (mon[th)."

מערב can also be found in the medieval astrological text *Baraita de-Mazzalot*, *The External Tractate of the Zodiacal Signs* (ברייתא דמזלות), see for example in section two (S.A. Wertheimer and A.J. Wertheimer [eds.], *Batei Midrashot: Twenty-Five Midrashim Published for the First Time from Manuscripts Discovered in the Genizoth of Jerusalem and Egypt* [2d enl. and am. ed.; vol. 2; Jerusalem: Ktab Wasepher, 1968], 13-14). This section deals with the parallel rising in the east and the setting in the west of the zodiacal signs, for example, "in the case of *Aries* rising from the east, *Libra* sets in the west" (בזמן שהטלה עלה מן המזרח מאזנים שקע מערב). It is, however, difficult to understand Carmignac's proposal of a dual form of עֶרֶב given the use of מערב in *Baraita de-Mazzalot*. Carmignac's interpretation, therefore, still lacks a clear astrological understanding.

confused" is clearly intended. The word is followed by the qualification that it is "not too much" (ולא שגיא), hardly said of something if the sense were "pleasing."[81] In *The Book of the Reading of the Hands by an Indian Sage* the use of ערב in the sense of "to be sweet, pleasing" is also attested. A person's voice can be sweet and pleasant (ואותו שקול ערב ומתוק).[82] Based on these two occurrences of ערב in physiognomic texts, I take the sense of מעורבים in 4Q186 1 i 9 to be "mixed." 4Q186 1 i 9 probably continues the description of a certain feature or part of the head as being of mixed character, but at the same time stresses the limited extent of this; the rest of the person's head should not show this characteristic.

"The Second Column"

In 4Q186 1 ii 6 the phrase והואה מן העמוד השני ("And he is from the second column") occurs,[83] the exact sense of which is difficult to ascertain. The subject of הואה ("he") is the described person in the preceding physiognomic account, referred to by the suffixes attached to the different parts of the body.[84] No new subject has been introduced. It is, therefore, probable that it is the described individual who is "from the second column," whatever the meaning of this sentence might be. Furthermore, it is evident that a new element is introduced following the physiognomic description. The conjunctive-*waw* in והואה ("And he") makes this clear.[85]

The sense of העמוד השני ("the second column") cannot be reconciled with any of the various meanings of עמוד ("column") in the Hebrew Bible, the Dead Sea Scrolls, or rabbinic literature.[86] Most scholars translate עמוד with "column," but there is great divergence between the various interpretations regarding the specific sense of the terminology used in *4QZodiacal Physiognomy*. A distinction must be made between astrological and non-astrological interpretations. There are basically three astrological explana-

[81] See 4Q561 1 i 1.

[82] Scholem, "Physiognomy," 491.9. But the first interpretation is confirmed also by this same text, and, even more, in exactly the same form as in 4Q186 1 i 9 (in another manuscript of this text Scholem notes the reading מעורבים). It is said that the lines on the palm of the hand are mixed, that they are intertwined with each other (וכל אשר יהיה כפו מעורבבים אלו עם אלו), see Scholem, "Physiognomy," 489.12.

[83] Another small fragment, 4Q186 4, bearing]עֹמוד השני has been wrongly joined by Allegro (DJD 5.90-91) to 4Q186 2 i.

[84] 4Q186 1 ii 5: ושוקיו ("his thighs") and ואצבעות רגליו ("his toes").

[85] In 4Q186 1 iii 8 and 4Q186 2 i 6 another element subsequent to the physiognomic parts is introduced in the same manner by the use of conjunctive-*waw* in ורוח לו ("and there is a spirit for him"). At the same time, these latter two examples show that the words העמוד השני do not occupy a fixed position in the sequence in *4QZodiacal Physiognomy*.

[86] See the notes and comments on 4Q186 1 ii in Appendix I.

tions and one non-astrological understanding for the terminology העמוד השני.

עמוד as a Reference to a Zodiacal Sign

The most widely held interpretation is to connect העמוד השני ("the second column") somehow with the zodiacal sign *Taurus*. The term עמוד ("column") is then a reference either to a sign of the zodiac or to the space between the zodiacal signs in the form of a column or pillar.[87]

It is evident that this interpretation of עמוד ("column") is dependent on the relationship assumed between the ordinal שני ("second") and the zodiacal sign *Taurus*. The most common starting point of the zodiacal circle in antiquity seems to have been *Aries*.[88] In all sorts of astrological lists the sequence starts with *Aries*.[89] Ptolemy explains that:

[87] Although Carmignac, "Les Horoscopes," 203, is not able to explain the exact sense of עמוד, he suggests that the use of the adjective שני must relate to the fact that the zodiacal sign *Taurus* (שור), mentioned in 4Q186 1 ii 9, is the second sign of the zodiac circle. According to A. Dupont-Sommer, "Deux documents horoscopiques esséniens découverts à Qoumrân, près de la Mer Morte," CRAI (1965), 239-53, at 241-42, the sense of עמוד in *4QZodiacal Physiognomy* is that it designates a division of the celestial circle referring either to a zodiacal sign or one of the thirty-six decans. He comes to the conclusion, like Carmignac, that the use of the adjective שני must be taken as a reference to *Taurus* as the second sign of the zodiac circle. Albani, "Horoscopes in the Qumran Scrolls," 284-85, 309, states that this zodiacal interpretation, according to which עמוד refers to one of the twelve positions of the zodiac circle, is convincing. Furthermore, Strugnell, "Notes en marge du volume V," 276, suggests that עמוד might have been a scribal error for עומד, an astronomical term for "station, position," cf. also Maier, *Texte vom Toten Meer*, 2:135-36; F. García Martínez, *The Dead Sea Scrolls Translated: The Qumran Texts in English* (2d ed.; trans. W.G.E. Watson; Leiden: E.J. Brill and Grand Rapids, Michigan: Eerdmans, 1996), 456.

See also Bergmeier, *Glaube als Gabe*, 78, 80, who, first, argues that *Taurus* is the second station (העומד השני) because the sun passes through the twelve signs of the zodiac in a year that begins with the zodiacal sign *Aries* in the month Nisan, and, second, proposes that the expression "the second station" can be compared with the phrase *secundae stationis* in Firmicus Maternus, *Mathesis* 5.4.10. The two expressions, however, cannot be compared. The terms *statio* and στηριγμός (see below) refer to the two points in the planetary motion in which a planet seems to stand still before beginning retrograde movement (for the first station, *primam stationem*, see also Firmicus Maternus, *Mathesis* 5.3.10 in connection with Saturn's position in *Gemini*). *Taurus* as "the second station" is a different concept not covered by the sense that *statio* has in Firmicus Maternus, where it relates to planetary motion.

Delcor, "Recherches sur un horoscope," 301-2, interprets עמוד as representing the dividing pillar or wall between the signs of the zodiac, visible on ancient iconography of the zodiacal circle. As *Taurus* is the individual's zodiacal sign in 4Q186 1 ii, this means that this person is part of the second pillar, the one separating *Taurus* from the first zodiacal sign *Aries*. Therefore, *4QZodiacal Physiognomy* can say that a certain person belongs to the second column or pillar. According to Delcor, this interpretation of עמוד as the dividing pillar between zodiacal signs would also explain the use of the particle מן in והואה מן העמוד השני as expressing origin. See also H. Lichtenberger, *Studien zum Menschenbild in Texten aus der Qumrangemeinde* (SUNT 15; Göttingen: Vandenhoeck & Ruprecht, 1980), 144-46.

[88] But in Egypt the sign of *Cancer*, marking the summer solstice, commenced the zodiacal circle, see A. Bouché-Leclercq, *L'astrologie grecque* (Paris: Ernest Leroux, 1899), 137 n. 2.

> although there is no natural beginning of the zodiac, since it is a circle, they assume that the sign which begins with the vernal equinox, that of Aries, is the starting-point of them all.[90]

If עמוד ("column") is taken to be a reference to a zodiacal sign, it is not likely that it is the same as a zodiacal sign in a direct manner,[91] but rather a

---

The details of the not yet satisfactorily resolved matter of the selenodromion in 4Q318 (*4QZodiology and Brontology ar*), beginning with *Taurus* as the first sign in the month of Nisan, are further ignored here.

Wise, "Thunder in Gemini," 39-48, suggests that 4Q318 reflects awareness of the concept of precession in that the text represents a *thema mundi* (horoscope of the world) considering the world to have been created some 4000 years previously, when the vernal equinox still occurred in the zodiacal sign *Taurus* and not in *Aries* (the second-century BCE Greek astronomer Hipparchus discovered that the equinoctial points are not fixed, but move to the west, opposite to the order of the zodiacal signs, at a rate of about 1° every century. Because astrologers work with a fixed zodiac, the actual constellations move away from the zodiacal signs to which they gave their name. See O. Neugebauer, *A History of Ancient Mathematical Astronomy* [3 vols; SHMPS 1; Berlin: Springer, 1975], 292-98; S.J. Tester, *A History of Western Astrology* [Woodbridge: Boydell, 1987], 71).

Albani, "Der Zodiakos in 4Q318," 27-35, argues that 4Q318 follows a Babylonian tradition transmitted in the astronomical compendium MUL.APIN ("Plough Star") in which, according to Tablet I iv 31–39, the monthly course of the moon begins with the *Pleiades* and *Taurus* (respectively the "Stars" [MUL.MUL, *Zappu*] and the "Bull of Heaven" [^{mul}GU$_4$.AN.NA, *Alû*]), which were later reduced to the one zodiacal constellation of *Taurus*. For the relevant passage, see H. Hunger and D. Pingree (eds.), *MUL.APIN: An Astronomical Compendium in Cuneiform* (AfO.B 24; Horn, Austria: Ferdinand Berger, 1989), 67-69, 144. The dating of this tradition is controversial, but the editors opt for a date around 1000 BCE, see Hunger and Pingree, *MUL.APIN*, 10-12, to which add the compilation date of around 700 BCE proposed by J. Koch, *Neue Untersuchungen zur Topographie des babylonischen Fixsternhimmels* (Wiesbaden: Otto Harrassowitz, 1989). Greenfield et al., DJD 36.265, find Albani's explanation convincing. Cf. also M.J. Geller, "New Documents from the Dead Sea: Babylonian Science in Aramaic," in *Boundaries of the Ancient Near Eastern World: A Tribute to Cyrus H. Gordon* (eds. M. Lubetski, C. Gottlieb and S. Keller; JSOTSup 273; Sheffield: Sheffield Academic Press, 1998), 224-29, at 224-27.

J. Ben-Dov, *Astronomy and Calendars at Qumran – Sources and Trends* (Diss. Jerusalem, 2005 [in Hebrew]), 213-14, disagrees with Wise's and Albani's interpretations. Regarding Wise's suggestion, Ben-Dov states that it is not reasonable to assume that knowledge of precession is already reflected in a first-century BCE Aramaic document such as 4Q318. Concerning Albani's proposal, Ben-Dov doubts that MUL.APIN dictated the beginning of the zodiac in *Taurus*. The list of seventeen constellations in MUL.APIN through which the moon passes every month does not begin with *Taurus*; it begins with the Pleiades to be followed by the "Bull of Heaven." Both constellations do indeed belong to the zodiacal constellation *Taurus*, but not until, several centuries after the compilation of MUL.APIN, the zodiac was reduced to consist of twelve zodiacal signs, schematically of 30° equal length.

89 The sequence of zodiacal signs is then *Aries*, *Taurus*, *Gemini*, *Cancer*, *Leo*, *Virgo*, *Libra*, *Scorpio*, *Sagittarius*, *Capricorn*, *Aquarius*, and *Pisces*.

90 Ptolemy, *Tetrabiblos* 1.10.2. Translation from F.E. Robbins, *Ptolemy: Tetrabiblos* (LCL 435; Cambridge, Massachusetts: Harvard University Press, 2001), 59-61. The numbering of books, chapters, and paragraphs of Ptolemy's *Tetrabiblos* throughout this study is according to the recent critical edition, W. Hübner (ed.), *Claudius Ptolemaeus: ΑΠΟΤΕΛΕΣΜΑΤΙΚΑ* (vol. 3.1; Teubner; Stuttgart: B.G. Teubner, 1998). See also *Tetrabiblos* 2.11.2–3. Cf. Hephaestion, *Apotelesmatica* 1.1.2.

reference to a position within the zodiac.[92] The use of עמוד in *4QZodiacal Physiognomy* could be understood as another, somewhat complicated, way of saying that someone is under a particular position within the circle of the zodiac beginning with *Aries*.

עמוד as a Reference to a Zodiacal Quadrant

According to the second astrological interpretation, עמוד ("column") is a reference to a zodiacal quadrant.

Schmidt considers it unlikely that *4QZodiacal Physiognomy* would refer to the same astrological phenomenon, i.e., a zodiacal sign, with two different technical terms, i.e. העמוד השני ("the second column") and שור ("*Taurus*").[93] In addition, since the phrase is used at least one more time and probably in another account (4Q186 4), he objects that it cannot refer to the zodiacal sign *Taurus*. The reason is that such an equation implies that different physiognomic types are described under the same zodiacal sign, whereas one would expect differentiation.[94]

Schmidt, therefore, suggests that the term עמוד ("column") refers to a unit of three zodiacal signs, a quadrant. In Schmidt's hypothesis the four quadrants correspond with the four seasons, the first quadrant beginning with the zodiacal sign *Capricorn* (the winter solstice). Consequently, "the second column" (העמוד השני) denotes the second quadrant containing the zodiacal sign *Taurus*. This interpretation answers Schmidt's second objection, because עמוד includes in this sense three signs of the zodiac, which means that different physiognomic types can be discussed with reference to

---

91 It seems more likely that בהמה in 4Q186 1 ii 9, a possible equivalent of the Greek ζῴδιον and the Latin *signum*, represents an attempt to translate the concept "zodiacal sign" directly into Hebrew. See the excursus on בהמה in Chapter Two.

92 Delcor, "Recherches sur un horoscope," 302, refers to the use of *manzāzu* (KI.GUB), "station, position," in Babylonian astronomy, but this does not present an exact analogy for the proposed sense of עמוד in *4QZodiacal Physiognomy*. The term *manzāzu* does not denote one of the twelve positions of the zodiacal circle. It is used to refer to a planetary or stellar position when first seen on a particular night, or, more precisely, to the place on the horizon above which a planet or star rises or sets, see e.g. E. Reiner and D. Pingree, *Babylonian Planetary Omens Part Three* (CM 11; Groningen: STYX, 1998), 18. Cf. also D. Brown, *Mesopotamian Planetary Astronomy-Astrology* (CM 18; Groningen: STYX, 2000), 65, 69, 235.

93 Schmidt, "Astrologie juive ancienne," 128. It is not clear whether Schmidt might mean בהמה ("animal") as the second technical term instead of שור ("*Taurus*").

94 Bergmeier and Albani support the interpretation of העמוד השני as a reference to *Taurus*, but provide another solution to the problem that different physiognomic types were discussed under the same sign. Their solution is based on their interpretation of the words ברגל השור, "in the foot of *Taurus*" (4Q186 1 ii 9), as a specific ecliptical part of the zodiacal sign *Taurus*, suggesting that *4QZodiacal Physiognomy* presupposes a division of the zodiacal signs in which case each particular partition of the zodiacal sign could be linked to a different physiognomic type, thereby creating the possibility for different physiognomic characters to be assigned to the same sign of the zodiac. See Chapter Three.

העמוד השני, while an actual sign can be explicitly mentioned, such as *Taurus* in 4Q186 1 ii 9.[95]

In ancient astrology all sorts of divisions, classifications, and qualities of the zodiac circle were conceived. One of these subdivisions of the zodiac circle grouped the zodiacal signs into quadrants. The quadrants can be connected again with certain qualities, such as the four elements: air, fire, earth, and water, or the ages of life: childhood, youth, adult life, and old age. The quadrants themselves were connected to the seasons: spring, summer, autumn, and winter.[96]

Contrary to the most common arrangement of quadrants beginning with *Aries*, Schmidt's interpretation of "the second column" (העמוד השני) entails that the zodiacal signs from *Aries* to *Gemini* be assigned to the second quadrant. This does not render it necessarily impossible that העמוד השני is a reference to the second quadrant,[97] but it does make it less likely.[98]

[95] Schmidt, "Astrologie juive ancienne," 127-28, 132-33, 136-38. According to this reconstruction the remaining text successively traces the physical and spiritual portrait of individuals conceived in the three decans belonging respectively to the zodiacal signs *Aries*, *Taurus*, and *Gemini*.

[96] Taking *Aries* as the beginning of the zodiacal circle, the first quadrant commenced in spring with *Aries*, *Taurus*, and *Gemini*. Continuing with summer (*Cancer*, *Leo*, and *Virgo*), autumn (*Libra*, *Scorpio*, and *Sagittarius*), and winter (*Capricorn*, *Aquarius*, and *Pisces*) as respectively the second, third, and fourth quadrants. Cf. Manilius, *Astronomica* 2.841–855; Ptolemy, *Tetrabiblos* 1.10.2. See W. Hübner, *Die Eigenschaften der Tierkreiszeichen in der Antike: Ihre Darstellung und Verwendung unter besonderer Berücksichtigung des Manilius* (SA.B 22; Wiesbaden: Franz Steiner, 1982), 82-87, 241-43, 251, 259-60.

Apart from this fixed system of quadrants, there is another arrangement, variable regarding the zodiacal circle, that is based on the ascendant. From the ascendant (ὡροσκόπος) the quadrants are defined successively to each cardinal point (κέντρον), which are mid-heaven (μεσουράνημα, *medium caeli*), setting (δῦνον, descendant), and the nadir or lower mid-heaven (ἀντιμεσουράνημα, *imum medium caelum*). The first quadrant (τεταρτημόριον) is that from the ascendant to mid-heaven, the second is from mid-heaven to setting, the third is from setting to lower mid-heaven, and the fourth is that from lower mid-heaven to the ascendant. Depending on the circumstance of which part of the zodiacal circle, i.e. which zodiacal sign, is ascending at a particular time, the assignment of each of the zodiacal signs to a certain quadrant will be made successively. See Ptolemy, *Tetrabiblos* 1.6.2; 1.13.4; Sextus Empiricus, *Against the Professors* 5.12; H. Gundel and R. Böker, "Zodiakos: Der Tierkreis in der Antike," *RE* 2.19 (1972), 462-709, at 476, 478; Tester, *A History of Western Astrology*, 71-72; H.G. Gundel and A. Kehl, "Horoskop," *RAC* 16 (1994), 597-662, at 601-2.

[97] Ptolemy, *Tetrabiblos* 2.11.3–4, explains, although in the context of mundane not individual astrology, that there are those who, in order to make investigations which deal with the year, have made use of each of the four tropical points, i.e. vernal and autumnal equinoxes and summer and winter solstices, as the starting point of the zodiacal circle and the new year. Perhaps *4QZodiacal Physiognomy* belongs to a tradition according to which the zodiacal circle begins with the winter solstice in *Capricorn*?

[98] Cf. also Albani, "Horoscopes in the Qumran Scrolls," 328.

עמוד as a Reference to One of the Phases of the Moon

The third astrological explanation specifically connects עמוד ("column") with one of the phases of the moon. It is, however, only mentioned in passing and has not been fully worked out.

Wise translates the sentence והואה מן העמוד השני as "and who was born during the second phase of the moon." This interpretation is suggested by "similar phrasing in Ptolemy's *Tetrabiblos*, where he is describing phases or 'stations' of the moon (i.e., the places where it 'stands')."[99] However, Wise's interpretative translation "born during the second phase of the moon" is not supported by the source he adduces.[100] Also, the date and origin of the concept of lunar stations is not clear.[101]

[99] Wise, "Horoscope Written in Code," 277.

[100] Unfortunately, Wise does not provide any further explanation regarding the way the astrological system in *4QZodiacal Physiognomy* exactly worked. Nor does he give references in Ptolemy's *Tetrabiblos*, nor make clear whether he supports Strugnell's emendation of עומד ("station, position") for עמוד. The following remarks, therefore, are of a limited and preliminary nature. Ptolemy, *Tetrabiblos* 1.8 describes the four phases of the moon with regard to the various effects that the moon causes during these phases, which are new moon, first quarter, full moon, and last quarter. The second and fourth phases are referred to by the word διχότομος, which cannot be interpreted as the place where the moon "stands." It literally means, "cut in half, divided equally" (LSJ s.v. διχότομος). Ptolemy refers to the second phase of the moon, being the first quarter, as πρώτης διχοτόμου and to the fourth phase, which is the last quarter, as δευτέρας διχοτόμου. In the same section Ptolemy also discusses the effects that the planets exercise on their courses through the sky. The two points at which the planets seem to stand still in their motion before beginning retrograde movement Ptolemy calls "station" (στηριγμός, from στηρίζω, literally "being fixed, standing still," LSJ s.v. στηριγμός) explaining the (irregular) planetary courses by the theory of epicycles. Ptolemy distinguishes between the point of rising to the first station (πρώτου στηριγμοῦ), from the first station to evening rising, from evening rising to the second station (δευτέρου στηριγμοῦ), and from the second station to setting. Cf. Bouché-Leclercq, *L'astrologie*, 111-23; O. Neugebauer, *The Exact Sciences in Antiquity* (2d ed.; New York: Dover Publications, 1969), 122-26, 198-201; Neugebauer, *HAMA*, 190-206, 267-73; W. Hübner, "The Ptolemaic View of the Universe," *GRBS* 41 (2000): 59-93, at 62.

Notwithstanding the difficulty of fixing the exact moment of planetary stations (Ptolemy, *Almagest* 9.2), examples of ancient Greek horoscopes note whether a planet is in its first or second station (στηριγμός). See the examples for the years 81, 137, and 497 CE in Neugebauer and van Hoesen, *Greek Horoscopes*, 22-23, 39-40, 155. See also P. Oxy. 2555 9.12 in Baccani, *Oroscopi Greci*, 83. Also, in Babylonian horoscopes reference is made to a planet's stationary point (UŠ or *nenmudu*), see AO 17649 rev. 1, 4 and MLC 2190 obv. 10 in Rochberg, *Babylonian Horoscopes*, 51-55, 83-85.

Regarding the language used (i.e., "station"), Ptolemy's description of the two points in the planetary movement at which they appear to stand still seems to fit better with Wise's suggestion than Ptolemy's account of the second and fourth phases of the moon. But the terminology used refers to a specific aspect of planetary motion, and not that of the moon. It seems, however, unlikely that עמוד should be interpreted as a reference to a planet's stationary point at the time of birth, since no planet to which it might apply is mentioned in the text.

[101] Gundel and Böker, "Zodiakos," 509-22: "Die 27 bzw. 28 Mondstationen waren in hellenistischer Zeit im östlichen Mittelmeerraum nicht unbekannt, haben aber in der antiken Astronomie und Astrologie wenig Beachtung gefunden" (509); Neugebauer, *HAMA*, 6, 1073 n. 2, states that this concept has no Mesopotamian or Greek background, but represents a

עמוד *as a Reference to a Column of a (Heavenly) Scroll*

Dismissing the possibility of a convincing astrological (or for that matter physiognomic) understanding of העמוד השני ("the second column"), Alexander suggests seeking the proper context of this phrase in the early Jewish tradition of heavenly books in which the history of the world is written down and in which the names of mankind are recorded.[102] He argues that עמוד is used as a reference to a "column" or a "list" in a scroll. Only two such columns were imagined in *4QZodiacal Physiognomy*. One column or list would have been for the righteous, the other for the wicked.[103] Alexander suggests that people's place in this book could be discerned through their physiognomies. He distinguishes between two notions of heavenly books in early Judaism. First, heavenly records are said to list everyone's deeds. These records will be consulted during a final judgment that will pronounce the rewards and punishments for everybody.[104] Secondly, it is also said that heavenly records contain everything that will befall the world and every individual. They have the predetermined history of the world written down in them.[105] *4QZodiacal Physiognomy* reflects the second idea of heavenly books: "whether one is good or bad has been foreordained and

---

contribution from Indian astrological tradition; according to Tester, *History of Western Astrology*, 81-84, a Babylonian origin for the lunar mansions is fairly clear (82).

102 Alexander, "Physiognonomy," 388-89.

103 Alexander assumes that the second column is the list in which the righteous is inscribed. He argues that the partitioning of light and darkness on a nine-point scale excludes the possibility of an equal division. One is either for the most part good or for the most part bad. The individual portrayed in 4Q186 1 ii has six parts in the house of light and three parts in the pit/house of darkness and belongs to the second column. Therefore, the second column is the one in which righteous people are listed.

104 Alexander, "Physiognonomy," 388 n. 8, refers to Dan 7:10 and Rev 20:12, where a heavenly court sits in judgment and books (ספרין, βιβλία) are opened (see J.J. Collins, *Daniel: A Commentary on the Book of Daniel* [Hermeneia; Minneapolis: Fortress, 1993], 109, 303; D.E. Aune, *Revelation 17-22* [WBC 52C; Dallas, Texas: Word Books, 1998], 1102). See also *1 En.* 81:4; 89:61–64.68.70–71.76; 90:17.20; 98:7–8; 100:10; 104:7; *Jub.* 19:9; *m. ʾAbot* 1:18; 2:1; 3:2 (quoting Mal 3:16).

105 Alexander, "Physiognonomy," 388 n. 8, refers to Rev 5:1 (see D.E. Aune, *Revelation 1-5* [WBC 52A; Dallas, Texas: Word Books, 1997], 341-46), and to *3 En.* 45:1, where the heavenly curtain (פרגוד) is said to contain everything done or to be done by all the generations of the world until the last generation, see P. Alexander, "3 (Hebrew Apocalypse of) Enoch," *OTP* 1: 223-315, at 296. Alexander, "3 Enoch," 296, also mentions *Hekhalot Rabbati* 6:3, where it is said that an angel opens tablets (פנקסין/הפנקסים) for R. Ishmael in which certain events that will happen to Israel are already recorded, see §122 in P. Schäfer et al. (eds.), *Synopse zur Hekhalot-Literatur* (TSAJ 2; Tübingen: J.C.B. Mohr [Paul Siebeck], 1981), 60-61, and P. Schäfer et al. (eds.), *Übersetzung der Hekhalot-Literatur II §§81-334* (TSAJ 17; Tübingen: J.C.B. Mohr [Paul Siebeck], 1987), 52. See also *1 En.* 81:1–2; 93:1–3; *Jub.* 16:3.9; 24:33; 31:32; 32:21; 4Q180 1 3. Alexander's second category seems also to be expressed in Ps 139:16. Although the textual evidence is not without difficulties (cf. e.g. MT with 11QPs^{a} 22:7–9), the sense seems to be that divine foreknowledge of things to be is contained in a book or books.

depends on the column in which one's name is inscribed in the heavenly record."[106]

References to heavenly books and tablets as such are not restricted to the apocryphal and pseudepigraphic literature of early Judaism. The idea of heavenly tablets already appears in Sumerian texts, and they are also mentioned a few times in the Hebrew Bible. The notion of heavenly books and tablets is complex and can be divided into more than just two categories. Heavenly records play a special role in beliefs about a final judgment.[107]

Alexander's interpretation is unlikely for two reasons. First, because it invokes the concept of heavenly books for which there is no other evidence in the text (unlike the astrological interpretations that can refer to another astrological notion, namely that of the zodiacal sign *Taurus* in 4Q186 1 ii

[106] Alexander, "Physiognonomy," 388-89.

[107] For scholarly treatments of the notion of heavenly records, see e.g. R. Eppel, "Les tables de la Loi et les tables célestes," *RHPR* 17 (1937): 401-12; H. Bietenhard, *Die himmlische Welt im Urchristentum und Spätjudentum* (WUNT 2; Tübingen: J.C.B. Mohr [Paul Siebeck], 1951), 231-54; L. Koep, *Das himmlische Buch in Antike und Christentum: Eine religionsgeschichtliche Untersuchung zur altchristlichen Bildersprache* (Theophaneia 8; Bonn: Peter Hanstein, 1952); S.M. Paul, "Heavenly Books and the Book of Life," *JANESCU* 5 (1973): 345-53; E. Rau, *Kosmologie, Eschatologie und die Lehrautorität Henochs: Traditions- und formgeschichtliche Untersuchungen zum äth. Henochbuch und zu verwandten Schriften* (Diss. Hamburg, 1974), 345-98; F. García Martínez, "The Heavenly Tablets in the Book of Jubilees," in *Studies in the Book of Jubilees* (eds. M. Albani, J. Frey and A. Lange; TSAJ 65; Tübingen: Mohr Siebeck, 1997), 243-60; M. Himmelfarb, "Torah, Testimony, and Heavenly Tablets: The Claim to Authority of the *Book of Jubilees*," in *A Multiform Heritage: Studies on Early Judaism and Christianity in Honor of Robert A. Kraft* (ed. B.G. Wright; HS 24; Atlanta, Georgia: Scholars Press, 1999), 19-29; G.W.E. Nickelsburg, *1 Enoch 1: A Commentary on the Book of 1 Enoch, Chapters 1-36; 81-108* (Hermeneia; Minneapolis: Fortress, 2001), 478-80.

Regarding Alexander's proposal, it is interesting to observe that the idea of two heavenly books, one that records the deeds of the wicked and the other that lists the deeds of the righteous, is attested in Jewish tradition. See *Apoc. Zeph.* 3:5–9; *2 Bar.* 24:1; *Jub.* 30:19–22; 36:10; *Pss. Sol.* 13:10-11; *T. Ab.* 12:8.12; 13:9 (see D.C. Allison, Jr., *Testament of Abraham* (CEJL; Berlin: Walter de Gruyter, 2003), 264-65, 269); *b. ʿArak.* 10b. Cf. also Aune, *Revelation 1-5*, 224 and Aune, *Revelation 17-22*, 1102 (in *b. Roš Haš.* 16b three books are mentioned, "one for the thoroughly wicked, one for the thoroughly righteous, and one for the intermediate").

The concept of heavenly books is also attested in the Dead Sea Scrolls. In *4QInstruction*[c] Malachi's book of remembrance (Mal 3:16), meant for the benefit of those who fear the Lord and think on his name, has been paired with an engraving that contains the punishments for the iniquities of the sons of שות (4Q417 1 i 14–16), see J. Strugnell, D.J. Harrington and T. Elgvin (eds.), *Qumran Cave 4.XXIV: Sapiential Texts, Part 2: 4QInstruction (Mûsār lĕ Mēvîn): 4Q415 ff. with a Re-edition of 1Q26 by John Strugnell and Daniel J. Harrington, S.J. and an Edition of 4Q423 by Torleif Elgvin, in Consultation with Joseph A. Fitzmyer, S.J.* (DJD 34; Oxford: Clarendon, 1999), 162-63. Also, in CD 20:19–20 the quote from Mal 3:16 is expanded with the statement that things are written in the book of remembrance "until salvation and justice are revealed to those who fear God." Cf. F. Nötscher, "Himmlische Bücher und Schicksalsglaube in Qumran," *RevQ* 1/3 (1959): 405-11; Lange, *Weisheit und Prädestination*, 69-83. See also the possible references in 4Q381 31 8, 4Q541 2 i 6–7, 4Q541 7 4, and 11Q18 19 5–6.

9). Second, because the term עמוד ("column") is not attested this early in such a sense and other terms were available.

The terminology used in *4QZodiacal Physiognomy* does not provide any evidence that the concept of a heavenly record is being referred to. There is no mention of a book, tablet or writing (ספר, לוח, כתב), but one might object that this may be due to the fragmentary state of the manuscript. One should perhaps allow for the possibility that in some section of the text a book was specifically mentioned in which one column is preserved for the wicked and another for the righteous. But, as Alexander points out himself, his proposed interpretation of עמוד as a "column" or "list" in a scroll is not attested this early and he can only appeal to the intrinsic probability of his explanation.[108] Other terms for referring to a column on a scroll were available, however.[109] This lessens the probability of Alexander's interpretation of עמוד ("column"). One might modify Alexander's proposal by interpreting עמוד in the sense of a stick around which a scroll is rolled as a *pars pro toto* for an entire scroll, one scroll for the wicked and one scroll for the righteous.[110] It is, however, unlikely that עמוד is used in this sense in *4QZodiacal Physiognomy*, especially in the context of the phrase והואה מן העמוד השני ("And he is from the second column").

Assessing the Meaning and Significance of העמוד השני

It is obvious that the main problem with the interpretations that have been put forward is that neither supports the specific sense of עמוד ("column") outside *4QZodiacal Physiognomy*. This means that *4QZodiacal Physiognomy* presents a terminology and a concept otherwise unknown to us in Hebrew. If one assumes that עמוד is in some way a meaningful reference to a concept, the question is, of course, which concept. Is it a reference to a zodiacal position, a zodiacal quadrant, a phase of the moon, a column on a scroll, or something else? Furthermore, the meaning and significance of the phrase העמוד השני ("the second column") cannot be determined in isolation to the, albeit fragmentary, context of the rest of *4QZodiacal Physiognomy*.

108 Alexander, "Physiognonomy," 388 n. 7. Cf. E. Ben Yehuda, *Dictionary and Thesaurus of the Hebrew Language* (vol. 5; New York: Thomas Yoseloff, 1960), s.v. עַמּוּד, 4555-57.

109 In Jer 36:23 the word דלת, in the form of the *hapax legomenon* דְּלָתוֹת, is clearly used as a reference to the columns on a leather sheet of a scroll. This is translated with σελίδας (from σελίς, "plank, leaf of papyrus") in the LXX, a word Alexander, "Physiognonomy," 388 n. 7, refers to in support of his proposal for עמוד. See e.g. R. Lansing Hicks, "*Delet* and *Megillah*: A Fresh Approach to Jeremiah xxxvi," *VT* 33 (1983): 46-66; Tov, *Scribal Practices*, 82. In rabbinic literature the word דף is employed for referring to columns on a scroll, such as, for example, in *y. Meg.* 1.71c and *b. Menaḥ.* 29b–30b.

110 In rabbinic literature עמוד or עמודים occur as references to a roller or wooden bar attached to the uninscribed area of one, or, in the case of Torah scrolls, to either end of a scroll (*y. Meg.* 1.71d; *b. B. Bat.* 13b–14a). And at Qumran evidence of such a stick has been preserved in the case of *11QapocrPs*. Cf. Tov, *Scribal Practices*, 115-18.

How is the concept to which these words refer related to three other elements in the text?

In 4Q186 1 ii the phrase והואה מן העמוד השני ("And he is from the second column") is positioned between (1) the physiognomic description, (2) the numbers concerning the "house of light" and the "house of darkness," and (3) the zodiacal data with regard to the moment of birth of the individual. Considering the first element, is the idea that people are "from the second column" a consequence of their physiognomies, and, if so, in what way? With regard to the second element, is the information provided by the words העמוד השני ("the second column") necessary for understanding the numbers allotted to the "house of light" and the "house of darkness?" Regarding the last element, is knowledge that someone belongs to "the second column" superfluous or basic to understanding the zodiacal information provided for his birth?

Of course, one also has to bear in mind that the term עמוד ("column"), and consequently the concept it refers to, has a variable position in the sequence of elements in *4QZodiacal Physiognomy*. The enigmatic words העמוד השני ("the second column") occur at least twice in the extant text of *4QZodiacal Physiognomy*: 4Q186 1 ii 6; 4 1 (cf. 4Q186 6 2). In 4Q186 1 ii 6 העמוד השני follows as a new element in the sequence after the bodily features, clearly marked by the use of a conjunctive-*waw* (והואה מן, "And he is from..."). This is followed by the next element, introduced by the words רוח לו ("There is a spirit for him"), concerning the different numbers in the "house of light" and the "house of darkness." This sequence, however, seems not to be the general rule. There are two indications. First, in 4Q186 1 iii 8 and 2 i 6 עמוד is not mentioned subsequent to the physiognomic description and before the phrase רוח לו. The words ורוח לו ("And there is a spirit for him") introduce the element following immediately after the physical features have been described. This is also evidently marked by the use of a conjunctive-*waw*. It is evident that in 4Q186 1 iii 8–9, immediately following the physiognomic element, the next part concerning the "house of light" and the "house of darkness" is introduced by the words ורוח לו. It is clear, therefore, that in two out of three accounts the term עמוד does not follow the physiognomic descriptions, whereas the reference to the "house of light" and the "house of darkness" does. Apparently, the latter concept is more important than the one referred to by עמוד. Second, 4Q186 4 1 shows that when a reference to עמוד occurs it is not necessarily followed by רוח לו. Here the phrase ה[עֿמוד השני is certainly not followed by רוח לו as in 4Q186 1 ii 6–7, but by a word beginning with *šin* (see also 4Q186 6 2).

The optional character of עמוד ("column") in *4QZodiacal Physiognomy* raises the question of its significance in the text. Perhaps a statement regarding the number of עמוד to which a person belongs could follow after the other three elements (physiognomy, light/darkness division, zodiacal horoscope) had been presented. One should allow for the possibility that the text was not consequent regarding the phrase העמוד השני ("the second column"). Equally possible is that it was left out altogether in some entries of the text. Unfortunately, the manuscript does not continue after 4Q186 1 iii 9 and is too fragmentary from 4Q186 2 i 6. Be that as it may, the optional, maybe even redundant, position of the phrase העמוד השני in the sequence of elements in *4QZodiacal Physiognomy* means that what it referred to was not intrinsically connected to the other three elements.

It is evident that all the proposals for the specific sense of העמוד השני ("the second column") in *4QZodiacal Physiognomy* can be no more than hypothetical. The simplest, but unsatisfactory, solution seems to be to take the phrase העמוד השני as a reference to the second position in the zodiacal circle. Only in 4Q186 1 ii does it occur in context, and then in relation to the zodiacal sign *Taurus*, which is the second sign of the zodiacal circle according to the most common view. If direct context determines the meaning of this phrase, and I do not see what else could do better, then this is the easiest solution. It is unsatisfactory, however, because it is completely superfluous and does not add any significant information to a text that is otherwise succinct and meaningful.

"AND THIS IS THE HOROSCOPE UNDER WHICH HE WAS BORN"

The noun מולד ("horoscope") does not occur in the Hebrew Bible.[111] In Rabbinic Hebrew מוֹלָד refers to "descendant," while מוֹלָד refers to the "birth-time," especially of the new moon.[112] In the Dead Sea Scrolls מולד is used in a general and, in some of the wisdom texts, more specific, astrological, sense.

In general, מולד refers to the circumstances surrounding people's birth. These need not necessarily be understood in terms of astrological circumstances. This depends on the context in which the term appears. In the *Instruction* texts the reader is encouraged to learn (לקח) or examine (דרש) people's מולד, both verbs being used synonymously.[113] One should allow

[111] Biblical Hebrew has the abstract noun מולדת in the sense of "relatives" or "offspring," and the *hi'pil* participle מוליד (Isa 66:9), מולדים (Jer 16:3) "to beget, deliver," but these occurrences do not shed light on the use of מולד in *4QZodiacal Physiognomy*.

[112] Cf. Jastrow 742; S. Gandz, "The Problem of the Molad," *Studies in Hebrew Astronomy and Mathematics* (S. Gandz; New York: Ktav, 1970), 120-49.

[113] 4Q415 11 11; 4Q416 2 iii 9 (=4Q418 9+ 8); 4Q416 2 iii 20 (=4Q418 10a, b 3); 4Q417 2 i 11 (=4Q416 2 i 6); 4Q418 202 1. Cf. Strugnell, Harrington and Elgvin, DJD 34.62, 117, 123, 182, 423, who do not think that the two verbs are always synonymous when used in combination with מולד in *Instruction*. See also A. Lange, "Kognitives *lqḥ* in Sap A, im T^{e}nak und Sir," *ZAH* 9 (1996): 190-95.

for the possibility that the addressee is in some of these passages being encouraged to determine people's "horoscope."[114]

In the case of the technical phrase בית מולדים, one should assume that it specifically refers to people's nativities, that is, the configuration of heavenly bodies in relation to the zodiacal circle at the moment of birth. It is not likely that such a technical phrase would have different meanings in different contexts in the Dead Sea Scrolls. The phrase בית מולדים appears four times in wisdom contexts.[115] Matthew Morgenstern has drawn attention to the Syriac parallel *bēṯ yaldā*.[116] This occurs, for example, in *The Book of the Laws of Countries* by Bardaiṣan of Edessa (154–222/3).[117] There it refers to the arrangement of planets vis-à-vis the zodiac at the time of birth, i.e. the nativity. It is probable that בית מולדים in the Dead Sea Scrolls has a sense similar to the Syriac *bēṯ yaldā*, and should be translated with "nativity." Following Han Drijvers's translation of *bēṯ yaldā* with "horoscope,"[118] Morgenstern translates בית מולדים with "horoscope." But such a translation seems not completely appropriate for an ancient text. The term "horoscope" is used in modern parlance for the configuration of planets in relation to the zodiacal signs, but in antiquity ὡροσκόπος was used only for the zodiacal sign rising in the east at the moment of birth, the ascen-

[114] Cf. A. Caquot, "Les textes de sagesse de Qoumrân (Aperçu préliminaire)," *RHPR* 76 (1996): 1-34, at 9; M. Morgenstern, "The Meaning of בית מולדים in the Qumran Wisdom Texts," *JJS* 51 (2000): 141-44; E.J.C. Tigchelaar, *To Increase Learning for the Understanding Ones: Reading and Reconstructing the Fragmentary Early Jewish Sapiential Text 4QInstruction* (STDJ 44; Leiden: Brill, 2001), 238; J.M. Baumgarten, "מינוח קומראני ואסטרולוגי ב'מוסר למבין' (Some Qumranic and Astrological Terms in *Mûsār lᵉ Mēvîn*)," *Tarbiz* 72 (2003): 321-28, at 324-26; M. Kister, "Wisdom Literature and Its Relation to Other Genres: From Ben Sira to *Mysteries*," in *Sapiential Perspectives: Wisdom Literature in Light of the Dead Sea Scrolls: Proceedings of the Sixth International Symposium of the Orion Center for the Study of the Dead Sea Scrolls and Associated Literature, 20-22 May, 2001* (eds. J.J. Collins and G.E. Sterling; STDJ 51; Leiden: Brill, 2004), 13-47, at 28, 34, 36-37; F. Schmidt, "'Recherche son thème de géniture dans le Mystère de ce qui doit être:' Astrologie et prédestination à Qoumrân," in *Qumrân et le judaïsme du tournant de notre ère* (eds. A. Lemaire and S. Mimouni; CREJ 39; Leuven: Peeters, 2006), *forthcoming*. I am grateful to Francis Schmidt for sending me his article prior to publication. B.G. Wold, *Women, Men and Angels* (WUNT 201; Tübingen: Mohr Siebeck, 2005), 187, however, considers the term in a non-astrological sense throughout the *Instruction* text.

In 4Q417 2 i 11 the phrase וקח מולדי ישׁע ("and comprehend the nativities of salvation") may have an eschatological meaning, referring possibly to people's horoscopes at the end of times.

[115] See *4QMysteries* (4Q299) 1 4; 3a ii–b 13; 5 5; *4QInstruction*[a] (4Q415) 2 ii 9. Cf. L. Schiffman, "4QMysteries[a]," in *Qumran Cave 4.XV: Sapiential Texts, Part 1* (eds. T. Elgvin et al.; DJD 20; Oxford: Clarendon, 1997), 33-97, at 37; Strugnell, Harrington and Elgvin, DJD 34.49; Morgenstern, "Meaning of בית מולדים."

[116] Morgenstern, "Meaning of בית מולדים," 142-43.

[117] See H.J.W. Drijvers, *The Book of the Laws of Countries: Dialogue on Fate of Bardaiṣan of Edessa* (STT 3; Assen: Van Gorcum, 1965), 32.16; 36.8; 40.2; 48.12; 52.3, 7, 9–10, 22; 54.4; 58.5. Cf. H.J.W. Drijvers, *Bardaiṣan of Edessa* (Assen: Van Gorcum, 1966), 86-88.

[118] But Drijvers uses "nativity" more often to translate *bēṯ yaldā*, see Drijvers, *Book of the Laws of Countries*, 49, 53, 55.

dant. In antiquity the word γένεσις, among other things, referred to the nativity of people.[119] Understanding the Syriac *bēṯ yaldā* as a reference to people's nativity is also borne out by Diodorus of Tarsus' *On Fate* (fourth century), transmitted by Photius of Constantinople (ninth century), who refers to it with the Greek γενεθλιαλογίας, "casting of a nativity."[120] The term "horoscope," then, for the phrase בית מולדים should be used with great care and awareness that it does not mean the same as the Greek term ὡροσκόπος, but is rather an equivalent of the Greek γένεσις.

In *4QZodiacal Physiognomy*, מולד ("horoscope") is a technical term that refers to the astrological sign under which people were born. In 4Q186 1 ii 9 the statement concerning the מולד under which the person was born is followed by a specific reference to a part of the zodiacal sign *Taurus* (ברגל השור, "in the foot of *Taurus*"). Such a technical context makes general understandings of the noun מולד, such as the time of birth,[121] the occasion of birth itself,[122] or to the one born,[123] improbable here. The astrological context in *4QZodiacal Physiognomy* warrants understanding מולד more specifically as a technical astrological term for the nativity of an individual, analogous to γένεσις and *genitura* in Greco-Roman astrology.[124]

Schmidt agrees that מולד ("horoscope") is a technical astrological term in *4QZodiacal Physiognomy*, but he suggests a particular interpretation of it that depends on his general interpretation of the text. He argues that *4QZodiacal Physiognomy* distinguishes between the moment of conception and that of birth. According to Schmidt the text is primarily interested in understanding the division between diurnal and nocturnal decans during pregnancy. He therefore suggests reading מולד in 4Q186 1 ii 8 as a passive participle *hop'al* הַמּוּלָד [sic!] referring to conception, and ילוד as a passive participle *qal* יָלוּד referring to birth: "And this is the conceived one because

[119] On γένεσις in an astrological sense, cf. J.N. Bremmer, "Foolish Egyptians: Apion and Anoubion in the *Pseudo-Clementines*," in *The Wisdom of Egypt: Jewish, Early Christian, and Gnostic Essays in Honour of Gerard P. Luttikhuizen* (eds. A. Hilhorst and G.H. van Kooten; AJEC 59; Leiden: Brill, 2005), 311-29, at 313 n. 10. For the meaning of the term ὡροσκόπος, see e.g. Bouché-Leclercq, *L'astrologie grecque*, 257-75; Gundel and Kehl, "Horoskop," 599-600; W. Hübner, "Zur Verwendung und Umschreibung des Terminus ὡροσκόπος in der astrologischen Lehrdichtung der Antike," *Mene* 1 (2001): 219-38, at 221.

[120] Cf. Drijvers, *Bardaiṣan*, 70.

[121] Allegro, DJD 5.89.

[122] Carmignac, "Les horoscopes," 203.

[123] Dupont-Sommer, "Deux documents horoscopiques," 242.

[124] Cf. Delcor, "Recherches sur un horoscope," 304-5; M. Hengel, *Judentum und Hellenismus: Studien zu ihrer Begegnung unter besonderer Berücksichtigung Palästinas bis zur Mitte des 2. Jh.s v. Chr.* (3d ed.; WUNT 10; Tübingen: J.C.B. Mohr [Paul Siebeck], 1988), 433-35; Bergmeier, *Glaube als Gabe*, 78; Albani, "Horoscopes in the Qumran Scrolls," 282, 286; Morgenstern, "Meaning of בית מולדים," 143.

of which he [that is, the individual portrayed] is born: in the foot of Taurus."[125]

This understanding of מולד ("horoscope") is not necessary, however, to make sense of the Hebrew in 4Q186 1 ii 8. Furthermore, the astrological system for determining the moment of conception that Schmidt argues in favor of is too hypothetical and does not correspond to any of the systems devised by ancient astrologers that are known to us.[126] The term מולד in *4QZodiacal Physiognomy* is best taken in the sense of "nativity," i.e. the constellation of planets vis-à-vis the zodiac at the moment of birth. In this case "nativity" is restricted to the zodiacal sign, or rather part of it, under which people were born.

The translation "horoscope" for מולד in *4QZodiacal Physiognomy* has been chosen for two reasons. First, because it is a better-known term in contemporary English than "nativity." Second, because the text demonstrates an interest in the ascendant. It is evident that מולד is a derivative of ילד, like γένεσις is of γίγνομαι, and that it, therefore, is an equivalent of the Greek γένεσις rather than of ὡροσκόπος. The interesting thing to point out, however, is that *4QZodiacal Physiognomy* does not refer to planets at all, but only to a specific part of a zodiacal sign that is the ascendant, literally the ὡροσκόπος. Because *4QZodiacal Physiognomy* is primarily interested in the ascendant, the translation "horoscope" is appropriate for מולד (with the caveat that the term itself is equivalent to γένεσις).[127]

4QZodiacal Physiognomy demonstrates that the appearance and shape of the body is believed to be due to the result of the position of the zodiacal sign under which a person was born. In 4Q186 2 i 3–4 the physiognomic description is directly related to the person's "horoscope." The person described in 4Q186 2 i is said to be neither tall nor short, "and that because of his horoscope" (והואה[]ממולדו).[128]

Magico-Medicinal Stones

Contrary to the otherwise inverted written character of the text, 4Q186 1 ii 2 has the words אבן צונם, a kind of granite, written in the normal order from

[125] Schmidt, "Astrologie juive ancienne," 132 ("Ancient Jewish Astrology," 196).

[126] For a fuller discussion, see Chapter Three.

[127] Cf. also the occurrence of מולדה, "his horoscope," in 4Q534 1 i 10; 1 ii+2 1–2, 6; 4Q535 2 1. For the astrological sense here, see already Starcky, "Un texte messianique araméen," 61-62.

[128] In 4Q186 4 2 מולדו ("his horoscope") occurs, probably equivalent to המולד in 4Q186 1 ii 8. The text is too fragmentary to reconstruct the exact reference.

right to left. Previously, no-one has commented in detail on the occurrence of this stone in the text.[129]

In late antique and medieval Jewish magical texts, the names of illnesses, demons, and gods were sometimes intentionally corrupted, and letters or words in incantation texts scrambled.[130] The same phenomena can be found in Greek curse tablets (*defixiones* or καταδέσμοι).[131] It was believed that "writing or reciting backwards was more effective than doing so in the normal manner," and that "reversing the order of letters or words led to reversing the natural order of things."[132] A magical power is also demonstrated by the use of different scripts in later Arabic magic.[133]

4QZodiacal Physiognomy, however, is not a magical text in which some words or phrases are written in reverse. As the entire text is in inverted writing, it is improbable that this was intended to make the text more effective as in magical formulae. Yet, it may suggest that the regular order of writing represented some form of magical power, although it is difficult to imagine what kind of power or how it may have worked. Perhaps אבן צונם represented some sort of magical stone, and maybe it was thought that the normal direction of writing had some sort of magical power in this otherwise inverted written text. Unfortunately, the text is too fragmentary to substantiate this suggestion.[134]

Apart from the way אבן צונם is written in *4QZodiacal Physiognomy*, the occurrence of a certain type of stone in a zodiacal physiognomic list should not surprise us. Already in Babylonian astrology certain stones, plants, and trees were associated with specific zodiacal signs and subdivisions of signs,

[129] Wise, "Horoscope Written in Code," 277, suggests that *4QZodiacal Physiognomy* incorporated ideas about birthstones.

[130] See J. Naveh, "Lamp Inscriptions and Inverted Writing," *IEJ* 38 (1988): 36-43. An example from the Dead Sea Scrolls is the Aramaic text 4Q560, see J. Naveh, "Fragments of an Aramaic Magic Book from Qumran," *IEJ* 48 (1998): 252-61, at 259, and J. Naveh, "Hebrew Graffiti from the First Temple Period," *IEJ* 51 (2001): 194-207, at 195. While discussing a deliberately distorted inscription from sixth-century BCE Khirbet Beit Lei, Joseph Naveh ("Hebrew Graffiti," 195 n. 9) refers in passing to *4QZodiacal Physiognomy*.

[131] Cf. J.G. Gager (ed.), *Curse Tablets and Binding Spells from the Ancient World* (New York: Oxford University Press, 1992), 5.

[132] Naveh, "Lamp Inscriptions," 41.

[133] Cf. D. De Smet, "L'alphabet secret des Ismaéliens ou la force magique de l'écriture," in *Charmes et sortilèges: Magie et magiciens* (ed. R. Gyselen; ResOr 14; Bures-sur-Yvette: Groupe pour l'Étude de la Civilisation du Moyen-Orient, 2002), 51-60.

[134] According to Carmignac, "Les Horoscopes," 202, the words אבן צונם represent the beginning of a new line and immediately follow on from the final word of the previous line (מבא according to Carmignac), because they are written in the regular manner from right to left. This would mean that the whole of l.2 was written in the normal way from right to left, but this need not be so. It is possible that the rest of l.2 was written in the inverted way typical of *4QZodiacal Physiognomy* and that only the last two words were written from right to left. The fact that the final words were written in the regular way does not necessarily mean that the entire line was written thus.

and also with certain magical operations.[135] And in Greco-Roman astrology lists were compiled that associated stones, colors, plants, animals, and letters of the alphabet with the signs of the zodiac.[136] These texts are examples of a tendency to bring together various branches of knowledge in one type of text, such as compendia or catalogues.[137]

It is possible that a compendium text such as *4QZodiacal Physiognomy* associates certain stones with different physiognomic types and zodiacal signs. If so, this text from Qumran is important evidence that in first-century BCE Jewish culture stones were related with signs of the zodiac.

Josephus ascribes divine communicative power to the twelve stones on the high priest's breastplate and assigns them to the twelve tribes of Israel,[138] but he also refers to a zodiacal interpretation for them.[139] Before him, Philo suggested a zodiacal interpretation for the twelve stones, and emphasized their different colors:

> Secondly, the stones at the breast, which are dissimilar in color, and are distributed in four rows of threes, what else should they signify but the zodiac circle? [...] It is an excellent and indeed a splendid point that the twelve stones are of different colors and none of them like to any other. For each of the signs of the zodiac also produces its own particular coloring in the air and earth and water and their phases, and also in the different kinds of animals and plants.[140]

Philo thus shows knowledge of the association between stones, colors, and zodiacal signs, and their various influences on living beings and plants.[141]

135 See e.g. Sachs, "Babylonian Horoscopes," 67, 72; E. Reiner, "The Uses of Astrology," *JAOS* 105 (1985): 589-95, at 592-93; E. Reiner, *Astral Magic in Babylonia* (TAPS 85/4; Philadelphia: American Philosophical Society, 1995), 25-42, 119-32; N.P Heeßel, "Stein, Pflanze und Holz: Ein neuer Text zur 'medizinischen Astrologie,'" *Or* 74 (2005): 1-22.

136 See e.g. Gundel and Böker, "Zodiakos," 575-80; Hübner, *Eigenschaften der Tierkreiszeichen*, 280; J. Scarborough, "The Pharmacology of Sacred Plants, Herbs, and Roots," in *Magika Hiera: Ancient Greek Magic and Religion* (eds. C.A. Faraone and D. Obbink; New York: Oxford University Press, 1991), 138-74, at 154-56; R. Gordon, "*Quaedam veritatis umbrae*: Hellenistic Magic and Astrology," in *Conventional Values of the Hellenistic Greeks* (eds. P. Bilde et al.; SHC 7; Aarhus: Aarhus University Press, 1997), 128-58, at 132.

137 Cf. Reiner, "Uses of Astrology," 592-93, quoting an unpublished paper by Leo Oppenheim.

138 Josephus, *Jewish War* 5.234; *Jewish Antiquities* 3.166–169, 216–218.

139 Josephus, *Jewish Antiquities* 3.186.

140 Philo, *On the Life of Moses* 2.124, 126. Translation from F.H. Colson, *Philo: VI* (LCL 289; Cambridge, Massachusetts: Harvard University Press, 1984), 508-11. For a zodiacal interpretation of the high priest's breastplate, cf. also *On the Special Laws* 1.87; *On Dreams* 1.214. I thank Jonathan Norton for bringing these latter two references to my attention. On both Philo and Josephus, see also N. Fernández Marcos, "Rewritten Bible or *Imitatio*?: The Vestments of the High-Priest," in *Studies in the Hebrew Bible, Qumran, and the Septuagint Presented to Eugene Ulrich* (eds. P.W. Flint, E. Tov and J.C. VanderKam; VTSup 101; Leiden: Brill, 2006), 321-36.

141 Cf. also Gundel and Böker, "Zodiakos," 577-78.

Josephus, furthermore, demonstrates that Jews were familiar with the medicinal properties of stones. He says of the Essenes that they examined medicinal roots and the properties of stones. The context here is clearly that of medicine. The Essenes' extraordinary interest in the writings of the ancients is particularly aimed, says Josephus, at those texts that are of use for soul and body:

> with the help of these, and with a view to the treatment of diseases, they make investigations into medicinal roots and the properties of stones.[142]

The medicinal function of plants and stones was no mystery for Second Temple period Jews. One should, therefore, allow for the possibility that *4QZodiacal Physiognomy* not only related stones with zodiacal signs and physiognomic types of people, but also that this association was of a magico-medicinal nature.[143]

Within a literary context that combines physiognomics and astrology, the element of stones, possibly of magico-medicinal nature, may shed further light on the nature of the list in *4QZodiacal Physiognomy*. It suggests that the catalogue related different things to each other: human body, zodiacal signs, and stones. This not only reveals a tendency to bring together various branches of knowledge in one type of text such as lists or catalogues, but it may also hint at the notion of cosmic sympathy, which is further explored in Chapter Five.

4QPHYSIOGNOMY AR (4Q561)

Name and Genre

Instead of the present name *4QPhysiognomy/Horoscope ar* for this Aramaic text,[144] the simple name *4QPhysiognomy ar* is more appropriate. The remaining fragments of this text contain purely physiognomic accounts.

[142] Josephus, *Jewish War* 2.136. Text and translation from H.St.J. Thackeray, *Josephus: The Jewish War, Books I-II* (LCL 203; Cambridge, Massachusetts: Harvard University Press, 1997), 374-75.

[143] See further Chapter Five. Whether 4Q186 1 ii 2 is part of the account in 4Q186 1 ii 5–9 or belongs to the end of a previous account cannot be ascertained, but the latter possibility seems more likely. If the text is structured according to physiognomic criteria, and if the text associates signs of the zodiac with various stones, it seems to make more sense to list the related stones subsequent to the physiognomic and zodiacal data.

[144] See Tov et al., DJD 39.74. Starcky named the text "Horoscopique araméen" (*Hor ar*), see Starcky, "Un texte messianique araméen," 51; *A Preliminary Concordance to the Hebrew and Aramaic Fragments from Qumrân Caves II-X, Including Especially the Unpublished Material from Cave IV, Printed from a Card Index Prepared by R.E. Brown, S.S., J.A. Fitzmyer, S.J., W.G. Oxtoby, and J. Teixidor, Prepared and Arranged for Printing by H.-P. Richter* (5 vols.; *editorum in usum*; Göttingen: Privately Printed, 1988), 4.

There are no clear and indisputable references to zodiacal signs or other astrological notions.[145] There is no reason whatsoever to suggest the genre horoscope for this text. The fragments represent the remains of a physiognomic catalogue that lists the bodily descriptions of different types of people.

Material Reconstruction, Columns, and Measurements

4QPhysiognomy ar consists of seven fragments,[146] some of which are collated.[147] For one of these seven fragments two columns can be reconstructed (4Q561 1 i and ii), and while it is probable that some of the other fragments should be placed near each other they cannot be joined on material grounds. The two reconstructed columns provide the only clue to measure the column width of this manuscript. It is, however, impossible to establish the number of columns that the original manuscript may have had.

From Starcky's transcriptions in the *Preliminary Concordance* and his preliminary translation it is clear that he understood 4Q561 1 i as preserving more or less the entire column width. For example, the eyes of the person described in the extant part of this column are between light and dark-colored (4Q561 1 i 1–2), and his nose is long and beautiful (4Q561 1 i 2–3).[148] The descriptions seem thus to continue from one line to the other.

4Q561 1 i has a top margin, measuring ca. 0.9 cm. From the amount of blank space following נגיד (4Q561 1 i 2), and a possible guide dot in l.3, it is clear that this fragment preserves the upper and left part of the column. A bottom margin is not extant.

The right margin cannot be ascertained materially, but can plausibly be assumed. First, the words בׄין (l.2), שׄפיר (l.3), דׄק (l.4), and [מ]מחקׄין (l.5) seem to begin on the right along the same vertical line. But, of course, this is not entirely certain as no beginning letter has been completely preserved. Secondly, the physiognomic descriptions seem to carry on from one line to another. It is unlikely that the descriptions were extremely detailed, and that

[145] Cf. also Alexander, "Physiognonomy," 393 n. 18; M.O. Wise, "An Aramaic Horoscope (4Q561)," in *The Dead Sea Scrolls: A New Translation*, 567-68, at 567; J. Zimmermann, *Messianische Texte aus Qumran: Königliche, priesterliche und prophetische Messiasvorstellungen in den Schriftfunden von Qumran* (WUNT 104; Tübingen: J.C.B. Mohr [Paul Siebeck], 1998), 193-94. Holst and Høgenhaven, "Physiognomy," 42.

[146] See PAM 41.944; 41.954; 42.438; 43.598.

[147] In the *Preliminary Concordance*, 4, it is noted that eight fragments have been catalogued (4Q561 1 i and ii are catalogued as *Hor ar* 1 and 3). In PAM 43.598 another fragment appears that has not been recorded in the actual concordance. This fragment appears upside down. It contains the remains of two lines, but only one word is still legible: תתחמדין or תתחמרין. The fragment does not belong to *4QPhysiognomy ar* because the handwriting is slightly different (cf. *mem* and *taw*). Cf. Wise, "4Q561," 230-31; Holst and Høgenhaven, "Physiognomy," 27-28.

[148] *Preliminary Concordance* 2238, 2262; Starcky, "Un texte messianique araméen," 64.

many more parts of the head were enumerated between the nose (l.2) and the teeth (l.3), such as ears or lips. This is illustrated, for example, by 4Q561 1 i 3: "and his teeth are even. And his beard" (ושנוהי שוין ודקנה). The portrayal is succinct, and quickly moves on to another part of the body.

If indeed 4Q561 1 i gives a good indication of the column width, and if this is representative for the entire manuscript, it means that *4QPhysiognomy ar* preserves one of the narrowest columns for a manuscript from the Dead Sea Scrolls. The column width of *4QPhysiognomy ar* is ca. 5.5-6.0 cm.[149] Such a narrow column width probably means that this Aramaic text cannot have been a large composition. But it is impossible to determine the length of the text because not enough material has been preserved.

4Q561 1 ii consists of two fragments joined together by Starcky. The first fragment has clearly preserved a right and top margin. The top margin measures ca. 1.0 cm. As for the right margin, from להוה (4Q561 1 ii 1) the leather extends ca. 1.2 cm to the right. Perhaps even it represents the beginning of a sheet. It seems as if some stitches have been preserved, but this is not entirely clear from the photograph.[150]

There are three reasons for the probability of the join between the two fragments.[151] First, they can be joined physically along the diagonal tear between ll.1 and 2. Second, the stroke to the left of the second *lamed* of ו[ל]מ[ת]ת֯מלי in l.2 might very well be the down stroke of *qop* continuing from l.1. Thirdly, the (partially reconstructed) beginning letters of the second fragment stand on one vertical line, in accordance with להוה in the first fragment in l.1. If the join is accepted, the consequence is that one can assume a right margin, too, for the second fragment, or, in this case, 4Q561 1 ii 2–8.

Furthermore, from PAM 43.598 it is apparent that Starcky joins 4Q561 1 i with 4Q561 1 ii.[152] This join is probable. One has to assume, however, that the leather of 4Q561 1 i extends a bit further to the left in order to have enough space for *yod* in עינוה[י]. This means that it is not possible to estab-

[149] Cf. 4Q394 1–2 i–v: 1.5-2 cm; 4Q448 B: 2.6 cm; 4Q298 3–4 ii: 6.5-7 cm; 4Q504 *recto*: 6.5-7 cm; 4Q511 63 iv: 2.5 cm (left edge of a sheet, including the end of a hymn and containing only three lines with three or four words each); 4Q521 2 ii: 7.5-8 cm; 4Q368 1 ii–iii: ca. 8 cm (the *tefilim* are not taken into account, while 4Q341 [ca. 5 cm] is a writing exercise on a scrap of leather). Cf. H. Stegemann, "Methods for the Reconstruction of Scrolls from Scattered Fragments," in *Archaeology and History in the Dead Sea Scrolls: The New York University Conference in Memory of Yigael Yadin* (ed. L.H. Schiffman; JSPSup 8; Sheffield: JSOT Press, 1990), 189-220, at 198 n. 78; Tov, *Scribal Practices*, 83. In Biblical manuscripts the phenomenon of narrow columns is more common, see E. Eshel, H. Eshel and A. Yardeni, "4QApocryphal Psalm and Prayer," in *Qumran Cave 4.VI: Poetical and Liturgical Texts, Part 1* (eds. E. Eshel et al.; DJD 11; Oxford: Clarendon, 1998), 403-25, at 404 n. 5.

[150] See PAM 43.598.

[151] Cf. PAM 43.598.

[152] Cf. also Wise, "4Q561," 228; Holst and Høgenhaven, "Physiognomy," 36.

lish an exact physical join. But assuming that the extant fragments of the original manuscript were in the vicinity of each other,[153] the join is likely. Both texts preserve a more or less similar top margin. In addition, 4Q561 1 i preserves the left part of the column, and 4Q561 1 ii has a clear right margin. Because of these material considerations the join is probable.

4Q561 2 consists of three fragments joined together by Starcky.[154] The join seems materially sound. The shape of the fragments fits neatly with each other, and in the case of *šin* (שקוהי, l.3), *reš* (רגלה, l.5), and *yod* (יפלֿ/קֿ, l.6) the letters fit all three fragments. No top, bottom, or side margins can be discerned.

It is possible that this fragment continues the physical description in 4Q561 1 i. 4Q561 1 i 4–5 describe a person's limbs, while 4Q561 2 3–5 describe someone's thighs and foot. Assuming that the *a capite ad calcem* principle guided the descriptions of the body in *4QPhysiognomy ar* 4Q561 2 could be placed beneath 4Q561 1 i as a continuation of that column, but this cannot be ascertained materially.[155]

Wise joins a fragment from another manuscript that Starcky coined *ar S* to the left of 4Q561 2.[156] This latter text consists of three fragments that appear at the top in PAM 43.598. The join is improbable for two reasons. First, there is no material join. Second, the scribal format is slightly different. The interlinear space is 0.4 cm between *ar S* 2 2 and *ar S* 2 3, but 0.3 cm between 4Q561 2 3 and 4Q561 2 4, while Wise suggests the first to be to a continuation of the latter. Starcky probably judged correctly when assigning the three fragments of *ar S* to another manuscript, although both *4QPhysiognomy ar* and *ar S* were perhaps copied by the same scribe.[157]

4Q561 3 clearly preserves the left part of the column. This is evident from the amount of blank space to the left of ll.1–3. No top or bottom margins have been preserved.

Wise places this fragment under 4Q561 1 i 5.[158] But this join is incorrect.[159] First, the state of the leather of 4Q561 3 seems different from that of 4Q561 1 i. From the photographs its surface seems less smooth. Second, if one were to follow Wise's reading of 4Q561 1 i 5, the possible *nun* in 4Q561 3 1 represents the continuation of final *nun* from דקין in 4Q561 1 i

153 Cf. Stegemann, "Methods for the Reconstruction," 206.

154 In PAM 41.954 only the right part appears, whereas in PAM 43.598 two more fragments are joined.

155 Cf. also García Martínez, *Dead Sea Scrolls Translated*, 456.

156 Wise, "4Q561," 228. Cf. also the other two fragments assigned to the same text, Wise, "4Q561," 230.

157 Holst and Høgenhaven, "Physiognomy," 39. *Qop* is written a bit differently in that the down stroke in *4QPhysiognomy ar* is straight, while that in *ar S* is curved ("s"-shaped), but this need not indicate another scribe.

158 Wise, "4Q561," 228-29.

159 See PAM 41.944; 43.598.

5, and a trace of the lower part of the reconstructed final *nun* from לעבי֯ן֯[in 4Q561 1 i 5 should also be visible in 4Q561 3 1 to the left of לה֯ from line 2. But this is not the case. For these reasons this fragment is best treated separately.

4Q561 4 is a small fragment that contains the left part of the column, as is shown by the blank space following מ֯מר[in l.1.

4Q561 5 is a bit more substantial than the previous fragment, but it has no indications of any margins. It cannot be determined whether this fragment preceding 4Q561 2 originally stood to the right of 4Q561 1 i as Wise suggests.[160]

From Starcky's transcriptions in the *Preliminary Concordance* one can infer that he considers 4Q561 6 to have preserved the right part of the column. This assumption seems reasonable, given the fact that the words in three lines begin exactly on the same vertical line.

Wise joins this fragment with 4Q561 5, because of the upper part of a final *nun* in 4Q561 5 4 and the lower part of a final *nun* in 4Q561 6 1.[161] But it is doubtful whether the shape of the leather of both fragments permits this join. Therefore, I treat both as separate fragments.

Finally, although Starcky assigns 4Q561 7 to *4QPhysiognomy ar*, I think it doubtful whether it belongs to the same manuscript. The distance between the two lines is 0.2 cm, whereas the line distance in *4QPhysiognomy ar* is ca. 0.3-0.4 cm.[162]

Paleography and Date

The script of *4QPhysiognomy ar* can be characterized, according to the Cross typology, as an early Herodian "Round" semiformal hand, with some features that are late Hasmonean.[163]

ʾAlep has a right arm thickened at the top, but this has not yet developed into a *serif*, and the left leg joins the oblique axis at the top and is bent to the right. *Bet* appears with an angular corner, although in some cases it still has something of its tick at the right, upper shoulder,[164] but less pronounced than in *4QZodiacal Physiognomy*. *Bet* does not yet appear with the baseline extending beyond the vertical right down stroke, the "tail." *Gimel* has a right down stroke that is gently curved, but it is not yet clearly bent to the right at the top.[165] The left leg is connected higher on the right down

[160] Wise, "4Q561," 228.

[161] Wise, "4Q561," 228.

[162] See PAM 41.944; 43.598. In PAM 43.598 another unidentified small fragment appears upside down. It probably does not belong to *4QPhysiognomy ar*, see n. 147 above.

[163] Cross, "Development of the Jewish Script," 166-81. See also the paleographic analysis of Holst and Høgenhaven, "Physiognomy," 30.

[164] 4Q561 6 4: רב; 7 2: ו֯י֯תעב.

[165] But see 4Q561 4 2: ת֯גרל; 5 3: וסגלגל.

stroke. *Dalet* has already an "s"-shaped right leg characteristic of the semiformal style.[166] *He* has a crossbar or roof that is thick and heavily shaded. *Waw* and *yod* are clearly distinguished. The head of *yod* is larger than that of *waw*.[167] *Ḥet* has a right leg curved inward.[168] *Ṭet* is relatively broad. Medial *kap* has a down stroke that is straight[169] or curves outward to the right.[170] *Lamed* has a curved upper arm, sometimes thickened at the top,[171] but its hook is not yet very well pronounced. The medial *mem* has two forms: an older form, following late Hasmonean style, in which the left oblique stroke is penned last,[172] and a younger form (early Herodian) in which the left oblique is made first.[173] Medial *nun* appears with a down stroke bent to the right and slightly thickened at the top, but without *serif*. Final *nun* has a bent or shaded head. *Samek* appears in two forms: an older form in which it is not yet fully closed,[174] and a younger one in which it is.[175] *ʿAyin* has a right down stroke that is sharply curved at the top, while the arm below the top tends to straighten, which according to Cross is characteristic of the transition between late Hasmonean and early Herodian.[176] Medial *pe* has a sharp head. Final *pe* has the tendency to curl the head under toward the right down stroke.[177] *Qop* is penned continuously, forming a slight loop,[178] but also, apparently, with two strokes of the pen.[179] *Reš* has some variation in the width of the head. *Šin* has a right arm that shows the similar tendency like the right arm of *ʿayin* to straighten and sharply curve upwards at the top.[180] The left down stroke does not continue below the right arm. *Taw* shows the development in late Hasmonean script from being a "kerned" letter[181] to having a straight left leg at the top.[182]

On paleographic grounds a date for the manuscript of *4QPhysiognomy ar* between ca. 50–25 BCE, according to the Cross dating, seems probable.

[166] 4Q561 1 i 3: ודקנה; 2 2: די.
[167] 4Q561 1 i 4: אברוהי.
[168] 4Q561 1 i 5: [מ]מחקי̊ן.
[169] 4Q561 1 ii 7: כעבין; 6 2: כתפה.
[170] 4Q561 1 i 2: לאכומן.
[171] 4Q561 1 i 5: לעבי̊[ן].
[172] 4Q561 1 i 1: מערבין; 1 ii 2: [ו]ל[מ]תמלי; 5 2: לשמקמוקי.
[173] 4Q561 1 i 2: לאכומן; 1 ii 8: לקומתה; 2 7: ל]מסף.
[174] 4Q451 5 3: וסגלגל.
[175] 4Q561 2 7: ל]מסף.
[176] 4Q561 1 i 1: עינוה[י]; 1 i 5: לעבי̊[ן]; 1 ii 7: כעבין; 3 3: עקא.
[177] 4Q561 2 4: כ̊ף.
[178] 4Q561 1 i 4: ד̊ק; 1 ii 8: לקומתה.
[179] 4Q561 2 3: שקוהי.
[180] 4Q561 1 i 3: ושנוהי and שוין; 2 3: שקוהי.
[181] 4Q561 2 3: פתין; 6 2: כתפה.
[182] 4Q561 1 ii 8: לקומתה.

The manuscript was copied perhaps a bit earlier than *4QZodiacal Physiognomy*, but it was done more or less in the same period.

Contents

The fragments of *4QPhysiognomy ar* are the remains of a physiognomic catalogue that, as far as can be determined, contained at least two elements.

First, it is evident that several physiognomic types are listed in the extant fragments. Similar to *4QZodiacal Physiognomy*, the descriptions of the body run from head to toe.[183] Wise reconstructs at least five such descriptions from the extant text.[184] This might be correct, but it is also possible that some of the fragments he lists as separate descriptions belong to others, which would diminish the number of extant accounts.

Second, there is some evidence that the text originally listed prognostics for each physiognomic type. In 4Q561 3 3 the word עקא, "trouble," appears at the end of the line. This may have been an indication that a person with a certain kind of physical shape and appearance would encounter some sort of trouble at a certain point of time during his life. The occurrence of ל]מסף, "to]come to an end," in 4Q561 2 7 perhaps indicated the number of years when the person's life would come to an end.[185] 4Q561 7 2 says that something will be done (ו֯יתעבד֯), but what that might have been is unknown. These examples can be taken as indications of the person's future state. If so, this means *4QPhysiognomy ar* did contain certain interpretations of the physiognomic data enumerated.[186]

In addition, one should allow for the possibility that *4QPhysiognomy ar* "établit un lien entre l'esprit et le corps."[187] Starcky reads ר]וח לה in 4Q561 3 2, but his reading and understanding are not so straightforward. If *4QPhysiognomy ar* establishes a link between (human) spirit and body, it cannot be determined how exactly this was expressed, due to the fragmentary state of the manuscript.

Transcription and Translation

For notes and comments on readings see Appendix I.

4Q561 1 i

1 [ ]נׄוׄה֯י מׄעׄרבין ולא שגיא עינוה[י]
2 בׄין אׄרין לאכומן אפה נגיד

[183] See 4Q561 1 i, 4Q561 1 ii, and 4Q561 2.
[184] See Wise, "Aramaic Horoscope," 567-68.
[185] See also the notes on readings in 4Q561 2 7 in Appendix I.
[186] See, however, Geller, "New Documents from the Dead Sea," 227-29, who argues that this is not the case.
[187] Starcky, "Un texte messianique araméen," 64.

3 שׁפיר ושנוהי שוין ודקנה ·
4 דק להוה[ו]לא שגיא אברוהי
5 [מ]מחקי[ן ו]בי[ן מג]דמין לעבי[ן]

1. his [...] will be mixed but not too much. Hi[s] eyes
2. are between light and dark-colored. His nose is long
3. <and> beautiful, and his teeth are even. And his beard
4. will be thin [but] not too much so. His limbs
5. [are s]mooth[and]be[tween st]umped and thic[k]

4Q561 1 ii

1 להוה קל[ה
2 [ו]ל[מ]תמלי ות[
3 (ו)ל]א אריך[]··ה/ח [
4 ושער דקנה שח[ם
5 להון בין עבין ל[דקין
6 ואנין קטינ[ן
7 כעבין טפרוה[י
8 לקומתה ו[
9]ל[

[(And) someone
1. [his] voice will be [
2. [and f]illed (?) ...[
3. [(and) no]t long ... [
4. and the hair of his beard is da[rk/bla[ck
5. will be between thick and [thin
6. and they are slender[
7. are somewhat thick. Hi[s] nails[
8. Regarding his height and[
9.]...[

4Q561 2

1]מו[
2]די אניין כוין [
3] פתין שקוהי[
4] לעבין כף רגלה[
5]·ג [פתי] לה רגלה[
6]לת יפל/ק[
7 ל]מסף · [
8]קדו/י[

1.]...[
2.]they have burn marks (?) [
3.] broad. And his thighs[

4.] and thick. The sole of his foot[
5.]… [broad (?)] is his foot [
6.]… …[
7. to]come to an end …[
8.]…[

4Q561 3

]ן֯[1
ר]וח לה/קר]יח לה 2
] עקא 3
]שׂערן עׂבות 4

1.]…[
2.] he has a [sp]irit (?) / bald/smooth] is his (?)
3.] trouble
4.]thick hairs

4Q561 4

]·[]מו 1
]תׄרגל 2
]עׄל 3

1.]…[]…
2.]wavy (?)
3.]upon

4Q561 5

]··[1
]לׄשמקמוקׄיׄ []·[2
ב]ררי וסגלגל להׄוׄ]ה 3
לב]ןׄ לה שׂער רשה] 4

1.]…[
2. (between)…]and red[ ]…[
3.] will b[e br]ight and round [
4. white (?)] is the hair of his head[

4Q561 6

]ןׄ [1
כתפה [2
להון עלׄ] 3
ולא רב] 4

1.]…[
2.] his shoulder [
3.] they will be upon[

4.] and not large[

4Q561 7

1 ושגיא̊ן]
2] [ו̊יתעבד̊ן]

1. and abundant [
2. [ ]and it will be done[

Body and Spirit?

There is no unequivocal evidence indicating that the physiognomics in *4QPhysiognomy ar* is aimed at discerning people's spirit or character.

Starcky reconstructs ר]וחלה in 4Q561 3 2, and understands this as parallel to רוח לו in *4QZodiacal Physiognomy*. He states that this proves that *4QPhysiognomy ar* establishes a link between body and spirit. Furthermore, on the basis of the perceived parallel with *4QZodiacal Physiognomy*, Starcky suggests that *4QPhysiognomy ar* is also astrological.[188] What this means is that *4QPhysiognomy ar* would also have mentioned the spirit being divided between the "house of light" and the "house of darkness." This would have been so because "le début de la formule caractéristique est: 'il a comme esprit tant de parts de lumière et tant de ténèbres.'"[189] This line of reasoning has been accepted by other scholars, who take the words רוח לה to mean "his spirit," while some also assume that *4QPhysiognomy ar* contains references to the "house of light" and the "house of darkness."[190]

Regarding the references to light and darkness, it is evident that these do not actually occur in the extant text. The assumption of such a reference hinges entirely on the words רוח לה. However, even if ר]וח לה were the correct reconstruction, it is very doubtful whether 4Q561 3 2–3 contained references to a division between the "house of light" and the "house of darkness," in the way that these occur in *4QZodiacal Physiognomy*. Such a reconstruction of 4Q561 3 2–3 is highly unlikely because of the column width of *4QPhysiognomy ar*, which probably measures no more than ca.

[188] Starcky, "Un texte messianique araméen," 64-65.

[189] Starcky, "Un texte messianique araméen," 62: "Elle (the introduction to the characteristic formula) se retrouve sur l'un des fragments de notre manuscrit *4QHor ar*: *r*]*wḥ lh* [."

[190] See the note on 4Q561 3 2 in Appendix I, and add Alexander, "Physiognonomy," 393: *4QPhysiognomy ar* "evinces concern about the proportions of light and darkness"; A. Lange, "The Essene Position on Magic and Divination," in *Legal Texts and Legal Issues: Proceedings of the Second Meeting of the International Organization for Qumran Studies Cambridge 1995: Published in Honour of Joseph M. Baumgarten* (eds. M. Bernstein, F. García Martínez and J. Kampen; STDJ 23, Leiden: E.J. Brill, 1997), 377-435, at 390: "the same view of human beings as composed of different parts of light and darkness is expressed in 1QS 4:15-18, 23-25 and the text preserved in manuscripts 4Q186 and 4Q561."

5.5-6.0 cm (see above). The words ר]וח לה stand at the end of the line in 4Q561 3 2. This means that references to the "house of light" and the "house of darkness" would have stood at the beginning of 4Q561 3 3. However, at the end of this line there is the word עקא, indicating some sort of trouble that may befall someone (presumably the type of individual whose body was described in the original text). The word עקא did not begin a phrase, but was probably introduced by something like להוה, "there will be trouble for him" (להוה לה]עקא). In any case, on the basis of the column width there is not enough space in 4Q561 3 3 previous to עקא for references to certain numbers in the "house of light" and the "house of darkness," as in *4QZodiacal Physiognomy*; it is materially impossible. One should perhaps allow for the possibility that the formula from *4QZodiacal Physiognomy* occurred in abbreviated form in *4QPhysiognomy ar*. It is, however, difficult to imagine how this was done because the formula in *4QZodiacal Physiognomy* is itself already elliptical. In conclusion, there is no evidence which suggests that *4QPhysiognomy ar* originally contained references to light and darkness.

Assuming that Starcky's reconstruction ר]וחלה is correct, this raises the question of what these words might mean. Scholars have simply translated these words with "his spirit." The construction רוחלה, however, is not easy to reconcile with "his spirit." For "his spirit" one would instead expect רוחה.[191] In the physiognomic descriptions in the rest of *4QPhysiognomy ar* the nouns referring to the different body parts have suffixed pronouns attached to them,[192] but this is not exactly the same as a noun followed by a particle with suffixed pronoun. To be more precise, the words רוח לה should be understood as "he has a spirit" or "there is a spirit for him";[193] a point acknowledged by Starcky's translation "il a comme esprit..." (see above). It is not clear what is meant by "he has a spirit" though. The word ר]וח is not further qualified by an adjective[194] or noun.[195] Perhaps a more elaborate qualification followed subsequently to ר]וח לה, but this cannot be determined. It is possible that this construction was used on purpose to accentuate the difference with the physiognomic descriptions, but this is not clear; nor, for that matter, is it evident that the construction רוח לה in fact appears in *4QPhysiognomy ar*.

[191] Cf. 4Q541 10 4; 4Q545 9 2; 4Q550 1 4.

[192] 4Q561 1 i 1: עינוה]י[; 1 i 2: אפה; 1 i 3: ושנוהי and ודקנה; 1 i 4: אברוהי; 1 ii 4: ושער דקנה; 1 ii 7: טפרוה]י; 1 ii 8: לקומתה; 2 3: שקוהי; 2 4: כף רגלה; 2 5: רגלה; 5 4: שער רשה; 6 2: כתפה.

[193] Cf. also 1Q20 22:31; 4Q196 6 11.

[194] See e.g. $1QH^a$ 8:19 (according to *DSSSE*); 4Q393 1 ii–2 5; 4Q511 15 7; 81 3; 4Q538 1–2 4; 4Q542 1 i 10; 11Q5 19:15; 11Q6 4–5 16. Note also 4Q230 1 1, see E.J.C. Tigchelaar, "'These Are the Names of the Spirits of...': A Preliminary Edition of *4QCatalogue of Spirits (4Q230)* and New Manuscript Evidence for the *Two Spirits Treatise (4Q257* and *1Q29a)*," *RevQ* 21/84 (2004): 529-47, at 531.

[195] See e.g. 4Q230 1 2–3.

As already said, Starcky's reading and understanding of ר[וח לה is based on the occurrence of רוח לו in *4QZodiacal Physiognomy*,[196] but would it also have been proposed if that text were not known to us?

On the basis of *4QPhysiognomy ar* and the description of Sarai in the *Genesis Apocryphon*, one would anticipate an adjective prior to לה in 4Q561 3 2. The construction ל with attached suffixed pronoun occurs two more times in *4QPhysiognomy ar*, but the context is too fragmentary to determine the exact sense in these cases. In 4Q561 2 5 it is possible that an adjective preceded לה in the lacuna and described what the foot looks like, perhaps something like "he has a broad foot" (פתי[לה רגלה). In 4Q561 5 4 an adjective describing the hair is possible, such as "he has white hair" (לב[ן̊ לה שׂער רשה). The complete construction is clearly preserved in 1QapGen ar 20:2–4. Subsequently to an adjective comes *lamed* with suffixed pronoun, which is followed by a suffixed noun referring to a part of Sarai's body: ושפיר לה צלם אנפיהא ("beautiful the form of her face"); רקיק לה שער ראישה ("soft the hair of her head"); יאאלה חדיה ("lovely her breast").[197]

In the case of 4Q561 3 2 the letter prior to *ḥet* is most likely *waw*, but *yod* cannot be excluded, which suggests as a plausible adjective, for example, קריח ("bald, smooth"). This would mean that this line of the fragment contains part of a physiognomic description, perhaps portraying the head or the legs, but due to the fragmentary state it is not possible to be more specific.

In conclusion, perhaps 4Q561 3 2 referred in some way to the spiritual state of the described individual. But what was said about it cannot be surmised from this text, or from its presumed identicalness with *4QZodiacal Physiognomy*. The manuscript is too fragmentary to determine its exact meaning or to relate it to רוח לו in *4QZodiacal Physiognomy*.

[196] It seems certain that the occurrence of רוח לו in *4QZodiacal Physiognomy* was known prior to that of in ר[וח לה in *4QPhysiognomy ar*. 4Q186 1 ii 7 appears for the first time in PAM 40.615, a photograph taken in May 1953 (the material was acquired on September 20, 1952), while 4Q561 3 2 appears for the first time in PAM 41.944, a photograph taken in February 1956. Allegro arrived in Jerusalem to work on the material in October 1953, while Starcky received his lot of manuscripts in January 1954, after Allegro had already deciphered part of *4QZodiacal Physiognomy*. Allegro worked on the material in late 1953 (see Brown, *John Marco Allegro*, 29-30), while J. Starcky, "Le travail d'édition des fragments manuscrits de Qumrân: Communication de J. Starcky," *RB* 63 (1956): 66-67, did not yet comment on *4QPhysiognomy ar*, which might have been expected if the perceived parallel with a text from Allegro's lot had been known.

[197] Cf. also 1QapGen ar 20:3–4.6: יאין להון לה עיניהא ("lovely her eyes"); רגג הוא לה אנפהא ("graceful her nose"); שפיר לה כול לבנהא ("beautiful all her whiteness"); שלמא להן לה שקיהא ("perfect her legs"). The latter example is slightly different in that it uses a participial form in an adjectival sense.

THE RELATIONSHIP BETWEEN *4QPHYSIOGNOMY AR* (4Q561) AND *4QZODIACAL PHYSIOGNOMY* (4Q186)

According to Starcky, *4QPhysiognomy ar* represents "la version (ou l'adaptation) araméenne" of the Hebrew text *4QZodiacal Physiognomy*, and he suggests it thus also contained certain astrological information.[198] Starcky does not indicate what sort of astrology he is thinking of, but it seems likely he had something of a similar zodiacal nature as *4QZodiacal Physiognomy* in mind.

4QPhysiognomy ar has in common with *4QZodiacal Physiognomy* the enumeration of physiognomic descriptions of evidently different types.[199] These descriptions follow a pattern running from head to toe (*a capite ad calcem*). Another element possibly common to both texts is that they offer interpretations of the physiognomic types with regard to a future situation of people.[200]

Being descriptions of the human body, it is not surprising to find terminological overlap between these accounts in both texts. There are, of course, parts of the body that are referred to.[201] Furthermore, equivalent or similar descriptions are used for the physical traits,[202] and also with regard to syntactical construction there is common ground.[203] The physiognomic descriptions are ordered according to the *a capite ad calcem* principle, but, at the same time, a comparison between the accounts in both texts shows

[198] Starcky, "Un texte messianique araméen," 51, 65. Cf. J. Starcky, "Les quatre étapes du messianisme à Qumrân," *RB* 70 (1963): 481-505, at 503 n. 66. Milik understood both texts as part of a Book of Noah, see J.T. Milik, "Problèmes de la littérature Hénochique à la lumière des fragments araméens de Qumrân," *HTR* 64 (1971): 333-78, at 366; J.T. Milik, *The Books of Enoch: Aramaic Fragments of Qumrân Cave 4* (Oxford: Clarendon, 1976), 56. See also García Martínez, *Dead Sea Scrolls Translated*, 513; Lange, "Essene Position on Magic," 387-90.

[199] Zimmermann, *Messianische Texte*, 194; Holst and Høgenhaven, "Physiognomy," 40.

[200] Cf. 4Q561 2 7: ל]מסף ("to]come to an end"); 4Q561 3 3: עקא ("trouble"), and possibly 4Q186 1 ii 9: עני יהיה ("he will be poor").

[201] See for the *eyes*: 4Q561 1 i 1: עינוה[י], and 4Q186 2 i 1: ועי]ניו; *teeth*: 4Q561 1 i 3: ושנוהי, and 4Q186 1 iii 6; 2 i 2: ושניו; *beard*: 4Q561 1 i 3; 1 ii 4: (ו)דקנה, and 4Q186 2 i 1: וזקנו; *voice*: 4Q561 1 ii 1: קל]ה, and 4Q186 2 i 2: ובת קולו; *thighs*: 4Q561 2 3: שקוהי, and 4Q186 1 ii 5; 1 iii 7; 2 i 5: ושוקיו; *foot sole(s)*: 4Q561 2 4: כף רגלה, and 4Q186 2 i 5: וכפות רגליו. Cf. also for the *height*: 4Q561 1 ii 8: לקומתה, and 4Q186 2 i 3-4: והואה לוא ארוך ולוא קצר.

[202] Cf. *mixed*: 4Q561 1 i 1: מערבין, and 4Q186 1 i 9: מעורבים; *beautiful*: 4Q561 1 i 3: שפיר, and 4Q186 3 3: יפי; *even/well ordered*: 4Q561 1 i 3: שוין, and 4Q186 2 i 3: ויושבות על סרכמה; *thin/sparse*: 4Q561 1 i 4: דק, and 4Q186 2 i 2: ממע]ט; *smooth*: 4Q561 1 i 5: מ]מחקין[, and 4Q186 2 i 5: חלקות; *thick*: 4Q561 1 i 5: לעבי]ן[; 1 ii 5: עבין;1 ii 7: כעבין; 2 4: לעבין, and 4Q186 1 iii 7-8: עבות; *long*: 4Q561 1 ii 3: אריך, and 4Q186 1 ii 5-6: (ו)ארוכות; 2 i 3: ארוך; *slender*: 4Q561 1 ii 6: קטינן, and 4Q186 1 ii 5-6; 2 i 4: (ו)דקות; *broad*: 4Q561 2 3: פתין, and 4Q186 1 i 8: רחבים; *wavy*: 4Q561 4 2: תגרל, and 4Q186 2 i 2: תרגל; *round*: 4Q561 5 3: וסגלגל, and 4Q186 1 i 8: סגלגלים.

[203] Cf. e.g. 4Q561 1 i 2: בין ארין לאכומן (*between* light *and* dark-colored), and 4Q186 2 i 1: בין שחורות ובנין]מנמריות (*between* black *and* speckled).

that there is no fixed sequence for enumerating the different parts of the body. For example, in 4Q561 1 i 3 the teeth are described prior to the beard, but in 4Q186 2 i 1–3 the description of the beard is followed by that of the teeth. Also, in 4Q561 1 ii 7–8 the height is mentioned subsequently to the nails, whereas in 4Q186 2 i 3–5, the fingers are described subsequently to the height.

Terminological overlaps, however, are inherent to the genre of physiognomic texts and, apart from illustrating this generic connection, do not warrant the conclusion that *4QPhysiognomy ar* and *4QZodiacal Physiognomy* are different manuscripts of the same composition. Apart from the physiognomic descriptions, it is far from clear that *4QPhysiognomy ar* is the same text as *4QZodiacal Physiognomy* containing the same elements, but then in Aramaic. This entire interpretation hinges on the reading and understanding of ר]וחלה, which is problematic.

There are no grounds for assuming that *4QPhysiognomy ar* and *4QZodiacal Physiognomy* represent the same literary composition. The similarities between both texts are satisfactorily explained against their common background in the physiognomic tradition.[204] To argue that *4QPhysiognomy ar* is "aussi astrologique" is not borne out by the textual evidence.[205]

[204] Alexander, "Physiognonomy," 393 n. 18; Holst and Høgenhaven, "Physiognomy," 42.

[205] Starcky, "Un texte messianique araméen," 65. One cannot import astrological elements from *4QZodiacal Physiognomy* into *4QPhysiognomy ar* on the basis of doubtful evidence such as a presumed equivalence between רוחלו and ר]וח לה.

CHAPTER II

TO READ STRANGE MATTERS FROM THE HUMAN BODY: PHYSIOGNOMICS IN BABYLONIAN AND GRECO-ROMAN CULTURE AND LITERATURE

INTRODUCTION

In the previous chapter, *4QZodiacal Physiognomy* and *4QPhysiognomy ar* were presented as physiognomic texts – in the case of *4QZodiacal Physiognomy* we possess, more precisely, the remains of a physiognomic-astrological list. How do these two texts relate to Babylonian and Greco-Roman physiognomic texts and can texts from these two cultural realms help to elucidate certain features of the Qumran texts?

What is perceived as the purpose of physiognomic inquiry in these traditions, and according to which principles and methods is the physiognomic art supposed to operate? In other words, what is signified by the human body and how is the relationship between signifier and signified rationalized? It has been argued that the Qumran texts resemble Babylonian omen lists, but closer scrutiny of the textual format in both traditions does not corroborate that observation. What does this mean for a possible Babylonian or Hellenistic origin for the Qumran physiognomic texts? If *4QZodiacal Physiognomy* is a physiognomic-astrological compendium, how does its combination of these two elements relate to Babylonian and Greco-Roman texts that also combine astrology and physiognomic learning?

From a comparative perspective, this chapter will discuss Babylonian and Greco-Roman physiognomic traditions and their cultural and social contexts. The textual evidence for physiognomic literature will be presented and the function of these texts assessed, as well as the people cultivating this knowledge. Who practiced the physiognomic art and for what purpose? Who had access to the technical physiognomic texts? How widespread was knowledge of the physiognomic art and on what level? Some of these issues will return in Chapter Five in relation to the physiognomic texts from Qumran.

ANCIENT REFLECTIONS ON THE PHYSIOGNOMIC ART AND ITS PURPOSE

There is no explicit reflection on physiognomics in cuneiform literature, but the so-called *Esagil-kīn-apli Catalogue* does provide a definition of the physiognomic omen series *Šumma alamdimmû* ("If the form").[1] This catalogue is extant in a Neo-Assyrian and a Neo-Babylonian copy, but probably dates to the eleventh century BCE. Esagil-kīn-apli was a Babylonian scholar from Borsippa active during the reign of the Babylonian king Adad-apla-iddina (1069–1048 BCE).[2] In the biographical section of the *Esagil-kīn-apli Catalogue* the reader finds the following statement:

> *Alamdimmû* (concerns) external form and appearance (and how they imply) the fate of man that Ea and Asalluḫi/Marduk(?) ordained in Heaven.[3]

This definition describes the subject matter of the omen series *Alamdimmû* to have been the shape and appearance of the human body and what these mean for a person's fate.[4] The definition presents Babylonian physiognomics as a divinatory art predicting people's futures. This impression is confirmed by the omens in *Alamdimmû*. Most omen apodoses give predictions concerning health, length of life, wealth, offspring, family, and death.

On a textual level, Babylonian physiognomics was, therefore, principally a divinatory art that predicted people's future situations on the basis of their physical characteristics. This judgment, however, needs some qualification because Babylonian physiognomics also seems to have been partially concerned with the discernment of character.[5] Alongside the overwhelming number of predictive apodoses there are a few that provide clues about someone's character, for example:

> If the hair on his head is red, (*variant*) he is trustworthy. (II:87)
> If there is on the right side (of his face) a *ṣiršu*-pimple, he flourishes, he is modest. (VIII:125)

[1] The learned word *alamdimmû* ("form, figure"), used in scholarly texts, is a late, rare loanword from Sumerian alan-dim$_2$. See F.R. Kraus, *Die physiognomischen Omina der Babylonier* (MVAG 40/2; Leipzig: J.C. Hinrichs, 1935), 1; CAD A/I 332b s.v. *alamdimmû*.

[2] See I.L. Finkel, "Adad-apla-iddina, Esagil-kīn-apli, and the Series SA.GIG," in *A Scientific Humanist: Studies in Memory of Abraham Sachs* (eds. E. Leichty, M. deJong Ellis and P. Gerardi; OPSNKF 9; Philadelphia: The University Museum, 1988), 143-59; B. Böck, *Die babylonisch-assyrische Morphoskopie* (AfO.B 27; Vienna: Institut für Orientalistik der Universität Wien, 2000), 14-15, 55; N.P. Heeßel, *Babylonisch-assyrische Diagnostik* (AOAT 43; Münster: Ugarit, 2000), 13-17, 104-5.

[3] A66-8/B29'-30', Finkel, "Adad-apla-iddina," 148-49.

[4] The *Esagil-kīn-apli Catalogue* is rather unique in this regard. In Babylonian divination literature one looks in vain for introductory remarks that explicitly define the subject matter of a particular omen series and explain and justify the procedure followed in it. There is no conscious reflection on the Babylonian scholarly endeavors extant in cuneiform sources.

[5] See Barton, *Power and Knowledge*, 100; Böck, *Die babylonisch-assyrische Morphoskopie*, 1, 29-36.

> If he is of evil heart, he is a liar. (X:41)[6]

Greco-Roman literature provides much more evidence for reflections, either favorable or unfavorable,[7] on the goal of physiognomic inquiry. The interest of Greco-Roman physiognomics was in the judgment of people's characters. The Anonymous Latin author of the *De physiognomonia liber* (fourth century CE) succinctly states that physiognomics claims "to consider and discern the character of the soul from the character of the body."[8] Aulus Gellius (second century CE) is more elaborate and says that physiognomics:

> means to inquire into the character and dispositions of men by an inference drawn from their facial appearance and expression, and from the form and bearing of their whole body.[9]

To practice physiognomics was to observe the human body in detail and from that to figure out people's characters. The Pseudo-Aristotelian author of the *Physiognomonica*, a treatise written at the end of the fourth or the beginning of the third century BCE, describes the objective of physiognomic inquiry as follows:

> The science of physiognomics, as its name implies, deals with the natural character traits of mental character, and with such acquired ones as on their occurrence modify the characteristic signs studied by the physiognomist.[10]

The Peripatetic author explicates upon which signs of the body the physiognomist based his judgment about someone's mental character. These types of signs were:

[6] Böck, *Die babylonisch-assyrische Morphoskopie*, 80-81, 114-15, 120-21.

[7] E.g. Sextus Empiricus, *Outlines of Pyrrhonism* 1.85 (favorable); Artemidorus, *The Interpretation of Dreams* 2.69; Clement of Alexandria, *Stromata* 1.135.2 (unfavorable, but cf. *The Instructor* 3.3.15 where Clement is not negative about metoposcopy, contrary to Hippolytus, *Refutation of All Heresies* 4.15.3–4). See Evans, *Physiognomics*, 50, 73-74, 78 n. 48; Barton, *Power and Knowledge*, 99; B. Isaac, *The Invention of Racism in Classical Antiquity* (Princeton: Princeton University Press, 2004), 155-56. See furthermore V. Tsouna, "Doubts about Other Minds and the Science of Physiognomics," *CQ* 48 (1998): 175-86.

[8] Anonymous Latin, *De physiognomonia liber* §2, see J. André, *Anonyme Latin: Traité de physiognomonie* (Budé; Paris: Les Belles Lettres, 1981), 50.

[9] Aulus Gellius, *Attic Nights* 1.9.2. Translation from J.C. Rolfe, *The Attic Nights of Aulus Gellius* (LCL 195; Cambridge, Massachusetts: Harvard University Press, 1954), 45-47. Standing in the Western physiognomic tradition, both descriptions concur with the modern definition that physiognomics is "the art of judging character and disposition from the features of the face or the form and lineaments of the body generally." See *OED*, s.v. physiognomy.

[10] Pseudo-Aristotle, *Physiognomonica* 806a 23–25. Text and translation, adapted, from W.S. Hett, *Aristotle XIV: Minor Works* (LCL 307; Cambridge, Massachusetts: Harvard University Press, 1980), 90-93, and T. Loveday and E.S. Forster, "Physiognomonics," in *The Complete Works of Aristotle: The Revised Oxford Translation* (ed. J. Barnes; vol. 1; Princeton: Princeton University Press, 1995), 1237-50, at 1239. All translations of the *Physiognomonica* are taken, sometimes in adapted form, from Hett, unless otherwise noted.

> movements, shapes and colors, and from habits as appearing in the face, from the growth of hair, from the smoothness of the skin, from voice, from the condition of the flesh, from parts of the body, and from the general character of the body.[11]

The assumption is that these types of signs and their appearances are mutually connected with the natural traits of mental character; through them mental character expresses and shows itself. If that is true then the types of bodily signs point the physiognomist to types of mental character. This reveals the basic premise of Greco-Roman physiognomics, namely the inherent sympathy between body and soul. The correspondence between both is the justification for practicing the science of physiognomics. This is clearly stated in the *Physiognomonica*:

> soul and body react on each other; when the character of the soul changes, it changes also the form of the body, and conversely, when the form of the body changes, it changes the character of the soul. [...] Now if this is true (and it is invariably so), then it should be possible to physiognomize.[12]

Greco-Roman physiognomics was by and large concerned with the discernment of people's characters, whereas the predictive function was minimal.

There are some anecdotes about predictions made by metoposcopes, but few predictions occur in the learned treatises.[13] As Tamsyn Barton points out, this focus of Greco-Roman physiognomics on character is "in contrast to the astrology of the period. Whereas today's clients of astrologers are interested in a character analysis, the ancients were far more eager for predictions, preferably of success in all areas of life."[14] In the Greek *zodiologia* both aspects of physiognomics and astrology were combined. These extensive lists mention the various psychological and physical characteristics attributed to those born under each zodiacal sign, as well as predictions concerning their future welfare.

In general, therefore, one may say that the interest of Babylonian physiognomics was primarily directed at predicting people's fate, while Greco-Roman physiognomics was mainly concerned with revealing people's character. Ancient definitions of physiognomics demonstrate the different interests of physiognomic learning regarding what the human body signifies in these cultures.

[11] *Physiognomonica* 806a 27–34. Cf. A. Degkwitz, *Die pseudoaristotelischen 'Physiognomonica' Traktat A: Übersetzung und Kommentar* (Diss. Heidelberg, 1988), 66; Vogt, *Physiognomonica*, 309-10.

[12] *Physiognomonica* 808b 11–14; 805a 18, cf. also 805a 1–18. Cf. Degkwitz, *'Physiognomonica' Traktat A*, 56-58; Vogt, *Physiognomonica*, 394-96.

[13] See J. Mesk, "Die Beispiele in Polemons Physiognomonik," *WS* 50 (1932): 51-67, at 61-64; Barton, *Power and Knowledge*, 100 n. 37; M.W. Gleason, *Making Men: Sophists and Self-Presentation in Ancient Rome* (Princeton: Princeton University Press, 1995), 30, 48-51.

[14] Barton, *Power and Knowledge*, 100. Cf. also Evans, *Physiognomics*, 73.

BACKGROUNDS OF PHYSIOGNOMIC LEARNING AND LITERATURE: MESOPOTAMIA

In ancient Mesopotamia a scholarly literature of specialized omen series evolved in which physiognomic learning was handed down for almost two millennia.[15] Physiognomic omens were one of the many classes of Babylonian divination texts. Babylonian physiognomics should, therefore, first of all be understood against the general background of divination literature.[16]

Textual Evidence

The textual evidence for Babylonian physiognomics covers a wide range in time. The oldest evidence for physiognomic omens dates back to the Old Babylonian period (ca. 2000–1600 BCE). At the end of the second millennium the physiognomic material was collected in the standard series called *Šumma alamdimmû*.[17]

[15] In addition to the publications by Fritz Kraus, the most recent edition by Barbara Böck serves as the major reference to these texts in this study. See Kraus, *Die physiognomischen Omina der Babylonier*; F.R. Kraus, "Ein Sittenkanon in Omenform," *ZA* 43 (1936): 77-113; F.R. Kraus, "Babylonische Omina mit Ausdeutung der Begleiterscheinungen des Sprechens," *AfO* 11 (1936-1937): 219-30; F.R. Kraus, *Texte zur babylonischen Physiognomatik* (AfO.B 3; 1939; repr., Osnabrück: Biblio, 1967); Böck, *Die babylonisch-assyrische Morphoskopie*.

[16] See e.g. the treatments of Mesopotamian divination by J. Nougayrol, "La divination babylonienne," in *La divination: Tome Premier* (eds. A. Caquot and M. Leibovici; Paris: Presses Universitaires de France, 1968), 25-81; J. Bottéro, "Symptômes, signes, écritures: en Mésopotamie ancienne," in *Divination et Rationalité* (eds. J.P. Vernant et al.; Paris: Éditions du Seuil, 1974), 70-197; in summarized form J. Bottéro, *Mesopotamia: Writing, Reasoning, and the Gods* (Chicago: The University of Chicago Press, 1992), 125-37 (first published as *Mésopotamie: L'écriture, la raison et les dieux* [Paris: Gallimard, 1987], 157-69); A.L. Oppenheim, *Ancient Mesopotamia: Portrait of a Dead Civilization* (Revised Edition Completed by E. Reiner; Chicago: University of Chicago Press, 1977), 206-27; F.H. Cryer, *Divination in Ancient Israel and its Near Eastern Environment: A Socio-Historical Investigation* (JSOTSup 142; Sheffield: Sheffield Academic Press, 1994), 124-215 (good surveys of Ancient Near Eastern data, but conceptually disputable); A.K. Guinan, "Divination," in *The Context of Scripture: Volume I Canonical Compositions from the Biblical World* (eds. W.W. Hallo et al.; Leiden: Brill, 1997), 421-26; S.M. Maul, "Omina und Orakel. A. Mesopotamien," *RlA* 10.1/2 (2003), 45-88; Rochberg, *The Heavenly Writing*, 44-97.

[17] Before the physiognomic material was gathered together in *Alamdimmû*, there must have been numerous texts. Unfortunately, only four tablets are known thus far from the Old Babylonian period. Nevertheless, it is clear that the material incorporated in *Alamdimmû* derived from these tablets written in the Old Babylonian period, although they were compiled in an adapted form. Apart from the standard series *Alamdimmû*, there were also extra-serial physiognomic omens, such as the so-called *aḫû*-texts, and commentaries on the compendium. See Böck, *Die babylonisch-assyrische Morphoskopie*, 9-14, 19-23. For a discussion of the genre of *aḫû*-texts, see e.g. F. Rochberg-Halton, "The Assumed 29th *aḫû*-Tablet of *Enūma Anu Enlil*," in *Language, Literature, and History: Philological and Historical Studies Presented to Erica Reiner* (ed. F. Rochberg-Halton; AOS 67; New Haven, Connecticut: American Oriental Society, 1987), 327-50; S.J. Lieberman, "Canonical and Official Cuneiform Texts: Towards an Understanding of Assurbanipal's Personal Tablet Collection," in *Lingering over Words: Studies in Ancient Near Eastern Literature in Honor of William L. Moran* (eds. T.

Alamdimmû consists of twenty-seven tablets, some of which contain more than two hundred omens.[18] These deal with different parts of the human body in the order from head to toe. The omens devote attention to, for example, the color, quality and shape of the hair, the color and quality of the skin, and the position and color of different sorts of body marks. Furthermore, *Alamdimmû* lists omens that characterize ways of speech and describe the consequences of utterances and habitual, involuntary, movements.

Although the name *Šumma alamdimmû* is used to denote the entire collection of twenty-seven tablets, it actually only belongs to the first twelve tablets. There are four subseries that cover Tablets XIII–XXVII. After *Alamdimmû* proper, the next series numbers two tablets, of which only a small fragment has survived. It was named *Šumma nigdimdimmû* ("If the shape"), but unfortunately its fragmentary state does not enable one to assess its content. The second series, titled *Šumma kataduggû* ("If the utterance"), describes in one tablet the consequences of utterances and habitual conduct, showing correspondences with omens from the terrestrial series *Šumma ālu* ("If a city").[19] The appearance of women is described in two tablets bearing the name *Šumma sinništu qaqqada rabât* ("If the head of a woman is big"). Finally, the fourth sub-series *Šumma liptu* ("If the mole") probably numbered nine tablets, of which eight tablets describe the position of different sorts of body marks on the male body, while one tablet is devoted to the female body. This is followed by one tablet, *Šumma šer'ān pūt imittišu ittenebbi* ("If the vein on the right side of his forehead throbs"), that categorizes involuntary muscle movements.[20]

Abusch, J. Huehnergard and P. Steinkeller; HSS 37; Atlanta, Georgia: Scholars Press, 1990), 305-36, at 308.

[18] The complete text of the series *Alamdimmû* has not survived, and some tablets are not extant at all. Fortunately, the *Esagil-kīn-apli Catalogue* makes it possible to reconstruct the sequence of the entire series and some of its contents. Cf. Böck, *Die babylonisch-assyrische Morphoskopie*, 14-18. For the Neo-Assyrian copy of the catalogue, see J.V. Kinnier Wilson, "Two Medical Texts from Nimrud," *Iraq* 18 (1956): 130-46; J.V. Kinnier Wilson, "The Nimrud Catalogue of Medical and Physiognomic Omina," *Iraq* 24 (1962): 52-62. This copy of the catalogue was found in the Nabû temple library, which was in continuous use from 800 BCE until its destruction around 616 BCE. For an autograph of the Nimrud Catalogue see D.J. Wiseman and J.A. Black, *Literary Texts from the Temple of Nabû* (CTN 4; British School of Archaeology in Iraq, 1996), plates 44-45. For the Neo-Babylonian duplicate, see Finkel, "Adad-apla-iddina."

[19] Cf. Kraus, "Sittenkanon"; Böck, *Die babylonisch-assyrische Morphoskopie*, 40-42. For the series *Šumma ālu* see e.g. S.M. Freedman, *If a City is Set on a Height: The Akkadian Omen Series Šumma Alu ina Mēlê šakin: Volume 1: Tablets 1-21* (OPSNKF 17; Philadelphia: The University of Pennsylvania Museum, 1998); A.K. Guinan, "A Severed Head Laughed: Stories of Divinatory Interpretation," in *Magic and Divination in the Ancient World* (eds. L. Ciraolo and J. Seidel; AMD 2, Leiden: Brill/STYX, 2002), 7-40; Maul, "Omina und Orakel," 58-62.

[20] According to Böck, *Die babylonisch-assyrische Morphoskopie*, 1, the omens of the first twelve tablets refer to the male body. This is especially clear in the case of the different aspects of the male organ that are described in the tenth tablet (X:64–125). Furthermore,

The greatest part of the standard series *Alamdimmû* that has survived derives from the royal library of the Neo-Assyrian king Assurbanipal (668–631/627? BCE) in Nineveh. However, texts of *Alamdimmû* are not limited to the Neo-Assyrian period. Copies have also been unearthed from the residential area of Seleucid Uruk.[21] This demonstrates that the Babylonian physiognomic tradition was transmitted into the Hellenistic period. In addition to this continuing transmission of the standard series *Alamdimmû*, physiognomic learning is also attested in combination with astrology in other texts during the Hellenistic period.[22] Moreover, it is perfectly possible that in the Near East, Babylonian physiognomic learning was transmitted in a different form and in combination with astrology during late antique and even early medieval times, such as in the Mandean *Book of the Zodiac* (*Sfar Malwašia*).[23]

Alamdimmû's third subseries is explicitly devoted to the physiognomy of the female body. However, in the third tablet of *Alamdimmû* proper, one of the omens describing the occurrence of cuneiform signs on the forehead refers to the female: "If the TAB or UB sign is written (on the forehead), one who is barren will bear children, (and) one who has difficult labors will labor well" (III:97). This allows for the possibility that other omens in *Alamdimmû* could also be applied to females, and should be understood in a more general sense as omens on the human body. I thank Eckart Frahm for bringing this to my attention.

21 Böck, *Die babylonisch-assyrische Morphoskopie*, 7-9. The tablets from the private house in Seleucid Uruk belonged to the library of a family whose members held the title *āšipu*, "magician-exorcist," cf. O. Pedersén, *Archives and Libraries in the Ancient Near East 1500-300 B.C.* (Bethesda, Maryland: CDL, 1998), 212-13. I owe this reference to Geert De Breucker.

22 S. Langdon, "Miscellanea Assyriaca IV," *Babyloniaca* 7 (1913-1923): 230-36; R.D. Biggs, "An Esoteric Babylonian Commentary," *RA* 62 (1968): 51-58; J. Oelsner, *Materialien zur babylonischen Gesellschaft und Kultur in Hellenistischer Zeit* (Assyriologia 7; Budapest: Eötvös University, 1986), 175; Reiner, *Astral Magic*, 77-79; Böck, *Die babylonisch-assyrische Morphoskopie*, 8, 276-79; B. Böck, "'An Esoteric Babylonian Commentary' Revisited," *JAOS* 120 (2000): 615-20; E. Reiner, "Early Zodiologia and Related Matters," in *Wisdom, Gods and Literature: Studies in Assyriology in Honour of W.G. Lambert* (eds. A.R. George and I.L. Finkel; Winona Lake, Indiana: Eisenbrauns, 2000), 421-27.

23 Regarding the transmission of certain forms of ancient Mesopotamian magic and celestial divination in Mandean tradition, see C. Müller-Kessler, "Aramäische Beschwörungen und astronomische Omina in nachbabylonischer Zeit: Das Fortleben mesopotamischer Kultur im Vorderen Orient," in *Babylon: Focus mesopotamischer Geschichte, Wiege früher Gelehrsamkeit, Mythos in der Moderne: 2. Internationales Colloquium der Deutschen Orient-Gesellschaft, 24.-26. März 1998 in Berlin* (ed. J. Renger; CDOG 2; Saarbrücken: SDV, 1999), 427-43; F. Rochberg, "The Babylonian Origins of the Mandaean Book of the Zodiac," *ARAM* 11-12 (1999-2000): 237-47; C. Müller-Kessler, "A Charm Against Demons of Time," in *Mining the Archives: Festschrift for Christopher Walker on the Occasion of His 60th Birthday* (ed. C. Wunsch; BA 1; Dresden: ISLET, 2002), 183-89; Rochberg, *The Heavenly Writing*, 298-99. The Mandean *Book of the Zodiac* merges both Babylonian and Greek material, see nn. 204, 205 below.

Regarding the survival of Babylonian medical learning, see M.J. Geller, "The Survival of Babylonian Wissenschaft in Later Tradition," in *The Heirs of Assyria: Proceedings of the Opening Symposium of the Assyrian and Babylonian Intellectual Heritage Project Held in Tvärminne, Finland, October 8-11, 1998* (eds. S. Aro and R.M. Whiting; MS 1; Helsinki: The Neo-Assyrian Text Corpus Project, 2000), 1-6; M.J. Geller, "An Akkadian Vademecum in

Authorship

The compilation of the standard series *Alamdimmû* was probably made by the Babylonian scholar Esagil-kīn-apli in the eleventh century BCE. This is based on the *Esagil-kīn-apli Catalogue*. The biographical section of this catalogue opens with an explanation for the new edition of the diagnostic compendium SA.GIG (*Sakikkû*) and the physiognomic series *Alamdimmû*.[24] After elaborately introducing himself, Esagil-kīn-apli says that he has produced these standard series: "(Regarding) the twin series, their arrangement is one."[25] He emphasizes that he has ordered its material according to the principal "from head to foot" (*ištu muḫḫi adi šēpē*).[26] It is in light of this editorial work of systematically arranging the material by means of an *a capite ad calcem* classification that the claim of the catalogue to have produced a new standard series should be understood. It is in this sense that one can ascribe to Esagil-kīn-apli the editorial work of an authorized, new standard edition of *Alamdimmû* and *Sakikkû*, which he perhaps executed as head of a scribal school that collected and copied texts.[27]

Although the actual compilation of *Alamdimmû* was the work of Esagil-kīn-apli, there was also a tradition that ascribed authorship of the physiognomic omens to Ea, the god of wisdom and skills. The Neo-Assyrian *Catalogue of Texts and Authors* credits Ea with the authorship of works

---

the Babylonian Talmud," in *From Athens to Jerusalem: Medicine in Hellenized Jewish Lore and in Early Christian Literature* (eds. S. Kottek et al.; Pantaleon 33; Rotterdam: Erasmus, 2000), 13-32. I was unable to see M.J. Geller, "Akkadian Healing Therapies in the Babylonian Talmud," *Max-Planck-Institut für Wissenschaftsgeschichte Preprint* 259 (2004): 1-60.

24 In addition to *Alamdimmû* and *Sakikkû*, it is possible that Esagil-kīn-apli was also responsible for the ordering of "the confused and contradictory state of the astrological omen tradition around the mid-second millennium" into the astrological omen series *Enūma Anu Enlil*. Cf. U. Koch-Westenholz, *Mesopotamian Astrology: An Introduction to Babylonian and Assyrian Celestial Divination* (CNIP 19; Copenhagen: Museum Tusculanum, 1995), 42-43, 74-76.

In the Neo-Assyrian *Exorcist's Manual*, Esagil-kīn-apli is credited not only with *Alamdimmû* but with numerous compositions ranging from ritual and liturgical incantations and medical texts to astronomical and terrestrial omen series. See H. Zimmer, "Zu den 'Keilinschriften aus Assur religiösen Inhalts,'" *ZA* 30 (1915/1916): 184-229, at 204-29. For the most recent edition of the *Exorcist's Manual*, based on four new duplicates from the British Museum that provide additional bibliographic information about Esagil-kīn-apli, see M.J. Geller, "Incipits and Rubrics," in *Wisdom, Gods and Literature: Studies in Assyriology in Honour of W.G. Lambert* (eds. A.R. George and I.L. Finkel; Winona Lake, Indiana: Eisenbrauns, 2000), 225-58, at 242-54, 256-58.

25 A67-8/B30', Finkel, "Adad-apla-iddina," 149. I agree with Heeßel, *Babylonisch-assyrische Diagnostik*, 105 n. 38, that this remark refers to the *a capite ad calcem* principle that underlies the structuring of both series, and that it is not a reference to the fact that both series have been organized by the catalogue as Finkel, "Adad-apla-iddina," 149 n. 64, proposes.

26 A61-2/B25', Finkel, "Adad-apla-iddina," 149.

27 M.J. Geller, "Astronomy and Authorship," *BSOAS* 53 (1990): 209-13, at 212 n. 25; Böck, *Die babylonisch-assyrische Morphoskopie*, 14.

that belong to the so-called exorcism corpus (*āšipūtum*), saying that "[these] are by Ea." They include the astrological omen series *Enūma Anu Enlil*, the physiognomic series *Alamdimmû*, and the diagnostic omen compendium *Sakikkû*.[28]

The attribution of divine authorship is understandable. Babylonian tradition regarded divinatory knowledge as of divine origin. In addition, the role of the king as an intermediary between the gods and his scholars was stressed. The Neo-Assyrian *Enmeduranki-text* tells that the gods Šamaš and Adad taught Enmeduranki, the antediluvian king of Sippar, the practice of oil and liver divination. Enmeduranki then transmits the secrets that Šamaš and Adad taught him to the men of Nippur, Sippar, and Babylon, presumably the few learned ones.[29] The text presents knowledge of divination practices as originating with the gods and being transmitted to the scholars through the king. The cultivated image of the king as the guardian of divine knowledge is particularly clear in the case of Assurbanipal. He is presented as actively supporting the collection of divinatory texts for his own library, and also energetically involving himself with his diviners and their reports to him.[30]

Babylonian Scholars and Physiognomic Divination

The different forms of Babylonian divination were the domains of different types of scholars.[31] Following the differentiation between artificial and natural divination in Cicero's *On Divination*, scholars distinguish between

[28] See W.G. Lambert, "A Catalogue of Texts and Authors," *JCS* 16 (1962): 59-77, at 64-65, 72. Ea was also referred to in incantations, thereby providing the conjurations with working power. It was believed that the divine origin of the incantations rendered them effective, and therefore this origin was stressed. Cf. Lambert, "Catalogue," 72-73; B. Pongratz-Leisten, *Herrschaftswissen in Mesopotamien: Formen der Kommunikation zwischen Gott und König im 2. und 1. Jahrtausend v. Chr.* (SAAS 10; Helsinki: The Neo-Assyrian Text Corpus Project, 1999), 293-95.

[29] After this point the text is somewhat confused. It repeats the things Enmeduranki taught the learned men of the three cities, adding, however, astrology and related mathematics. See W.G. Lambert, "Enmeduranki and Related Matters," *JCS* 21 (1967): 126-38; W.G. Lambert, "The Qualifications of Babylonian Diviners," in *Festschrift für Rykle Borger zu seinem 65. Geburtstag am 24. Mai 1994: tikip santakki mala bašmu...* (ed. S.M. Maul; CM 10; Groningen: Styx, 1998), 141-58.

[30] See e.g. U. Jeyes, "Assurbanipal's *bārûtu*," in *Assyrien im Wandel der Zeiten: XXXIXe Rencontre Assyriologique Internationale Heidelberg 6.-10. Juli 1992* (eds. H. Waetzoldt and H. Hauptman; HSAO 6; Heidelberg: Heidelberger Orientverlag, 1997), 61-65; Pongratz-Leisten, *Herrschaftswissen in Mesopotamien*, 26-46, 286-320.

[31] The divisions in the corpus of divination texts correspond to those between types of scholars, cf. Rochberg, *The Heavenly Writing*, 94-95. It was, however, also possible that people held more than one scholarly title, see e.g. S. Parpola, "Mesopotamian Astrology and Astronomy as Domains of the Mesopotamian 'Wisdom,'" in *Die Rolle der Astronomie in den Kulturen Mesopotamiens: Beiträge zum 3. Grazer Morgenländischen Symposion (23.-27. September 1991)* (ed. H.D. Galter; GMS 3; Graz: GrazKult, 1993), 47-59; Pongratz-Leisten, *Herrschaftswissen in Mesopotamien*, 18.

provoked and unprovoked omens.[32] In general, the first category, containing *inter alia* extispicy, was the domain of the *bārû* ("haruspex"), while the latter belonged to the *āšipu* ("magician-exorcist").[33] The Neo-Assyrian *Exorcist's Manual*, from the collection of the *āšipu* Kiṣar-Nabû from Assur, assigns knowledge of the physiognomic series *Alamdimmû* to the *āšipu*.[34] Surveying the numerous descriptions of the human body and observing the detail with which this is done in *Alamdimmû*, it is evident that the study of physiognomics demanded a thorough knowledge of human anatomy. Learning human anatomy began with the study of lexical lists in school, such as the Sumerian lexical list *Ugu-mu* ("My skull").[35] Those who pursued their studies to become an *āšipu* had *Alamdimmû* in their curriculum.[36]

The divination scholars performed services for the temple and the palace.[37] Before the Hellenistic period they were especially closely connected with the palace. This is particularly clear in the case of the Neo-Assyrian celestial diviners, who sent reports to the king from various observatories throughout Mesopotamia and were dependent upon him for financial support.[38] However, a shift in cultural locus concerning celestial divination seems to have taken place sometime during the Persian period. The evidence for the intense involvement of the king with the diviners appears to decrease. During the Hellenistic period the context of celestial divination changed from the palace to the temple. The Late Babylonian temple was the

32 *On Divination* 1.12, 34, 72. This distinction by Quintus has been modified because both categories, according to Quintus' definition in *On Divination*, fall within the artificial form of divination since they are dependent on reflection and interpretation. Cf. Rochberg, *The Heavenly Writing*, 47-48.

33 Cf. Bottéro, "Symptômes, signes, écritures," 86-124; Oppenheim, *Ancient Mesopotamia*, 207-10; Koch-Westenholz, *Mesopotamian Astrology*, 9-10; Böck, *Die babylonisch-assyrische Morphoskopie*, 3; Rochberg, *The Heavenly Writing*, 47-49.

34 KAR 44 obv. 6. Cf. Zimmer, "Zu den 'Keilinschriften aus Assur religiösen Inhalts,'" 206-7; Böck, *Die babylonisch-assyrische Morphoskopie*, 3.

35 Böck, *Die babylonisch-assyrische Morphoskopie*, 45-54. For *Ugu-mu* see B. Landsberger and M. Civil (eds.), *The Series ḪAR-ra = ḫubullu: Tablet XV and Related Texts* (MSL 9; Rome: Pontificium Institutum Biblicum, 1967), 51-73. See furthermore *Sig$_7$-alan* (*Nabnītu*) in I.L. Finkel (ed. with the collaboration of M. Civil), *The Series SIG$_7$.ALAN = Nabnītu* (MSL 16; Rome: Pontificium Institutum Biblicum, 1982). See also A. Cavigneaux, "Lexikalische Listen," *RlA* 6 (1980-83), 609-41, at 630.

36 Böck, *Die babylonisch-assyrische Morphoskopie*, 4.

37 They did not, however, actually belong to the cultic or royal staff. See G.J.P. McEwan, *Priest and Temple in Hellenistic Babylonia* (FAS 4; Wiesbaden: Franz Steiner, 1981), 15-24; U. Koch-Westenholz, "Old Babylonian Extispicy Reports," in *Mining the Archives*, 131-45, at 140.

38 Cf. A.L. Oppenheim, "Divination and Celestial Observation in the Last Assyrian Empire," *Centaurus* 14 (1969): 97-135; H. Hunger, *Astrological Reports to Assyrian Kings* (SAA 8; Helsinki: Helsinki University Press, 1992); Koch-Westenholz, *Mesopotamian Astrology*, 56-73.

principal institution supporting the celestial diviners.[39] This probably also applies to other forms of divination that were the domain of the *āšipu*, such as physiognomics. In the Neo-Assyrian period the *āšipu* presumably performed certain advisory services for the court.[40] Later, however, in Seleucid Babylonia, the *āšipu* scholar functioned within the temple and was supported by temple finances.[41]

The Babylonian temples remained in function during the Seleucid and Arsacid periods and were responsible for the transmission and continuation of Mesopotamian culture during this time.[42] That the temple was a center of knowledge and learning in the Hellenistic period is illustrated by the Babylonian priest Berossos who wrote a history of his nation and culture around 300 BCE that was probably called *Babyloniaca*. Berossos' title "priest" does not imply that he held a cultic or religious function, but that he was connected to the Esagila, which was the main temple of Babylon. When writing his *Babyloniaca*, Berossos used native, cuneiform sources from the old archives, or copies of them, to which he would have had access as a member of the temple community.[43]

[39] Cf. F. Rochberg, "The Cultural Locus of Astronomy in Late Babylonia," in *Rolle der Astronomie*, 31-45; Rochberg, *The Heavenly Writing*, 209-36; T. Boiy, *Late Achaemenid and Hellenistic Babylon* (OLA 136; Leuven: Peeters, 2004), 272-3, 297.

[40] The biographical section of the *Esagil-kīn-apli Catalogue* ends with an exhortation to the *āšipu* to inform the king of his investigations on the basis of the diagnostic compendium *Sakikkû* and the physiognomic series *Alamdimmû*, see Finkel, "Adad-apla-iddina," 150.

[41] Rochberg, *The Heavenly Writing*, 95.

[42] Intriguing evidence for the survival of the scribal tradition at the Late Babylonian temple is provided, for example, by the Graeco-Babyloniaca tablets. These tablets contain cuneiform on one side and transcriptions in Greek letters on the other side. See e.g. M.J. Geller, "The Last Wedge," *ZA* 87 (1997): 43-95; M.J. Geller, "Graeco-Babyloniaca in Babylon," in *Babylon: Focus mesopotamischer Geschichte*, 377-83; Boiy, *Late Achaemenid and Hellenistic Babylon*, 41-43.

[43] G. De Breucker, "Berossos and the Mesopotamian Temple as Centre of Knowledge during the Hellenistic Period," in *Learned Antiquity: Scholarship and Society in the Near-East, the Greco-Roman World, and the Early Medieval West* (eds. A.A. MacDonald, M.W. Twomey and G.J. Reinink; GSCC 5; Leuven: Peeters, 2003), 13-23.

Despite the shift in cultural locus of the diviners away from the palace, one should allow for the possibility that the ideology of a close relationship between the king and his scholars remained an important element of Mesopotamian culture. This is demonstrated by the transmission of the *List of Sages and Scholars*, enumerating scholars in relation to kings, from the Rēš temple in Seleucid Uruk (see J. van Dijk, "Die Inschriftenfunde," in UVB 18 [Berlin: Gebr. Mann, 1962], 39-62, at 44-52, cf. De Breucker, "Berossos and the Mesopotamian Temple," 15). The temples were important instruments in the hands of the Seleucid government, and there is evidence that the Seleucid kings took an interest in the welfare of the temples, cf. S. Dalley et al., *The Legacy of Mesopotamia* (Oxford: Oxford University Press, 1998), 40-42; Rochberg, *The Heavenly Writing*, 231-33.

It is possible that royal interest in the learning of the Babylonian scholars did not cease completely and that Mesopotamian scholars continued to serve the Seleucid and Arsacid kings in one form or another, but evidence for their political advisory role is not traceable in these periods. The appearance of a Chaldean in the party of the Parthian envoy who predicts

Form and Principles of Babylonian Physiognomics

Babylonian physiognomics was transmitted in the form of omens. A general definition of an omen is an event or phenomenon that is regarded as a portent of something good or evil. The form of Babylonian omens is that of a conditional sentence. The introductory clause (*protasis*) begins with *šumma*, "if, provided that," in which an event or a phenomenon taken to be the sign is described. The consequent clause (*apodosis*), the beginning of which is not signaled by a particle, mentions that to which the sign refers (a future event or situation). In general, the protasis uses the preterite (past sense) for the verb, while the apodosis has the durative (present/future sense). This is the formal criterion to distinguish between the two parts of the omen. At the same time it also reveals the worldview, because it implies that there is an interval between observation (protasis) and prediction (apodosis), which allows for measurements to be taken to avoid what is stated in the apodosis setting in (see below on the *namburbi* rituals).

The relationship between the two parts of the conditional sentence could be based on a binary system of analogy or on an association of words or ideas.[44] Regarding the analogies, in general the leading principle is that right is positive (*pars familiaris*) and left is negative (*pars hostilis*). In the case of the physiognomic series *Alamdimmû*, this rule is easily made clear by those omens in which the position and color of different sorts of body marks, such as moles and pimples, are described, especially in its fifth subseries *Šumma liptu* ("If the mole"). It also shows that left as negative is both absolute and relative. The body marks themselves possess either a negative or a positive quality. According to their position and quality, the outcome described in the apodoses is either positive or negative. If a mark with a positive quality is positioned on the left side of a body part, the conclusion is negative. If the same mark appears on the right side, the result is positive. Consequently, if a mark with a negative value appears on the left side of the body, the outcome is positive, and if positioned on the right side, the conclusion is negative.[45]

With regard to the association based on the words used, the following example may suffice:

Sulla's fate from the form and appearance of his face is suggestive but inconclusive evidence, see n. 103 below.

[44] Böck, *Die babylonisch-assyrische Morphoskopie*, 38-40; B. Böck, "Babylonische Divination und Magie als Ausdruck der Denkstrukturen des altmesopotamischen Menschen," in *Babylon: Focus mesopotamischer Geschichte*, 409-25.

[45] See Böck, *Die babylonisch-assyrische Morphoskopie*, 39, 174-233.

> If on the back of the head (*kutallu*), on the right side, [an *umṣatu*-mole] is positioned, the person who supports him (*mukīl kutalšu*) will die, (his) heart will swell with misery.[46]

Finally, the next cases may illustrate the use of associative ideas:

> If the hair of his head is very thick, he will have satisfaction. If the hair of his head is thin, he will have dissatisfaction.[47]

Omens can be regarded as signs that establish links between what is potentially observed in the present and what occurs in the future. They do not seem to express causal relationships.[48] The omens contained signs of the gods communicating their messages to human beings; the protasis containing the divine sign, the apodosis the divine message (or the announcement of an inevitable future; inevitable, that is, if one did not pay heed to the prediction and take sufficient countermeasures, see below). Concerning physiognomics this would mean that the Babylonians probably did not think of any intrinsic relationship between the form and appearance of the human body and a person's future (or in some cases character). The human body was just one of the many places where the gods left their messages.

The notion of writing is paramount in all this. The gods were asked literally to place their answers, or, in other words, to write their signs, for example, in the entrails of animals to be deciphered by the *bārû*.[49] Although the metaphor "heavenly writing" (*šiṭir šamê* or *šiṭirti šamāmī*) is not used explicitly for astrology or celestial divination, "the notion of the stars as a heavenly script implies their capacity to be read and interpreted."[50] In *Alamdimmû* the writing of the gods is perhaps implied in a section that deals with the occurrence of signs on people's foreheads.[51]

46 *Šumma umṣatu* I.6.

47 *Alamdimmû* II:96–97. See also Geller, "New Documents from the Dead Sea," 228.

48 In the case of celestial divination, however, this perhaps does not describe entirely correctly the attitude towards the heavenly bodies. These were perhaps not just signs but also exerted influence on certain affairs. The issue is disputed. For example, David Brown maintains that the Babylonians did not ascribe a direct influence to planets and stars upon earthly matters, but rather saw the heavenly bodies as signs communicating divine messages (review of Pongratz-Leisten, *Herrschaftswissen in Mesopotamien*, *ZA* 94 [2004]: 112-21, at 114-15). Erica Reiner, however, has shown that there were areas in which the Babylonians acknowledged the influence of the stars (Reiner, *Astral Magic*).

49 See H.L.J. Vanstiphout, "*Miḫiltum*, or the Image of Cuneiform Writing," in *The Image in Writing* (VisRel 6; Leiden: E.J. Brill, 1988), 152-68, at 160; I. Starr, *Queries to the Sungod: Divination and Politics in Sargonid Assyria* (SAA 4; Helsinki: Helsinki University Press, 1990); U. Jeyes, *Old Babylonian Extispicy: Omen Texts in the British Museum* (UNHAII 64; Istanbul: Nederlands Historisch-Archeologisch Instituut, 1989); P. Steinkeller, "Of Stars and Men: The Conceptual and Mythological Setup of Babylonian Extispicy," in *Biblical and Oriental Essays in Memory of William L. Moran* (ed. A. Gianto; BibOr 48; Rome: Pontificio Istituto Biblico, 2005), 11-47.

50 Rochberg, *The Heavenly Writing*, 1. Cf. Reiner, *Astral Magic*, 9.

51 See *Alamdimmû* III:76–121.

The entire universe could potentially carry divine messages. The Babylonian divination literature consists of various celestial and terrestrial collections covering many domains of potential signifiers. These realms were congruent. Heavenly signs were thought to signify the same things as terrestrial ones. In the words of a Babylonian diviner's manual:

> The signs on earth just as those in the sky give us signals. Sky and earth both produce portents though appearing separately, they are not separate (because) sky and earth are related.[52]

To read, understand, and interpret these divine messages was the task of the Babylonian divination scholars. Understanding the divine messages was not just for the sake of knowing the future. Omens did not signal inevitable events. By means of countermeasures, such as the so-called *namburbi* rituals, an omen's negative outcome could be averted if known in time.[53] Regarding negative physiognomic omens, however, there is little evidence for apotropaic measures.[54]

Accessibility and Exclusivity of Physiognomic Learning

By means of certain terminology and formulas, Mesopotamian scholars expressed their concern for limiting the accessibility to divinatory knowledge and learning to their own circle. The *Enmeduranki-text* gives instructions to the master-initiate for the education of his son into the knowledge and rites of the diviner:

> the learned savant, who guards the secret of the great gods (*pirišti ilāni rabûti*), will bind by oath before Šamaš and Adad by tablet and stylus the son whom he loves and will teach him.[55]

The practice of labeling a text secret is well known from other examples from the Neo-Assyrian period until the Seleucid period.[56] The colophons sometimes contain a warning to keep the content of the text secret from the uninitiated. The typical formula used was: "The initiate (*mūdû*) may show the initiate, the uninitiated (*lā mūdû*) may not see." The Neo-Assyrian copy of the *Esagil-kīn-apli Catalogue* from Nimrud is labeled "secret of Ezida"

[52] A.L. Oppenheim, "A Babylonian Diviner's Manual," *JNES* 33 (1974): 197-220, at 204.

[53] See S.M. Maul, *Zukunftsbewältigung: Eine Untersuchung altorientalischen Denkens anhand der babylonisch-assyrischen Löserituale (Namburbi)* (BaF 18; Mainz: Philipp von Zabern, 1994).

[54] Cf. B. Böck, "Physiognomie und Schicksal? Oder wie der altmesopotamische Mensch mit einem durch ein physiognomisches Omen angekundigtes Unheil umgegangen sein mag," *Sefarad* 62 (2002): 241-57.

[55] Lambert, "Qualifications of Babylonian Diviners," 152. See also H.L.J. Vanstiphout and N. Veldhuis, "Ṭuppi ilāni takāltu pirišti šamê u erṣetim," *AIUON* 55 (1995): 30-32.

[56] Cf. R. Borger, "Geheimwissen," *RlA* 3 (1957-1971), 188-91; P.-A. Beaulieu, "New Light on Secret Knowledge in Late Babylonian Culture," *ZA* 82 (1992): 98-111, at 98-99, 110-11.

(*niṣirti E*[*zida*]).[57] In the biographical section, Esagil-kīn-apli explains that he established the collections of *Sakikkû* and *Alamdimmû* for knowledge, but that there are restrictions on the use of this learning:

A62/B26' [...] Take care! Pay [attention!]
A63-4/B27' Do not neglect your knowledge! He who does not attain(?) knowledge must not speak aloud the SA.GIG omens,
A65-6/B28' nor must he pronounce out loud Alamdimmû![58]

The reader is urged to be careful with the learned knowledge. Someone who is negligent or not sufficiently trained in these omen series must not speak them out aloud. This suggests that only those who had acquired the appropriate level of learning were suitable to make use and speak of the physiognomic knowledge in *Alamdimmû*.[59]

There is some ambivalence among Assyriologists regarding the notion of secrecy and exclusivity implied by the cuneiform formulas. Some argue that there was no genre of esoteric texts and that such a characterization is mainly based on our own inability to understand these texts.[60] Others point out that neither secrecy or esotericism imply incomprehensibility, only exclusivity.[61]

The latter position aptly captures the concept of secrecy. For those claiming to have knowledge of it, the exclusive, "secret" nature of learning functions as cultural capital and bestows status and prestige on them. In

57 A93, Kinnier Wilson, "Two Medical Texts," 139-40; Finkel, "Adad-apla-iddina," 152.

58 Finkel, "Adad-apla-iddina," 149.

59 Cf. also M.J. Geller, "West Meets East: Early Greek and Babylonian Diagnosis," *AfO* 48/49 (2001/2002): 50-75, at 51.

60 Cf. Böck, *Die babylonisch-assyrische Morphoskopie*, 43-44, 68-69. Borger, "Geheimwissen," argued that the secrecy formulas were not applied consistently. According to A. Livingstone, *Mystical and Mythological Explanatory Works of Assyrian and Babylonian Scholars* (Oxford: Clarendon, 1986), 1, the secrecy formulas and additional references "may simply indicate a scholarly pride in the value of literature and knowledge. While one need not doubt that certain texts or doctrines were only understood or held by a select few, it seems probable that many texts which appear 'esoteric' to a modern reader were readily understood by ancient scholars." Cf. also Reiner, *Astral Magic*, 33; H.L.J. Vanstiphout, "The nth Degree of Writing at Nineveh," *Iraq* 66 (2004): 51-54, at 53-54.

61 Cf. Rochberg, *The Heavenly Writing*, 212-18. In addition, it is possible that in the context of royal ideology, stressing the divine origin of knowledge and the function of the king as intermediary, kings such as Assurbanipal wielded a politics of secrecy with regard to scholarly learning in order to check and control knowledge and those who had it, the scholars, and thereby to affirm their power. See Pongratz-Leisten, *Herrschaftswissen in Mesopotamien*, 295-320 (cf. also the review by Brown [see n. 48 above] who disagrees that divination was used in maintaining royal hegemony). Beaulieu, "Secret Knowledge in Late Babylonian Culture," argues that the restriction alluded to in the so-called secrecy colophons might have existed in practice. His reading, however, of a Neo-Babylonian legal text to prove his argument has been refuted on grammatical grounds by M. Dietrich, "Babylonische Sklaven auf der Schreiberschule," in *Veenhof Anniversary Volume: Studies Presented to Klaas R. Veenhof on the Occasion of his Sixty-Fifth Birthday* (eds. W.H. van Soldt et al.; UNHAII 89; Leiden: Nederlands Instituut voor het Nabije Oosten, 2001), 67-81.

order to maintain the high social value of a body of knowledge, control over tradition, learning, and people is necessary. The concept of secrecy does not so much refer to the specific content of a body of knowledge or its comprehensibility. It is better understood as a means to organize the accessibility and availability of information and learning, and this in connection with the social status that it bestows on those possessing it. Secrecy can be described as a process in which the flow of information is suppressed across any boundaries that have been erected. The kind of information that is suppressed may vary, and the reason why does not have to be clear to an outsider. Strategies of secrecy and information control, however, can be made clear.[62] The words of warning in the *Esagil-kīn-apli Catalogue* reflect the importance of study and learning in gaining access to knowledge of physiognomic divination. Francesca Rochberg concludes that:

> the secrecy of the scholarly texts seems to be of the sort associated with trade knowledge. The scholars' knowledge was safeguarded and protected from the uninformed, and the integrity of the discipline was thereby maintained. [...] The interdiction against persons outside the circle of 'knowers' reflects the efforts of a particular scribal body to maintain control over its tradition and to protect a particular body of knowledge. The special status of the tradition in the view of the scribes, however, is expressed in the claim that the knowledge contained in the tablets was transmitted from a divine source.[63]

Functions of Physiognomic Divination

There is little evidence for the function(s) that Babylonian physiognomics may have had. Unlike other forms of divinatory learning, like extispicy or celestial divination, there is no evidence that the physiognomic omens in *Alamdimmû* were ever used in a divinatory practice. Regarding the relationship between omen literature and practice, scholars disagree on the character

[62] Cf. E.A. Brandt, "On Secrecy and the Control of Knowledge: Taos Pueblo," in *Secrecy: A Cross-Cultural Perspective* (ed. S.T. Tefft; New York: Human Sciences, 1980), 123-46; H.G. Kippenberg and G.G. Stroumsa (eds.), *Secrecy and Concealment: Studies in the History of Mediterranean and Near Eastern Religions* (SHR 65; Leiden: E.J. Brill, 1995); H.B. Urban, "Elitism and Esotericism: Strategies of Secrecy and Power in South Indian Tantra and French Masonry," *Numen* 44 (1997): 1-38; H.B. Urban, "The Torment of Secrecy: Ethical and Epistemological Problems in the Study of Esoteric Traditions," *HR* 37 (1998): 209-48; M. Douglas and G. Mars, "Terrorism: A Positive Feedback Game," *Human Relations* 56/7 (2003): 763-86; M. Popović, "Physiognomic Knowledge in Qumran and Babylonia: Form, Interdisciplinarity, and Secrecy," *DSD* 13 (2006): 150-76, at 166-76.

[63] Rochberg, *The Heavenly Writing*, 217. She rightly warns, however, against "too extreme a picture of an exclusive learned society," referring to an example of a fourth-century BCE scribe who participated in both learned scholarship and clerical record-keeping (217-18). On secrecy of craft knowledge in late antiquity, see P.O. Long, *Openness, Secrecy, Authorship: Technical Arts and the Culture of Knowledge from Antiquity to the Renaissance* (Baltimore: Johns Hopkins University Press, 2001).

and practical use of the omen series. Some argue that the standard series functioned as reference works for the diviner, while others say that the diviner would not have consulted them for decision-making purposes in his actual divinatory proceedings.[64] Regardless of the exact connection between the standard omen series and divinatory practice, there are no cuneiform references to physiognomic omens outside the standard and extra-serial texts that could throw light on the use and function of physiognomic knowledge.

Barbara Böck argues that Babylonian physiognomics was applied as an assessment tool for deciding which persons were eligible to enter the service of the royal court. The king would thus have been protected from the negative influence of people with bad omens.[65] Her argument is largely based on the ending of the biographical section of the *Esagil-kīn-apli Catalogue* where the *āšipu* is urged to report to the king:

A69/B31' [Let the *āšipu*] who makes the decisions, and who watches over people's lives,
A70-1/B32' who comprehensively knows SA.GIG and Alamdimmû, inspect (the patient) and check (the appropriate series),
A71/B33' [let him ponder], and let him put his diagnosis at the disposal of the king.[66]

Concerning the physiognomic omens with regard to women, however, Böck suggests their context was that of marital relations and childbirth.[67]

One should allow for the possibility that in Mesopotamia physiognomics functioned in a royal or cultic context. As the *Esagil-kīn-apli Catalogue* implies, the *āšipu* rendered certain physiognomic services to king and palace. The exact nature of these services, however, is not entirely clear and it

64 Cf. Koch-Westenholz, *Mesopotamian Astrology*, 13-19; Rochberg, *The Heavenly Writing*, 244-86. Perhaps omens were initially based on empirical observations that through non-empirical systematization were expanded to omen lists, with the observations contained therein forming the basis for the diviner's practice, see e.g. Bottéro, "Symptômes, signes, écritures." Another interpretation is that the omen compendia mainly represent purely theoretical and speculative knowledge. This knowledge evolved steadily out of a need to systematize every theoretical possibility of an event or phenomenon. But the actual occurrence in reality was often completely impossible. As such, this knowledge had no practical use for the diviner whatsoever, see e.g. N. Veldhuis, "Reading the Signs," in *All Those Nations...: Cultural Encounters within and with the Near East: Studies Presented to Han Drijvers at the Occasion of his Sixtyfifth Birthday by Colleagues and Students* (eds. H.L.J. Vanstiphout et al.; COMERS/ICOG Communications 2; Groningen: STYX, 1999), 161-74; Brown, *Mesopotamian Planetary Astronomy-Astrology*, 108-13. According to the latter position, this means that the occurrence of the same omens in standard series and, for example, scholarly reports does not imply that the diviner had consulted the series for reaching his conclusion. It merely demonstrates that what he knew was also to be found in the scholarly tradition, which he had learned as part of his scribal curriculum as a diviner. He need not have actually checked the tablets of a standard series.

65 Böck, *Die babylonisch-assyrische Morphoskopie*, 55-57.

66 Finkel, "Adad-apla-iddina," 150.

67 Böck, *Die babylonisch-assyrische Morphoskopie*, 58-59.

is not readily evident that most of the omens would have been of interest to the king. The predictions in *Alamdimmû* are primarily personal and concern the described subject. There are no omen apodoses that explicitly have consequences for king and country.[68] Some omens even predict that the king or palace will be harsh on a person.[69] The described subject seems, therefore, to be the party for whom most omen apodoses would have been of interest. Furthermore, the omens in *Alamdimmû* concern individual parts of the body. As Böck herself points out, there is nothing to suggest that a synthetic method was used for valuing and interpreting a group of signs from the body. From the few composite omens that have survived, it seems rather that no synthesis was attempted since no effort was made to bring contradictory apodoses in harmony with each other.[70] It is, therefore, not readily evident how the omens from *Alamdimmû* were used for screening court personnel.[71] All this also seems to apply to the cultic context that has been suggested. It has been argued that physiognomics was perhaps used in selecting candidates for religious positions as priests.[72] It is plausible that a physical examination was conducted as part of the admittance procedure for palace and temple. Although there is little evidence to support this, it is possible that the examination was some kind of divinatory screening by means of physiognomics, but the exact nature of it is not clear from the available sources.

Backgrounds of Physiognomic Learning and Literature: Greece and Rome

In the Greco-Roman world physiognomics was the domain of philosophers, physicians and rhetoricians. In their writings they dealt with the theoretical as well as the applied side of this art or science (τέχνη), while a physiognomic consciousness appears in literary works of history and biography,

68 See, however, *Alamdimmû* II:149: "If the head is wide, he will act loyal."

69 Böck, *Die babylonisch-assyrische Morphoskopie*, 32. Cf. the review by JoAnn Scurlock (*JAOS* 123 [2003]: 395-99, at 397).

70 Böck, *Die babylonisch-assyrische Morphoskopie*, 59-60.

71 There is a letter from a Babylonian scholar, called Marduk-šapik-zeri, in which he recommends himself and twenty of his students to the service of the Neo-Assyrian king (either Esarhaddon [680–669 BCE] or his son Assurbanipal [668–631/627? BCE]) on the basis of their knowledge of and proficiency in scientific literature, cf. #160 in S. Parpola, *Letters from Assyrian and Babylonian Scholars* (SAA 10; Helsinki: Helsinki University Press, 1993), 120-24. This letter suggests that learnedness was an important, if not the most important, criterion for entering into royal service. I owe this reference to Herman Vanstiphout.

72 Lambert, "Qualifications of Babylonian Diviners"; Veldhuis, "Reading the Signs," 169; Böck, *Die babylonisch-assyrische Morphoskopie*, 57-58.

drama, and satire.[73] One should, of course, also allow for the possibility that physiognomic ideas circulated on a more popular level, but there is very little evidence for such popular expressions.[74]

Textual Evidence

Four physiognomic treatises have been transmitted from antiquity. The most important one is the pseudo-Aristotelian *Physiognomonica*. From the Second Sophistic comes the Greek treatise of Polemo of Laodicea, now only extant in an Arabic translation. This work was paraphrased by Adamantius around 400, which represents the third physiognomic treatise. The fourth one is by an Anonymous Latin author, probably from the end of the fourth century.

The pseudo-Aristotelian work *Physiognomonica* is the first systematic treatment in the Greek world devoted to physiognomics.[75] The *Physiognomonica* clearly stands in the Aristotelian tradition with regard to the relationship between the body, the psyche, and the characteristics of both.[76] The connection, however, between the *Physiognomonica* and the name of Aristotle is a relatively late tradition,[77] and the Aristotelian authorship of

[73] The basic reference for physiognomic texts and passages is still R. Förster (ed.), *Scriptores Physiognomonici Graeci et Latini* (2 vols.; Leipzig: B.G. Teubner, 1893). See further e.g. Evans, *Physiognomics*; B.P. Reardon, *Courant littéraires grecs des IIe et IIIe siècles après J.-C.* (Paris: Les Belles Lettres, 1971), 243-48; André, *Anonyme Latin*; Degkwitz, *'Physiognomonica' Traktat A*; Barton, *Power and Knowledge*; Gleason, *Making Men*; F. Stok, "La fisiognomica fra teoria e pratica," in *Sciences exactes et sciences appliquées à Alexandrie (IIIe siècle av. J.-C. – Ier siècle ap. J.-C.): Actes du Colloque International de Saint-Étienne (6-8 juin 1996)* (eds. G. Argoud and J.-Y. Guillaumin; Mémoires 16; Saint-Étienne: Publications de l'Université de Saint-Étienne, 1998), 173-87; Vogt, *Physiognomonica*; A. Touwaide, "Physiognomik," *DNP* 9 (2000), 997-98; M.M. Sassi, *The Science of Man in Ancient Greece* (trans. P. Tucker; Chicago: University of Chicago Press, 2001); Isaac, *Invention of Racism*, 149-62.

[74] See Petronius, *Satyricon* 126.3, cf. Bollók, "Description of Paul," 12-15; Isaac, *Invention of Racism*, 155-56.

[75] See now the recent and thorough study by Sabine Vogt, *Physiognomonica*, which serves as the major reference in this study. The text is extant in medieval copies, but it has suffered some losses during its transmission, especially in its physiognomic catalogues. The amount of text included in the *Physiognomonica* and known to ancient authors, such as the fourth-century Anonymous Latin author, was greater than that which has come down to us through medieval copies. See Vogt, *Physiognomonica*, 197-227.

[76] Cf. Förster, *Scriptores Physiognomonici*, 1:xvi-xviii; Evans, *Physiognomics*, 6-10, 22-24; Barton, *Power and Knowledge*, 101; A. Degkwitz, "Die pseudoaristotelischen 'Physiognomonica,'" in *Geschichten der Physiognomik*, 23-44, at 29-31, 37-41; Vogt, *Physiognomonica*, 133-45. The secondary literature on Aristotle's position in general on the relationship between soul and body is vast, see e.g. P.J. van der Eijk, "Aristotle's Psycho-Physiological Account of the Soul-Body Relationship," in *Psyche and Soma: Physicians and Metaphysicians on the Mind-Body Problem from Antiquity to Enlightenment* (eds. J.P. Wright and P. Potter; Oxford: Clarendon, 2000), 57-77.

[77] There are just two concrete references from late antiquity to a book called *Physiognomonica* (Φυσιογνωμονικά) by Aristotle, see Vogt, *Physiognomonica*, 285. For example,

the text has been doubted by scholarship since the nineteenth century. Sabine Vogt does not commit herself in identifying the author (Aristotle or Pseudo-Aristotle) and labels the *Physiognomonica* as Aristotelian in content.[78]

The *Physiognomonica* is made up of two parts that each by itself represents a treatise on physiognomics (Tractate A and B).[79] Both treatises have an introductory part explaining the reasoning and methods of physiognomics. Although the theoretical section links up with notions set forth in Tractate A, Tractate B does not present a theoretical discussion but, in general, illustrates ideas by means of examples, thus complementing Tractate A. Following the introductory discourses, both treatises give a physiognomic catalogue. The two catalogues are structured differently. The first catalogue is basically ordered according to a list of character types ("the signified"), whereas the second one principally follows the parts of the human body ("the signifier"). Following the physiognomic catalogue, Tractate B has some final thoughts on physiognomic criteria and a hierarchy of signs.[80]

It is not necessary to assume that different authors, either belonging to or influenced by the Peripatetic school, wrote the treatises subsequently to each other, as most modern scholars do.[81] Vogt argues that the different emphasis and direction of both treatises are due to the fact that they were written for different purposes and different audiences. One and the same author (or perhaps two collaborating authors) could have done this.[82] Both parts of the *Physiognomonica* were written at the end of the fourth century or beginning of the third century BCE.[83] The two treatises differ from each

---

Diogenes Laertius, *Lives of Philosophers* 5.25, lists one book with that title (Φυσιογνωμονικὸν α΄) under Aristotle's writings, cf. Förster, *Scriptores Physiognomonici*, 1:xviii; Barton, *Power and Knowledge*, 101 n. 48. Other ancient authors, e.g. Pliny, *Natural History* 11.273-274, state more generally that Aristotle wrote on physiognomics, but they do not mention the title of the book. It is therefore not clear whether they are referring to the *Physiognomonica* as such.

[78] Vogt, *Physiognomonica*, 192-97.

[79] Somewhere in the middle (*Physiognomonica* 808b 11) the text begins to explain for a second time the definition and method of physiognomics.

[80] Cf. Vogt, *Physiognomonica*, 187, 285-87, 393-94.

[81] Cf. e.g. Evans, *Physiognomics*, 7; Degkwitz, *'Physiognomonica' Traktat A*, 1, 3-7; Barton, *Power and Knowledge*, 101; Degkwitz, "pseudoaristotelischen 'Physiognomonica,'" 27-29; Sassi, *Science of Man*, 35 n. 2.

[82] Vogt, *Physiognomonica*, 191: "Die unterschiedliche Auffassung der physiognomischen Methode in den beiden Traktaten kann demnach mit unterschiedlichen Absichten und Zielgruppen begründet werden. Während Traktat A sehr an der theoretischen Grundlegung des Faches interessiert ist und sie unter bewußtem Rückgriff auf Aristoteles' Diskussion der logischen Voraussetzungen des physiognomischen Zeichenschlusses erörtert, richtet sich Traktat B anscheinend an 'eingeweihte' Experten aus der Praxis, in der Erfahrung, Gesamteindruck und Tiervergleich eine große Rolle sowohl für die Legitimation als auch für die praktische Anwendung der Physiognomik spielen."

[83] Vogt, *Physiognomonica*, 197.

other, but there are no apparent contradictions between them, which is why it is possible to view them as complementary to each other, constituting one work.[84]

The Peripatetic *Physiognomonica* influenced in one way or another all further physiognomic writings in the Greco-Roman, and later Western, tradition.[85] Mention should be made of Loxus, an otherwise unknown physician probably from the third century BCE. The Anonymous Latin author says that he used his physiognomic work as a source, but the text of Loxus has not survived.[86]

The second physiognomic treatise to have been transmitted from antiquity is that of the physiognomist and rhetorician Polemo of Laodicea (ca. 88–145), representative of the city of Smyrna and beneficiary of the Emperor Hadrian.[87] Only one sentence by Polemo is extant in Greek. There is a Greek paraphrase by the iatrosophist Adamantius from ca. 400, and the text of his treatise is extant in an Arabic translation. The Latin translation made in 1884 is based on only one of the five Arabic manuscripts, one copied in Damascus in 1379.[88] A new edition is being prepared.[89] The third physiognomic writing to have survived is the paraphrase of Polemo by Adamantius (ca. 400).[90] A fourth treatise was thought to belong to Apuleius, but is now termed the Anonymous Latin author and dated to the end of the fourth century.[91] In the opening of his book, *De physiognomonia liber* §1, the author explicitly mentions the physician Loxus, the philosopher Aristotle, and the rhetorician Polemo as his sources.

Apart from these writings that deal entirely with physiognomics, there are many scattered references to physiognomics throughout Greek and Roman literature.[92] Together with the use of physiognomic notions in other forms of literature, these demonstrate how widespread and influential

[84] Cf. Degkwitz, *'Physiognomonica' Traktat A*, 4; Vogt, *Physiognomonica*, 192.

[85] Cf., for example, Degkwitz, "pseudoaristotelischen 'Physiognomonica,'" 41-43; M.M. Sassi, "Physiognomy," *OCD* 1181.

[86] On Loxus see e.g. G. Misener, "Loxus, Physician and Physiognomist," *CP* 18 (1923): 1-22; Evans, *Physiognomics*, 10-11; André, *Anonyme Latin*, 24-26; Vogt, *Physiognomonica*, 205.

[87] Förster, *Scriptores Physiognomonici*, 1:lxxv-lxxx, 98-294; Evans, *Physiognomics*, 11-15; Barton, *Power and Knowledge*, 102-31; Gleason, *Making Men*, esp. 21-81; Vogt, *Physiognomonica*, 202-3.

[88] Förster, *Scriptores Physiognomonici*, 1:lxxx gives 1356, but Barton, *Power and Knowledge*, 102 n. 52, rightly points out this does not correspond with the date of 757 after the Hegira, cf. Förster, *Scriptores Physiognomonici*, 1:96.

[89] This work is being done by Georges Boys-Stones, Peter Starr, and Simon Swain in Oxford. See Vogt, *Physiognomonica*, 202 n. 55.

[90] Förster, *Scriptores Physiognomonici*, 1:c-cix, 297-426; Evans, *Physiognomics*, 15-16; V. Nutton, "Adamantios," *DNP* 1 (1996), 103; Vogt, *Physiognomonica*, 203-4.

[91] Evans, *Physiognomics*, 16-17; André, *Anonyme Latin*, 31-39; Vogt, *Physiognomonica*, 205-8.

[92] Cf. Förster, *Scriptores Physiognomonici*, 1:237-352.

physiognomic ideas may have been, although physiognomics as found in the treatises is less culturally salient than astrology.[93] References will be made to the other physiognomic treatises, but the Peripatetic treatise *Physiognomonica* is singled out in this study because it set the example.

Highly important for this study are the texts that forge a connection between physiognomics and astrology. While physiognomics remained a distinct art unto itself, evident by the transmission of the treatises, it was at the same time, from the Hellenistic period onwards, incorporated by astrology. The connection between the planets and the signs of the zodiac on the one hand and the human person and body on the other hand served as another illustration of the sympathy between the macro and microcosmos.[94] The numerous astrological lists that have come down to us demonstrate that the idea that the zodiacal signs influenced the appearance and shape of the human body was widespread. This is most evident from the *zodiologia*, which enumerate all sorts of psychological and physical characteristics, prognostics, and more for types of people who were born under a certain zodiacal sign.

The Beginning and Origin of Greco-Roman Physiognomics

The basic notion of physiognomics, namely the signifying value of the human body, is so general that cultural influence from Mesopotamia need not have triggered this interest in the Greek world. Although in the case of astrology and astronomy some of the concepts in Hellenistic astrology clearly come from the East,[95] this cannot be determined in such a straightforward manner for physiognomics. Leaving aside the differences, the similarities are of too general a nature to warrant the conclusion that direct borrowing took place.[96]

Regarding the beginnings of physiognomics in the Greek world, there are different traditions crediting either the philosopher Pythagoras[97] or the

[93] Barton, *Power and Knowledge*, 96; Isaac, *Invention of Racism*, 150-51. For the importance of physiognomic knowledge in e.g. Apuleius' *Metamorphoses*, see most recently W. Keulen, "*Ad amussim congruentia*: Measuring the Intellectual in Apuleius," in *Lectiones Scrupulosae: Essays on the Text and Interpretation of Apuleius'* Metamorphoses (eds. W. Keulen, R. Nauta and S. Panayotakis; ANSupp 6; Groningen: Barkhuis, 2006), 168-202.

[94] Cf. J. Schmidt, "Physiognomik," *RE* 39 (1941), 1064-74, at 1066, 1070; Sassi, *Science of Man*, 161-87.

[95] See e.g. Neugebauer, *HAMA*, 589-614; F. Rochberg-Halton, "Elements of the Babylonian Contribution to Hellenistic Astrology," *JAOS* 108 (1988): 51-62; Koch-Westenholz, *Mesopotamian Astrology*, 177-79.

[96] Cf. also Kraus, *Die physiognomischen Omina der Babylonier*, 15-18; Böck, *Die babylonisch-assyrische Morphoskopie*, 61.

[97] See Aulus Gellius, *Attic Nights* 1.9.1–2; Hippolytus, *Refutation of All Heresies* 1.2.5; Iamblichus, *On the Pythagorean Life* 17.71, 74; Porphyrius, *Life of Pythagoras* 13, 54. Cf. C. Riedweg, *Pythagoras: Leben, Lehre, Nachwirkung; Eine Einführung* (Munich: C.H. Beck,

physician Hippocrates[98] as the discoverer of physiognomics.[99] There are also Greco-Roman traditions that seem to suggest a Babylonian origin.[100] Cicero tells of a physiognomist named Zopyrus who physiognomized Socrates as stupid, slow of wit, and a womanizer.[101] The name Zopyrus is of Persian origin. This anecdote is perhaps related to another one. Diogenes Laertius says that Aristotle tells of an unnamed *magus* who came from Syria and foretold Socrates a violent death.[102] Such a prediction would fit the interest of Babylonian physiognomics. Babylonian physiognomists also seem to have been active during the Roman period. Plutarch tells of a Chaldean who predicted a great future for Sulla on the basis of his face.[103] But this passage does not provide evidence of a cultural exchange of physiognomic ideas. The references to Pythagoras and Hippocrates seem to imply that physiognomic learning was not deemed to be of divine origin as it was in Babylonian culture.[104] But such a general observation perhaps ignores or downplays statements such as that by Adamantius who implies that the art is "God-given."[105]

It is in the writings of Hippocrates and his followers that the first examples of physiognomic learning appeared in the Greek world.[106] The Hippo-

---

2002), 35. These texts, however, do not explicitly mention Pythagoras as the "inventor" of the physiognomic art, but as one who practiced it.

98 Galen, *Quod animi mores corporis temperamenta sequantur* 7 = Kühn IV 797-98. Unfortunately, the recent study by Jacques Jouanna (*Hippocrates* [trans. M.B. DeBevoise; Medicine and Culture; Baltimore, Maryland: Johns Hopkins University Press, 1999]), does not discuss the direct attribution of this skill by later tradition to the figure of Hippocrates.

99 Cf. Förster, *Scriptores Physiognomonici*, 1:xiii-xv; Evans, *Physiognomics*, 5, 19, 27-28; Degkwitz, *'Physiognomonica' Traktat A*, 9 n. 43; Barton, *Power and Knowledge*, 98, 101; Vogt, *Physiognomonica*, 118, 201.

100 Cf. Förster, *Scriptores Physiognomonici*, 1:vii-xiii; Kraus, *Die physiognomischen Omina der Babylonier*, 14-15; Evans, *Physiognomics*, 6, 10, 42-43; Degkwitz, *'Physiognomonica' Traktat A*, 13-15; Barton, *Power and Knowledge*, 100; Böck, *Die babylonisch-assyrische Morphoskopie*, 61-62; Vogt, *Physiognomonica*, 114-17.

101 Cicero, *On Fate* 10; *Tusculan Disputations* 4.80.

102 Diogenes Laertius, *Lives of Philosophers* 2.45.

103 Plutarch, *Sulla* 5.5–6. Although the designation "Chaldean" need not necessarily indicate ethnic origin – it could also be a general indication for astrologers – here it probably refers to someone of Babylonian origin since the Chaldean came with the Parthian envoy. See Kraus, *Die physiognomischen Omina der Babylonier*, 14. See also n. 43 above.

104 Galen's mention of "the divine Hippocrates" (τὸν θεῖον Ἱπποκράτη), *Quod animi mores corporis temperamenta sequantur* 7 = Kühn IV 798, does not imply the divine origin of his knowledge. Cf. Jouanna, *Hippocrates*, 10-12, 37-38; D. Zeller, "The θεῖα φῦσις of Hippocrates and of Other 'Divine Men,'" in *Neues Testament und hellenistische Umwelt* (D. Zeller; BBB 150; Hamburg: Philo, 2006), 129-40. I owe the latter reference to Marlies Schipperheijn. The epithet "divine" can simply indicate high-level achievements (cf. the reference to divine men discovering the art of physiognomics in Adamantius' paraphrase, see n. 167 below), or Hephaestion of Thebes' reference to the method of "the divine Ptolemy" (see n. 236 below).

105 See n. 167 below.

106 This is true for the more technical, learned expressions of physiognomic knowledge. But "physiognomic consciousness" began much earlier with Homer and continued in the

cratic text *Airs, Waters, and Places* from around 400 BCE deals with ethnological physiognomics. It describes the influence of geographical conditions and climate on the nature, appearance, and health of entire peoples.[107] Another work from the Hippocratic corpus, *Epidemics*, exhibits in some instances the kind of learning found in the later physiognomic treatises:

> Those with ruddy complexion, flat nose, large eyes, are good.
> If the head is large and the eyes small, if they are stammerers, they are quick to anger.
> Those with large head, large dark eyes, thick, blunt nose, are good.
> Those with small head, thin neck, narrow chest, are equable.[108]

As a diagnostic method physiognomics was closely related to medicine.[109]

The earliest appearances of the Greek word for physiognomics, φυσιογνωμονία, are from the fourth century BCE with Demosthenes and Aristotle. In a speech against Aristogeiton ascribed to the Athenian politician Demosthenes, Demosthenes exhorts his fellow citizens to choose wisely because when they leave the courthouse those outside "will scan each one as he appears, and detect by their looks (φυσιογνωμονήσουσι) those who have voted for acquittal."[110] Although the use of the term seems non-technical, it remains its earliest occurrence. Aristotle has scattered remarks in his writings about physiognomics.[111] He mentions, for example, in *Generation of Animals* "a physiognomist (φυσιογνώμων) who in his lectures used to show how all people's faces could be reduced to those of two or

poetry, prose, and drama of classical authors. See Evans, *Physiognomics*, 33-35, 58-62, 67-68.

[107] Cf. Evans, *Physiognomics*, 19-20; Vogt, *Physiognomonica*, 113-14; Jouanna, *Hippocrates*, 210-32; B. Gundert, "Soma and Psyche in Hippocratic Medicine," in *Psyche and Soma*, 13-35, at 19-20; Sassi, *Science of Man*, 105-11; Isaac, *Invention of Racism*, 60-69.

[108] *Epidemics* 2.5.1; 2.6.1. Translation from W.D. Smith, *Hippocrates: Volume VII* (LCL 477; Cambridge, Massachusetts: Harvard University Press, 1994), 75, 81. See also Evans, *Physiognomics*, 20 n. 20; Barton, *Power and Knowledge*, 101.

[109] Cf. André, *Anonyme Latin*, 17-18; Barton, *Power and Knowledge*, 97-99. In this light it is interesting to observe that the same connection is made between physiognomics and medicine in Babylonian scholarship. The *Esagil-kīn-apli Catalogue* ascribes both the physiognomic series *Alamdimmû* and the diagnostic series *Sakikkû* to the expertise of the *āšipu*. On the medical expertise, see E.K. Ritter, "Magical-Expert (=*āšipu*) and Physician (=*asû*): Notes on Two Complementary Professions in Babylonian Medicine," in *Studies in Honor of Benno Landsberger on his Seventy-Fifth Birthday, April 21, 1965* (eds. H.G. Güterbock and T. Jacobsen; AS 16; Chicago: University of Chicago Press, 1965), 299-321.

[110] Demosthenes, *Orations* 25:98. Text and translation from J.H. Vince, *Demosthenes III* (LCL 299; Cambridge, Massachusetts: Harvard University Press, 1986), 574-75.

[111] In addition to the references cited in n. 76 above, cf. also Sassi, *Science of Man*, 43-47.

three animals."[112] And in *Prior Analytics* Aristotle presents physiognomics as a form of logical inference:

> It is possible to judge men's character from their physical appearance (τὸ δὲ φυσιογνωμονεῖν δυνατόν ἐστιν), if one grants that body and soul change together in all natural character traits.[113]

The obviousness with which Demosthenes and Aristotle use the derivatives of the term physiognomics implies that, by their time, it was a well-known concept to their audiences, which needed no further introduction.[114]

The focus on character links Greek physiognomics with rhetoric's interest in stereotype character types. From Aristotle onwards there is a connection between rhetoric and the study of characters and passions. For example, Theophrastus (ca. 370–285 BCE), who was Aristotle's successor as head of his school in Athens, wrote a small book with thirty characters. This work was written slightly earlier than the Peripatetic *Physiognomonica* with which it shares some formal correspondences, but Theophrastus did not give physical descriptions of these types in his *Characters*. Instead he focused on the psychological signs of stock characters. Both writings can be seen as different expressions of a similar interest, namely the classification and recognition of individual set character types.[115] The concrete elaboration of this interest seems to have commenced with Aristotle and his school at the beginning of the Hellenistic period. There are references to earlier physiognomists, but not to physiognomic treatises. The Peripatetic *Physiognomonica* probably represents the first effort to put physiognomics systematically in writing.[116]

Principles and Methods of Physiognomics

The *Physiognomonica* and the Anonymous Latin author devote attention to the principles and methods of physiognomic inquiry, but our ancient sources do not provide much information about the practical application of these methods.[117]

[112] Aristotle, *Generation of Animals* 769b 20–21. Text and translation from A.L. Peck, *Aristotle: Generation of Animals* (LCL 366; Cambridge, Massachusetts: Harvard University Press, 1963), 418-19. Cf. Vogt, *Physiognomonica*, 117-18.

[113] Aristotle, *Prior Analytics* 70b 6–7. Text and translation, slightly adapted, from H. Tredennick, *Aristotle: Prior Analytics* (LCL 325; Cambridge, Massachusetts: Harvard University Press, 1962), 526-27.

[114] Cf. Vogt, *Physiognomonica*, 37 n. 3.

[115] See Barton, *Power and Knowledge*, 110-11; Vogt, *Physiognomonica*, 88-107; Sassi, *Science of Man*, 76-77; Isaac, *Invention of Racism*, 162-63.

[116] Cf. Vogt, *Physiognomonica*, 119.

[117] Cf. Vogt, *Physiognomonica*, 145-46. Plutarch, *Sulla* 5.5–6, says that the Chaldean who examined Sulla did so according to the principles of the discipline (τὰς τῆς τέχνης ὑποθέσεις), but these are not explicated. Zopyrus is said to have based his judgment of Socrates on the latter not having a curved collarbone (Cicero, *On Fate* 10). The Stoic Cleanthes,

The *Physiognomonica* starts by simply stating that soul and body are mutually influenced, but it does not explain why this is so:

> Mental characteristics follow bodily ones and are not in themselves unaffected by bodily impulses. [...] Conversely, [...] the body suffers sympathetically with affections of the soul [...].[118]

To this the author of Tractate B adds the synchronicity of the reciprocal influence between body and soul; the one follows the other immediately (808b 20–28).[119]

Body, Character, and Humors

The reason and cause of this relationship between mental and physical character are not explained in the *Physiognomonica*.[120] Tractate B mentions briefly that there are forms of the body that are influenced by heat and cold (809a 6). This reference to a humoral basis for the relationship between mind and body is not elaborated upon in the *Physiognomonica*, but the theory of humors clearly lies behind some of the descriptions,[121] and is already found in Aristotle.[122] The theory of humors did not originate with Aristotle, but was developed before him.[123]

---

finally, recognized a kinaedic type of man by his sneeze (Diogenes Laertius, *Lives of Philosophers* 7.173). Polemo gives examples of characterizations and predictions listing the indicative bodily features (cf. Mesk, "Beispiele in Polemons Physiognomonik").

118 *Physiognomonica* 805a 1–6, cf. 808b 11–14. For a discussion of the terminology used to refer to character, body, and soul in the *Physiognomonica*, see Degkwitz, *'Physiognomonica' Traktat A*, 17-18, 56-58, 64-66; Vogt, *Physiognomonica*, 288-94, 304-5. This claim of a reciprocal influence between body and character is not really explained but it is justified by three empirical examples, which are, however, somewhat problematic, see *Physiognomonica* 805a 3–17. Cf. Degkwitz, *'Physiognomonica' Traktat A*, 25; Vogt, *Physiognomonica*, 287-88.

119 He takes madness as an example of a mental state that is, however, treated by physicians by purging the body with drugs. The form of the body was influenced by the disease, but changed as a result of the treatments that simultaneously freed the soul from its madness. Cf. Vogt, *Physiognomonica*, 398.

120 Galen, *Mixtures* 2.6 = Kühn I 624, remarks that those who attempt the art of physiognomics do not add the reason for connections like a large amount of hair on the chest and a spirited character, or, if on the thighs, a lustful character. The Anonymous Latin author follows Loxus in taking the location of the soul in the blood and understanding the state of the blood as influencing the form and appearance of the body, cf. *De physiognomonia liber* §§2, 12.

121 See e.g. the descriptions of the brave and cowardly types of men (807a 31–807b 12), cf. Vogt, *Physiognomonica*, 339-51.

122 See n. 76 above. In *On the Movement of Animals* (701b 27–33) Aristotle says that because of heat or cold or a similar affection a change is caused in the region of the heart that greatly affects the body, causing redness or paleness, shivering or trembling or the opposites of these. For example, living beings that consist of a great deal of fluid have straight hair, because the fluid advances like a stream in their hair. For the same reason, namely an abundance of moisture, the Scythians by the Black Sea and the Thracians have straight hair; both their constitutions and the surrounding air are moist. Ethiopians and other people who live in hot regions, however, have curly hair since both their brains and the surrounding air are dry,

Regarding physiognomics, the physician Galen (129–ca. 216) is considered to be the one who brought together fully the different strands of previous thinking concerning people, environment, and the humors that, affected by environment, shape the form and appearance of the body.[124] He devoted a small book about physiognomics called *That the Faculties of the Soul Follow the Mixtures of the Body* to explaining why soul and body are mutually influenced. The mixture (κρᾶσις) of the four fluids constitutes not only the shape and appearance of the body, but also the soul. This would explain why a certain character would have been recognizable by the shape and appearance of the body. Galen was the first who "skillfully combined the Aristotelian parallelisms of men and animals in the study of physiognomy with the theory of the humours circulating in the body."[125]

Philosophers and physicians in antiquity debated about the nature of the soul, whether it was a bodily or non-bodily substance, as well as its relationship to the body.[126] Galen admits his ignorance of the nature of the soul's substance, even after years of study, but he has difficulty accepting it as non-bodily, and states that even if this were the case, one would have to admit "at least that it is slave to the mixtures of the body."[127] For many there was not a strict division between body and soul, and physical and psychological states were seen as interrelated. So much so that people believed physiognomics was able to interpret the visible signs on the body as

says Aristotle in *Generation of Animals* (782b 31–37). The constitution not only affected the outward appearance of people's bodies, but also their characters. In *On the Soul* Aristotle argues that the affections of the psyche are inseparable from the body and always associated with it (*On the Soul* 403a 15–18). While discussing the cause for melancholy among outstanding men like philosophers, statesmen, poets, and artists, the pseudo-Aristotelian *Problems* 30 describes the influence of black bile on the forming of character as being a natural process (953a 10–955a 40).

123 See e.g. G.E.R. Lloyd, "The Hot, the Cold, the Dry and the Wet in Greek Philosophy," *JHS* 84 (1964): 92-106; R. Klibansky, E. Panofsky and F. Saxl, *Saturn and Melancholy: Studies in the History of Natural Philosophy, Religion, and Art* (London: Thomas Nelson, 1964; repr., Nendeln, Liechtenstein: Kraus, 1979), 3-66; I.W. Müller, *Humoralmedizin: Physiologische, pathologische und therapeutische Grundlagen der galenistischen Heilkunst* (Heidelberg: Karl F. Haug, 1993); V. Nutton, "Säftelehre," *DNP* 10 (2001), 1208-10; Sassi, *Science of Man*, 152-60. The notion that illness results from an imbalance of the humors is alien to the Babylonian concept of disease, cf. n. 188 below. Nevertheless, some earlier stages of Greek medicine, especially those termed "Cnidian," may have been influenced by Babylonian medicine, cf. Geller, "West Meets East."

124 Slightly earlier, the astrologer Ptolemy of Alexandria expressed his astrological climate theory against the background of the influence of environment on bodily characteristics based on the humors. See the recent discussion by Isaac, *Invention of Racism*, 55-109, esp. 99-101.

125 E.C. Evans, "Galen the Physician as Physiognomist," *TAPA* 76 (1945): 287-98, at 298. See also H. von Staden, "Body, Soul, and Nerves: Epicurus, Herophilus, Erasistratus, the Stoics, and Galen," in *Psyche and Soma*, 79-116, at 105-16.

126 See e.g. some of the articles in Wright and Potter, *Psyche and Soma*.

127 Galen, *Quod animi mores corporis temperamenta sequantur* 3 = Kühn IV 776-79. Translation from Singer, *Galen: Selected Works*, 155.

signifiers of the invisible character in the soul,[128] thus reflecting the absence of a firm boundary between the inner and outer body.[129]

Physiognomic Reasoning

The conceptual context of Greco-Roman physiognomics may have been based on the physiological theory of humors, but the *Physiognomonica* is more concerned with the correct logical procedures to draw conclusions concerning people's characters from the signs of the body.[130] The interest is primarily semiotic and closely follows Aristotle's reasoning at the end of *Prior Analytics* where he presents physiognomics as a form of logical inference called *enthymeme*, used to denote syllogisms that leave out one of the premises for rhetorical purposes.[131]

Adopting one of Carlo Ginzburg's theses, Maria Sassi has characterized the sort of reasoning in physiognomics as abductive: a "semiotic mode of inference from effect to cause where an observed fact is assumed to be the *result* (or effect) of some general *rule* (or cause), of which it constitutes a *case*."[132] According to Sassi the following chain of reasoning is implied in physiognomics:

Result	Animal C has physical trait A.
Case	Animal C has psychological trait B.
Rule	All animals with psychological trait B have physical trait A.

The procedure of abduction is described as a creative process. The conclusion (*case*) remains conjectural and can only be validated externally. The semiotic problem is that a one-on-one correspondence (*rule*) between sign (physical trait A) and signified (psychological trait B) is hypothetical. Thus, "it is not necessarily the case that all, *and only*, the animals with psychological trait B show physical trait A, and it is thus equally not the case that B as well as A may be predicated of C."[133]

[128] Cf. Iamblichus, *On the Pythagorean Life* 17.71. Cf. also this notion with regard to medicine in Arctinus, *Sack of Troy* fr. 1: "recognize the invisible and cure the incurable." Cf. V. Nutton, *Ancient Medicine* (London: Routledge, 2004), 45 n. 59, for references in the Hippocratic Corpus.

[129] Martin, *The Corinthian Body*, 18-20.

[130] Other physiognomic writings hardly bother about the physiological basis for the correspondence between body and soul either. Cf. Polemo 1.210.18–23F(örster), see Barton, *Power and Knowledge*, 104. See also the Anonymous Latin author in n. 120 above.

[131] Cf. R. Smith, *Aristotle: Prior Analytics: Translated, with Introduction, Notes, and Commentary* (Indianapolis: Hackett, 1989), 226-28; Barton, *Power and Knowledge*, 104-7; Vogt, *Physiognomonica*, 120-33.

[132] Sassi, *Science of Man*, xvi-xvii.63-76, at 70 (cf. Barton, *Power and Knowledge*, 134-36); C. Ginzburg, "Morelli, Freud, and Sherlock Holmes: Clues and Scientific Method," in *The Sign of Three: Dupin, Holmes, Peirce* (eds. U. Eco and T.A. Sebeok; Bloomington: Indiana University Press, 1983), 81-118.

[133] Sassi, *Science of Man*, 71. Aristotle attempts to validate physiognomic reasoning by granting certain assumptions (*Prior Analytics* 70b 7–38), basically transforming it into a first-

Tractate A of the *Physiognomonica* introduces three methods that were practiced by physiognomists: a zoological method, an ethnological method, and a method based on analogy of characteristic facial expressions.[134] The zoological method is based on comparing human beings with animals and their characters, and assumes that a correspondence in physique entails one in character.[135] The ethnological method is rooted in the idea that peoples are classified according to physical and psychological characteristics, and that individuals can be compared to these set types.[136] The method based on analogy of characteristic facial expressions takes the expressions of temporary emotions as indicative of permanent ones.[137] The author of Tractate A admits on the one hand that each of these three methods makes it possible to practice physiognomics, but on the other hand expresses his criticism of them.[138]

An important distinction between both treatises of the *Physiognomonica* is the methodological procedure demanded in Tractate B to distinguish between male and female characteristics. The divisions are made in conjunction with animal comparisons. Thus, sex difference and animal comparisons are the main methods used to justify physiognomic inferences in Tractate B.[139] The male form functions as the standard from which the female forms and characteristics deviate:

---

figure inference that is always valid. The assumptions by which physiognomizing (φυσιογνωμονεῖν) is possible are: (1) body and soul both change under influence of natural affections, (2) there is one sign (physical trait) of one thing (psychological trait), and (3) it is possible to know the affections and signs of each class of animals. This means that a particular affect that applies to any indivisible class must have a corresponding sign with it, which enables one to deduce by the outward appearance an affect of the psyche, reasoning from effect to cause.

[134] Cf. A.MacC. Armstrong, "The Methods of the Greek Physiognomists," *GR* 5 (1958): 52-56; André, *Anonyme Latin*, 12-15; Vogt, *Physiognomonica*, 151-53.

[135] *Physiognomonica* 805a 21–24. Cf. the unnamed physiognomist in Aristotle, *Generation of Animals* 769b 20–21. The comparison with animals also appears in Babylonian physiognomics, see Böck, *Die babylonisch-assyrische Morphoskopie*, 40.

[136] *Physiognomonica* 805a 26–28. Ethnological physiognomics already appears in the Hippocratic text *Airs, Waters, and Places* (cf. n. 107 above).

[137] *Physiognomonica* 805a 28–31.

[138] The third method that deduces character from affections on the basis of similar facial expressions is briefly criticized (805a 33–805b 10), cf. Vogt, *Physiognomonica*, 300-1. The zoological method receives elaborate discussion, as well as a suggested solution for the criticism expressed (805b 10–27), cf. Vogt, *Physiognomonica*, 129-30, 301-7. The ethnological method is more or less ignored in the *Physiognomonica* after this. It is only referred to in two short remarks: 806b 15–18: people living in the north are brave and stiff haired, whereas people in the south are cowardly and have soft hair; 808a 31: the little-minded man looks like someone from Corinth or Leucadia. Cf. Vogt, *Physiognomonica*, 296-98, 385.

[139] *Physiognomonica* 809a 26–810a 13. Tractate A refers only once explicitly to the sex difference (806b 31–34), but never relates to this distinction in its catalogue. Vogt, *Physiognomonica*, 153-63, 407-17. See also Gleason, *Making Men*, 55-81; Barton, *Power and Knowledge*, 115-18; Sassi, *Science of Man*, 82-139.

> Of all the animals that we attempt to breed the females are tamer and gentler in soul than the males, but less powerful, and more susceptible to rearing and handling. This being their character, they have less spirit than the males. [...] But it seems to me that females have a more evil disposition than males, are more forward and less courageous. Women and the female animals bred by us are evidently so. [...] Moreover, this is also obvious, that in each class each female has a smaller head, a narrower face and a more slender neck than the male, as well as a weaker chest and smaller ribs, and that the loins and thighs are more covered with flesh than in the males, that the female has knock-knees and spindly calves, neater feet, and the whole shape of the body built for charm rather than for nobility, with less strong sinews and with softer, moister flesh. The males are in every respect opposite to this; their nature is as a class braver and more honest, that of the female being more cowardly and less honest.[140]

In addition to discussing previous methods, the *Physiognomonica* expresses methodological concerns for practicing physiognomics. Four basic rules of thumb can be distilled.[141]

First, bodily characteristics are valid signs only if there is an exclusive connection to permanent mental characteristics. The theoretical character of this rule is made clear by the fact that even the *Physiognomonica* does not live up to it, for mental characters have several bodily signs and one bodily trait can signify various character features.[142]

Therefore, Tractate B formulates a second rule directed at a more practical level: to judge someone on the basis of his general appearance (ἀπὸ τῆς ἐπιπρεπείας) by that bodily sign that stands out and strikes the eye. The point is to grasp that aspect typical of people's general appearance and draw a physiognomic conclusion.

The third principle is closely related to the second, because it formulates how the reader is to proceed in finding the significant feature typical of people's appearance by introducing a hierarchy of signs:

> In all selection of signs some give a much clearer demonstration of the subject than others. Clearest of all are those that appear in the most favorable position. The most favorable part for examination is the region round the eyes, forehead, head and face; secondly, the region of the breast and shoulders, and lastly that of the legs and feet; the parts about the belly are of least importance. Generally speaking, these regions supply the clearest signs, in which there is greatest evidence of intelligence.[143]

Finally, the author of Tractate A states that the physiognomist cannot rely on a single sign, but must base his judgment on the agreement of several

140 *Physiognomonica* 809a 30–809b 13.

141 The following is based on Vogt, *Physiognomonica*, 146-50.

142 Cf. Vogt, *Physiognomonica*, 463-80.

143 *Physiognomonica* 814a 9–814b 9. Cf. also 806b 34-37; Polemo 1.168.1–9F; Anonymous Latin, *De physiognomonia liber* §§10–11. See Barton, *Power and Knowledge*, 109-10.

signs. If several of the signs (for a particular character) coincide in one individual the probability of the inference drawn is greater (806b 37–807a 3).[144]

Notwithstanding these concerns with correct reasoning, it remains important to realize that "their efficacy derives, not from an exhaustive amassing of empirical data, but rather from a classification of the world oriented – and guaranteed – by ideological values."[145] For example, regarding the second rule of *epiprepeia* Sassi argues that its use:

> highlights certain tautological convolutions, indicative of a conscious attempt to codify notions already defined at the level of a collectively shared framework. Thus a bodily feature is judged according to the general impression produced by the individual, who is in turn influenced by that same feature and by the meaning it carries in a context clearly structured by a scale of social values [...]. [...] In other words, the social behavior of men and women, of slaves and the free, and so forth is so standardized as to become self-evident and 'natural' and to impose a norm.[146]

Although the physical descriptions seem to evolve into ever more complex and nuanced distinctions, the characterization of people stays broadly within familiar stereotypes, as known from, for example, Theophrastus. It has been argued that physiognomic literature is another reflection of the notion that the ideal human being was Greek, male, and free. This type of person is never explicitly described, but he forms the implicit point of reference against which the others, i.e. women, barbarians, and animals, are characterized as inferior. This schema structured a large part of elite thinking, and, for example, surfaces clearly in ancient rhetoric where the woman, the foreigner, and the animal present important *topoi* in invective speeches as the monstrous other.[147]

Contexts and Functions of Physiognomic Texts and Learning

As in the case of Babylonian physiognomic texts and practice, one needs to distinguish between the context and function of Greco-Roman physiognomic texts and physiognomic practice.

Physiognomic texts and catalogues may have been transmitted, read, and studied for the mere knowledge of their content. Thus, the Anonymous Latin author's book intends to transmit Greek sources on physiognomics for his Latin readers.[148] Apart from transmitting this learning and providing

[144] Cf. Vogt, *Physiognomonica*, 329-30. Vogt points out that this is "eine pragmatische Anweisung, die nichts mehr mit den theoretischeren Darlegungen zur logischen Signifikanz in früheren Abschnitten zu tun hat, sondern direkt aus der Praxis zu stammen scheint." (329)

[145] Sassi, *Science of Man*, 75-76.

[146] Sassi, *Science of Man*, 53-55. Cf. Vogt, *Physiognomonica*, 402-5. With regard to Polemo, see Gleason, *Making Men*, 34-37.

[147] Cf. Barton, *Power and Knowledge*; Gleason, *Making Men*; Sassi, *Science of Man*.

[148] Anonymous Latin, *De physiognomonia liber* §1.

information about it, the book need not necessarily have served another, more practical, purpose.

Polemo, however, seems to instruct his readers on the practical use of his book. He makes clear that no one could include all physiognomic material in a book, but at the same time he assures his reader that his book will enable him to make progress by himself in the art of physiognomics:

> But for you, after you have learnt all the signs and made trial of all according to what has been described to you, this will be the base for physiognomics.[149]

Polemo seems to have intended his book as a guide and tool in the study and practice of physiognomics:

> For you to master this science, it will be enough that you learn thoroughly what I have described to you and then apply what you have learned. For not even I have attained mastery of these things without much study and lengthy observation.[150]

He also says that for those who really try, the process of learning the art of physiognomics does not seem very long.[151] But, as Barton comments, these "are no more than the usual encouraging remarks. Polemo will keep his monopoly of knowledge of a τέχνη."[152] It has been argued that a physiognomic text like that by Polemo embodied in the first place his effort to present himself as a physiognomist. The point of this treatise was to elevate his status as one who had mastered full control of the art and to boast of his achievements. The examples of concrete cases were intended to support and illustrate this claim, as well as to inspire the reader with awe for his successes as a physiognomist. The minute subdivisions, especially with regard to the eye, are the mark of the τέχνη and intend to convey the high level of learning. The claim to educate the reader should not be taken at face value. In the case of Polemo the physiognomic text suits primarily the purpose of self-presentation.[153]

The treatises alone would probably not have been enough for someone to become a physiognomist.[154] Studying a text like the *Physiognomonica* did not yet make the reader a physiognomist:

> One needs great familiarity with all the facts, if one hopes to be competent to discuss all these things in detail.[155]

[149] Polemo 1.120.9–12F, cf. 1.158.22–160.2F. References and translation from Barton, *Power and Knowledge*, 108.

[150] Polemo 1.192.14–18F. Translation from Gleason, *Making Men*, 38-39.

[151] Polemo 1.292.24–27F. Anonymous Latin, *De physiognomonia liber* §3, is more straightforward in admitting that the road of study is long.

[152] Barton, *Power and Knowledge*, 109.

[153] Cf. Barton, *Power and Knowledge*, 107-8; Gleason, *Making Men*, 38-52.

[154] Cf. Barton, *Power and Knowledge*, 110; Gleason, *Making Men*, 38.

This familiarity, the text says, could only be attained by much practice. This would most likely have happened under the guidance of a master physiognomist, for example someone like Polemo. Physiognomics purported to be an art (τέχνη) that presupposed much training and experience. Concrete information, however, about teachers, students, and the appropriation of the physiognomic art (τέχνη) is lacking.

In addition, the treatises themselves did not give clear instructions on the use of the information amassed in the catalogues. The *Physiognomonica*, for example, does not make clear when each of the four rules is to be applied.[156] Perhaps readers who practiced physiognomics simply took from the catalogues what suited their purposes. But since much of physiognomic knowledge received its credibility against a background of shared social values about types of people, it is not necessary to assume that the physiognomic treatises were used as tools of reference in actual practice.[157] Because of the importance of social values for the credibility of the art, the texts may be regarded as attempts to codify such social presuppositions.

Regarding the availability of the physiognomic treatises and the people who bought and read them, there is not much evidence. Of course, the audience for such technical writings written in an elite milieu would have been the elite itself, but there is no reason to assume that the circulation of this sort of learning was restricted on purpose. Unlike astrological treatises, which, in words reminiscent of the Babylonian secrecy formulas, have prohibitions against divulging astrological learning to the uninitiated,[158] physiognomic writings do not impose such explicit limitations on their dissemination.

Obviously, people like Polemo, Adamantius, and the Anonymous Latin author were familiar with physiognomic writings. Perhaps they owned such writings themselves, just as Galen collected medical writings,[159] or they had access to them by other means, such as a library. Perhaps excerpts from the physiognomic treatises or from their catalogues were also available to a larger audience through shops that sold *mirabilia* literature. At one point Gellius tells of his time in the port of Brundisium where he bought a bundle of Greek books "filled with marvelous tales, things unheard of, incredi-

[155] *Physiognomonica* 809a 1–3. Cf. Vogt, *Physiognomonica*, 401 and n. 82 above on the possible practical background to Tractate B.

[156] Vogt, *Physiognomonica*, 148, 150.

[157] Although Suetonius shows awareness of physiognomic principles in his portrayal of emperors in *Lives of the Caesars*, there is no reason to assume direct dependence on physiognomic treatises, cf. Evans, *Physiognomics*, 51-56.

[158] See Barton, *Power and Knowledge*, 82-85. Such statements were probably a common *topos* used to enhance the status of the text and its author, cf., for example, the ending of Gellius' preface to his *Attic Nights*.

[159] Cf. Nutton, *Ancient Medicine*, 4.

ble."[160] Some of the tales concerned barbarians, magic, and the transformation of women into men; stories also told by the marvels collector Phlegon of Tralles in his *On Marvels*. It has been noted that this contemporary of Polemo in some ways matches the latter's inquisitiveness for out of the ordinary cases.[161] Whether this points to a shared audience for physiognomic and *mirabilia* literature in this period cannot be ascertained, but it is possible and certainly suggestive. Another way in which the greater public could have become acquainted with physiognomics was through hearing speeches such as those by Polemo, but, although they are said to be on occasion attended by many, the audience for such speeches "remains a shadowy crowd, whose lower social limits are uncertain."[162]

Besides being embodied in treatises, physiognomics could be put to use as a tool for measuring people in different contexts. Thus, it was applied in ancient rhetoric. The study of physiognomics not only instructed the orator about the correct deployment of his own body, mimicry, gestures, and voice during the delivery of a speech,[163] physiognomics also provided ammunition for speeches of invective by presenting set descriptions of stock characters who were easy targets of ridicule to be used in the slander of opponents, such as in the case of Polemo and Favorinus.[164] In a political context, physiognomics could function as a resource for attacking rivals and enemies since appearance was made to function as a moral indicator. Its positive use for praise (*encomium*) was less popular, or at least so it seems from our sources.[165]

Physiognomics as an instrument for measuring people can be, and has been, used to exercise various forms of social control.[166] It was claimed to

160 Aulus Gellius, *Attic Nights* 9.4.3. Translation from J.C. Rolfe, *The Attic Nights of Aulus Gellius* (LCL 200; Cambridge, Massachusetts: Harvard University Press, 1968), 163. I owe this reference to Wytse Keulen.

161 Gleason, *Making Men*, 39-40.

162 Barton, *Power and Knowledge*, 99.

163 Cf. Evans, *Physiognomics*, 39-46; H.MacL. Currie, "Aristotle and Quintilian: Physiognomical Reflections," in *Aristotle on Nature and Living Things: Philosophical and Historical Studies Presented to David M. Balme on his Seventieth Birthday* (ed. A. Gotthelf; Pittsburgh, Pennsylvania: Mathesis and Bristol: Bristol Classical, 1985), 359-66; F. Graf, "Gestures and Conventions: The Gestures of Roman Actors and Orators," in *A Cultural History of Gesture: From Antiquity to the Present Day* (eds. J.N. Bremmer and H. Roodenburg; Cambridge: Polity, 1991), 36-58; Barton, *Power and Knowledge*, 103-4; Gleason, *Making Men*, 101-30; Vogt, *Physiognomonica*, 92-96.

164 See e.g. Gleason, *Making Men*, 55-81.

165 Cf. Barton, *Power and Knowledge*, 111-31.

166 Cf., for example, the importance of people's physiognomy in medieval Arab slave trade, in the classification of female hysterics at the Salpêtrière and criminals by Cesare Lombroso in the nineteenth century, and in Victorian art, see T. Fahd, *La divination arabe: Études religieuses, sociologiques et folkloriques sur le milieu natif de l'Islam* (Leiden: E.J. Brill, 1966), 387-88; M. Cowling, *The Artist as Anthropologist: The Representation of Type and Character in Victorian Art* (Cambridge: Cambridge University Press, 1989); M. Gibson,

help people in their social relationships and dealings with others, such as in the paraphrase of Adamantius, which offers a few words of introduction that probably derive from Polemo:

> If divine men have made any discovery that can be of truly immense benefit to those who study it, it is physiognomics. For nobody would deposit in trust his financial assets, his heirlooms, his wife, or his children – or enter into any sort of social relationship – with a person whose form foretells the signs of dishonesty, lechery, or double-dealing. As if by some God-given, inerrant, and prophetic art, the physiognomist understands the character and purposes, so to speak, of all men: how to choose associates only from those who are worthy, and how to guard against the evildoing of unprincipled people without having to experience it first. For this reason, wise men should apply themselves with all their strength to working through the signs of this art.[167]

Not only did physiognomics enforce social classifications already shared by an elite framework of reference, it could also function more directly to exercise control over the actual make-up of groups of people. Traditions about Pythagoras and his group of followers exemplify this application of physiognomics.[168] Later tradition ascribes to Pythagoras the use of physiognomics as a means to admit or reject new applicants to his community.[169] It was clearly believed that physiognomic judgment was applied, but how it would have exactly functioned is not made explicit. Gellius says that:

> the order and method followed by Pythagoras, and afterwards by his school and successors, in admitting and training their pupils were as follows: At the very outset he 'physiognomized' the young men who presented themselves for instruction. [...] Then, when he had thus examined a man and found him suitable, he at once gave orders that he should be admitted to the school [...].[170]

The biographers Porphyrius (ca. 234–ca. 301) and Iamblichus (ca. 250–ca. 330) also attest that Pythagoras applied a physiognomic test in order to determine people's natures. The signs by which Pythagoras recognized the invisible character of the soul were a person's shape and form, walking, and

---

Born to Crime: Cesare Lombroso and the Origins of Biological Criminology (Westport, Connecticut: Praeger, 2002); G. Didi-Huberman, *Invention of Hysteria: Charcot and the Photographic Iconography of the Salpêtrière* (trans. A. Hartz; Cambridge, Massachusetts: MIT Press, 2003); D.G. Horn, *The Criminal Body: Lombroso and the Anatomy of Deviance* (New York: Routledge, 2003).

167 Adamantius, *Physiognomonikon* 1.298–299. Translation, slightly adapted, from Gleason, *Making Men*, 37-38. Cf. also n. 104 above.

168 There is also a reference to Socrates discerning Plato's inner character through his exterior appearance before taking him as a student, see Apuleius, *On Plato* 1.1. Cf. Barton, *Power and Knowledge*, 100 n. 40.

169 Riedweg, *Pythagoras*, 58, 129-36, understands the physiognomic test as part of the specific structure of the Pythagorean sect.

170 Aulus Gellius, *Attic Nights* 1.9.1–3. Translation from Rolfe, *Attic Nights*, 45-47.

entire bodily movement. Porphyrius states that Pythagoras would never have made someone his friend or pupil before physiognomizing what sort of person he was.[171] But, except for enumerating the signs by which people were known physiognomically, no further details are provided. In his *The Life of Apollonius of Tyana*, Philostratus (ca. 170–205) recounts the procedures of admittance into the study of philosophy in India. One of the elements is a physical examination, suggesting the same sort of physiognomic test as ascribed to Pythagoras:

> The particulars of the youths themselves are duly learnt by inspection of them. For in many cases a man's eyes reveal the secrets of his character, and in many cases there is material for forming a judgment and appraising his value in his eyebrows and cheeks, for from these features the dispositions of people can be detected by wise and scientific men, as images are seen in a looking-glass.[172]

These examples demonstrate the belief that physiognomics could function as a tool for exercising social control, though they do not shed much light on the actual proceedings of such a physiognomic test. One should, nonetheless, allow for the possibility that physical, or more specifically physiognomic, examinations could be used by groups to control and maintain their boundaries for new members or other people; perhaps somewhat similar to medical examinations nowadays for certain professions.

Babylonian and Greco-Roman Physiognomics and the Dead Sea Scrolls: Comparative Issues[173]

Principles and Methods

From the Dead Sea Scrolls it appears that the human body was believed to signify different things. *4QPhysiognomy ar* suggests the predictive value of physiognomic observation, while *4QZodiacal Physiognomy* establishes a link between the human body and certain astrological elements and, in addition, perhaps characterization.[174] Regarding the semiotic relationship between signifier and signified that is suggested by the Qumran texts, some

[171] Iamblichus, *On the Pythagorean Life* 17.71, 74; Porphyrius, *Life of Pythagoras* 13, 54.

[172] Philostratus, *Life of Apollonius* 2.30. Translation from F.C. Conybeare, *Philostratus: The Life of Apollonius of Tyana, the Epistles of Apollonius and the Treatise of Eusebius* (LCL 16; Cambridge, Massachusetts: Harvard University Press, 1989), 197. Cf. J.-J. Flinterman, *Power,* Paideia, *and Pythagoreanism: Greek Identity, Conceptions of the Relationship between Philosophers and Monarchs, and Political Ideas in Philostratus'* Life of Apollonius (DMAHA 13; Amsterdam: J.C. Gieben, 1995), 105 n. 73. I owe this reference to Wytse Keulen.

[173] Cf. also Popović, "Physiognomic Knowledge in Qumran and Babylonia."

[174] 4Q186 1 ii 9: ענו יהיה, "he will be humble."

options need to be evaluated. As there is no explicit reflection on this issue in Jewish texts extant from this period, the considerations below remain speculative.

Divine Communication

It is possible that, somewhat similar to Babylonian physiognomic divination, the signs on the human body were understood to convey God's messages; through people's bodily shapes and appearance God communicated their future. That people believed that God communicated by means of portents is corroborated, for example, by certain passages in Josephus' writings.[175] It is, however, also possible that the predictive value of the human body was simply taken for granted, without any notion of divine signs.[176] The possible references to predictions in *4QPhysiognomy ar* are, unfortunately, too fragmentary to assess their relationship to the physiognomic descriptions.

Animal Comparisons

Another possibility is that the relationship between the body and what is signified by it was structured according to rules of inference reminiscent of Greco-Roman physiognomics. It has been argued that the mention of the person's animal being bull in 4Q186 1 ii 9 (וזה בהמתו שור) represents in some form a zoological method of animal comparisons linking certain human types with certain animals.[177] But this is far from clear, and the description of the body in 4Q186 1 ii is too fragmentary to assess whether this method of physiognomics is used here. Nevertheless, it seems unlikely that long and slender thighs and toes – attributed to the person whose animal is bull (4Q186 1 ii 5–6) – are to be expected for a bull-like type.[178]

[175] Josephus, *Jewish War* 6.288–310, discusses various portents prior to the First Revolt that were believed to be messages from God, and in *Jewish Antiquities* 15.144, some people believe the earthquake to have been a sign communicating God's wrath. See also Philo, *On the Creation of the Cosmos* §§58–59a, who says that "there are people who have based conjectures on heavenly movements and have given indications in advance that tremors and earthquakes would occur." Translation from D.T. Runia, *Philo of Alexandria: On the Creation of the Cosmos according to Moses* (PACS 1; Leiden: Brill, 2001), 60, cf. also 205-6.

[176] Nor is there any explanation of the predictive value of Greco-Roman physiognomics in ancient sources, cf. the references in n. 13 above.

[177] Alexander, "Physiognonomy," 389; Wise, "Horoscope Written in Code," 276. Cf. Albani, "Horoscopes in the Qumran Scrolls," 289.

[178] Cf. the descriptions of the bull and ox in physiognomic texts, Pseudo-Aristotle, *Physiognomonica* 807a 19; 810b 16; 811a 14; 811b 9.20.29.35 (cf. Vogt, *Physiognomonica*, 334-35.424-25); Anonymous Latin, *De physiognomonia liber* §120.

Excursus: בהמה *– Animal and Zodiacal Sign*

Most scholars render בהמה simply with "beast, animal," but one should allow for the possibility that the term attempts to translate a foreign word and conveys the astrological concept referred to by the Greek ζῴδιον or the Latin *animal.*[179] The word בהמה is not known from other Hebrew texts to be a *terminus technicus* for zodiacal sign. In later Hebrew texts a zodiacal sign is referred to with the word מזל.[180] But it is certainly possible that at times more terms were used to refer to this concept. Thus far, only one occurrence of the word מלויש, widely used in Syriac and Mandean sources,[181] is known from a Hebrew text from the Cairo Geniza in the sense of "sign of the zodiac."[182] And recently, it has been argued that the term is also used in this sense in Qumran Hebrew.[183] Since different terms were used over time, it is perfectly possible that בהמה has the meaning "zodiacal sign" in *4QZodiacal Physiognomy*.

179 Cf. Lichtenberger, *Studien zum Menschenbild*, 146; Bergmeier, *Glaube als Gabe*, 78 n. 430; cf. Maier, *Texte vom Toten Meer*, 2:136. For the Latin *animal* in its astrological sense, see A. Le Boeuffle, *Astronomie, astrologie: lexique latin* (Paris: Picard, 1987), 43; A. Le Boeuffle, *Les noms latins d'astres et de constellations* (Paris: Les Belles Lettres, 1977), 62. But Professor Wolfgang Hübner has pointed out to me that *animal* is rare and that *signum* is usually used to refer to zodiacal signs in Latin texts.

180 See e.g. the *Baraita de-Mazzalot* in Wertheimer and Wertheimer, *Batei Midrashot*, 13-37; the Cairo Genizah text T.-S. K 21.88 1/א 23; 2/א 9.22; 1/ב 11.26; 2/ב 12 in Gruenwald, "Jewish Physiognomic," 309-16.

181 For the Mandaic *maluaš(a)*, see e.g. the *Sfar Malwašia, Book of the Zodiac*, 147.20–148.9 in E.S. Drower, *The Book of the Zodiac* (OTF 36; London: Royal Asiatic Society, 1949), 93. Cf. also J.C. Greenfield and M. Sokoloff, "Astrological and Related Omen Texts in Jewish Palestinian Aramaic," *JNES* 48 (1989): 201-14, at 211.

182 See Schäfer, *Geniza Fragmente*, 137; Schäfer, "Metoposkopie und Chiromantik," 93. Perhaps this text (T.-S. K 21.95.L) makes a distinction between zodiacal *constellations*, referred to with מזלות (2b/13) and zodiacal *signs*, referred to with מלושיה{ם}(ן) (2b/14).

This interpretation is also proposed by K. von Stuckrad, *Das Ringen um die Astrologie: Jüdische und christliche Beiträge zum antiken Zeitverständnis* (RVV 49; Berlin: Walter de Gruyter, 2000), 195 n. 147, but rejected in favor of an interpretation according to which מלושיהן would refer specifically to the zodiacal sign of the ascendant that changes roughly every two hours (Von Stuckrad, *Ringen um die Astrologie*, 199-200), because of the statement in 2b/15–16 "when the child is born in these two hours" (כיון שהוא נודל הולד בב׳ שעות הללו). It seems, however, more likely that this latter statement should be understood with Schäfer, "Metoposkopie und Chiromantik," 94 n. 55, 95 n. 57, as an elaboration on the circumstance that the child is born under the zodiacal sign *Libra* on the first day under Jupiter or the moon (2b/14–15): הנולד במזל (ב)מאזנים ביום ראשון בצדק או בלבנה. It is probably a reference to the two hours of night and day during which Jupiter and the moon hold rulership. The statement "when the child is born in these two hours" is, therefore, not related to the sense of מלושיהן.

183 M. Kister, "Three Unknown Hebrew Words in Newly-Published Texts from Qumran," *Lešonénu* 63 (2000-2001): 35-40, at 35-36 (Hebrew). I owe this reference to Jonathan Ben-Dov.

"House of Light," "House of Darkness," and the Appearance of the Body
Finally, as has been noted by other scholars, with regard to *4QZodiacal Physiognomy* there seems to be a semiotic relationship between, on the one hand, the division of numbers in the "house of light" and the "house of darkness" and, on the other hand, the shape and appearance of the body as portrayed in the physiognomic descriptions. The text seems to imply that the more parts there are in the "house of light," the better someone looked. Those born at the moment when there are more parts of light have a more attractive appearance (4Q186 1 ii), than those born when there are more parts of darkness. These latter people look less attractive (4Q186 1 iii).[184] This suggests that the human body is related to the division of numbers; it signifies that division. The *Genesis Apocryphon* from Cave 1 provides information on the positive appreciation of certain physical characteristics of Sarai.[185] She is said to have a beautiful face and lovely eyes (1QapGen ar 20:2–3); the latter possibly contrasts with 4Q186 1 iii 6, if מיראות ("terrifying") refers to the eyes. Sarai's hands also have an attractive appearance; she has long (אריכן) and slender (וקטינן) fingers (1QapGen ar 20:5). This suggests that the long (ארוכות) and slender (דקות) thighs and toes of the type described in 4Q186 1 ii 5–6 may have been regarded as positive, attractive features, while, in contrast, the thick fingers of the type in 4Q186 1 iii 6–7 may have been seen as unattractive. Thus, the praising description of Sarai in the *Genesis Apocryphon* provides some evidence for a connection between the descriptions of the human body and the apportionment of numbers between the "house of light" and the "house of darkness."

The Qumran physiognomic texts themselves do not in any way suggest familiarity with the Greek notion of a humoral basis for the connection between the human body and the soul[186] or, for that matter, between the human body and astrological notions, since the heavenly bodies and zodiacal signs were believed to be made up of a mixture of the same four elements (see below). Nonetheless, it has been suggested that the Dead Sea Scrolls provide evidence for a disease etiology similar to certain Greek expressions of the theory of humors. Joseph Baumgarten has argued that Cave 4 fragments of the *Damascus Document* that deal with the treatment of skin diseases use the word רוח ("spirit") in a way similar to the use of πνεῦμα ("spirit") in some Greek medical writings.[187] If this interpretation were cor-

[184] Due to the impossibility of Allegro's reconstruction of 4Q186 2 i, there are no references to the division of numbers for the type of person described in that column. See Popović, "A Note," 638.

[185] See Appendix II.

[186] See Wise, "Horoscope Written in Code," 276.

[187] J.M. Baumgarten, "The 4Q Zadokite Fragments on Skin Disease," *JJS* 41 (1990): 153-65. Cf. Martin, *The Corinthian Body*, 166-67; C. Hempel, *The Laws of the Damascus*

rect, it would demonstrate Jewish familiarity with Greek physiological theories in Palestine during the Hellenistic-Early Roman period. However, this understanding is problematic. First, the parallels that Baumgarten adduces from Greek medicine are not that strong; πνεῦμα is a vital factor in human physiology, not the cause of disease, as is the case in the Cave 4 fragments of the *Damascus Document*. Second, the spirit (רוח) is said to enter, which suggests an external spirit entering the human body, an option Baumgarten also mentions. However, this is not a notion connected with πνεῦμα in Greek medicine in this way. This sort of external cause for disease seems very much in line with Babylonian medicine, but not with Greek medicine where illness has an internal cause resulting from an imbalance in the mixture of humors.[188]

Another possible understanding of the connection between body and soul could have been suggested by a reading of the second creation narrative where God after having formed man from the dust of the earth instills the breath of life in man by which he becomes a living being (Gen 2:7). Perhaps this verse prompted the theological justification given by the *Testament of Naphtali* for the correspondence between body and spirit as being created in resemblance to each other.[189]

Regarding the connection between the human body and astrological elements, *4QZodiacal Physiognomy* attests the notion of influence of the zodiacal signs on the shape and appearance of the human body.[190] Following Albani's ascendant interpretation (see Chapter Three), I understand the numbers assigned to the "house of light" and the "house of darkness" to be established by the position of the rising zodiacal sign above the eastern horizon at the moment of birth (i.e. the ascendant). The "house of light" refers to the area above the horizon and contains the parts of the zodiacal sign that have risen, while the "house of darkness" refers to the area below the earth where the parts of the sign not yet risen reside. The assumption is that the more parts of a zodiacal sign that have ascended into the light, the more powerful, and presumably beneficial, its radiating influence is on the body of the people born at that moment. The idea governing the organization of *4QZodiacal Physiognomy* is that, because of this close connection, in later life the appearance of the human body (signifier) can give one clues about the position of a person's horoscope, i.e. his ascendant zodiacal sign (the signified).

Document: Sources, Tradition and Redaction (STDJ 29; Leiden: Brill, 1998), 43-50; I. Werrett, *Ritual Purity and the Dead Sea Scrolls* (Diss. St Andrews, 2006), 31, 38-39.

188 Cf. J. Scurlock and B.R. Andersen, *Diagnoses in Assyrian and Babylonian Medicine: Ancient Sources, Translations, and Modern Medical Analyses* (Urbana: University of Illinois Press, 2005), 10-12.

189 *T. Naph.* 2:2. See Appendix II.

190 On the relationship between astrology and physiognomics, see below.

The Textual Format of Physiognomic Descriptions

In *4QZodiacal Physiognomy* the physiognomic descriptions stand at the beginning of each section in the list. They describe the human body from top to bottom. This same structure underlies the descriptions in *4QPhysiognomy ar*. As far as can be determined from the remaining amount of text, the physiognomic catalogues only contained descriptions of the entire body. They did not enumerate individual bodily signs directly followed by what was signified by them.

Recently, some scholars have noted a resemblance in form between the physiognomic texts from Qumran and Babylonian physiognomic omen lists.[191] But there is no evidence for the use of conditional sentences in the Qumran texts, which, if present, would imply the same form as the omens from Babylonian divination literature.

Furthermore, the structure of the Qumran texts differs from that of the Babylonian omen lists. Babylonian omen collections were methodically ordered and organized according to their protases, and this was also the case with *Alamdimmû*. This means that the different parts of the human body were the organizing principle behind the collection. For each body part various descriptions are given, almost exhaustively, before the list proceeds to another part of the body. The sequence of the descriptions follows the order from head to toe, an organizational principle mentioned explicitly in the *Esagil-kīn-apli Catalogue*.[192] The first twelve tablets of *Alamdimmû* describe the human body proceeding from head to toe. Here the reader finds in sequential order descriptive omens for the head, hair on the head, and various parts of the head, such as forehead, eyebrows, eyelids, eyes, nose, lips, tongue, teeth, jaw, cheeks, ears, chin, and finally the entire facial appearance. After the extensive treatment of the head and its various parts in the first eight tablets, the compendium moves in the ninth tablet downwards in its description of the human body, and lists accounts of the neck, and of the chest downwards to the genital area. The series continues with descriptions of buttocks and thighs, until it stops with the toes. The eleventh tablet ends with a section on ways of walking and comparisons with those of animals.

Neither the Hebrew *4QZodiacal Physiognomy* nor the Aramaic *4QPhysiognomy ar* are catalogues of the human body in the exact way that

[191] See Lange, "Essene Position on Magic," 387; Geller, "New Documents from the Dead Sea," 227-29. Geller limits himself to a comparison between the Aramaic text 4Q561 and the Babylonian physiognomic omens, as he thinks it likely that the Aramaic text has been influenced by this Babylonian tradition. Geller, "West Meets East," 70 n. 206, points out that the preservation of this text in Aramaic argues for a Mesopotamian origin in contrast with the Hebrew text 4Q186 which shows no Babylonian influence.

[192] A61-2/B25', Finkel, "Adad-apla-iddina," 148-49.

the Babylonian series *Alamdimmû* is. The Qumran texts are not structured and organized according to protases that describe a particular part of the body but contain overall descriptions of individual bodies. These descriptions themselves do follow the sequential principle *a capite ad calcem*, but this is not the same as the manner in which the entire lists of protases in *Alamdimmû* have been structured according to this rule.[193]

The twelfth and final tablet of the main series is not extant, but Böck suggests that it perhaps contained omens that listed several physical characteristics describing a person's entire bodily appearance. Their form may have been similar to some of the composite omens at the end of the subseries for the female.[194] At the end of one of the textual witnesses of this subseries, two compound descriptions of the female body are preserved, the better-preserved one reading:

> If the forehead glows, a tooth protrudes (*āṣītu*), the nose is straight, the lips are thin, the chin is flat, hands and feet are pointed, (and) she is covered with *ḫalû*- and *umṣatu*-marks, [...] this woman ..., she is near to the god, she is cheerful, barley and silver are set in place for her, she will acquire barley and silver, her days are long, (and) a basket she will not carry before her.[195]

The few remaining composite descriptions of the female body stand isolated within the corpus of omens in *Alamdimmû*. There are some omens that combine two, or three, descriptions, but this is rather different from composite descriptions of people's entire bodies from head to toe.[196] If indeed *Alamdimmû*'s twelfth tablet listed composite omens describing the entire body of certain types of individuals it would strengthen the argument that there is a resemblance in form between the physiognomic texts from Qumran and Babylonian physiognomic omen lists. This resemblance is nevertheless very limited. The greatest part of *Alamdimmû* concerns particular omens of individual parts of the human body.

Greco-Roman physiognomic texts also have descriptions of a particular part of the human body, followed directly by a characterization.[197] Overall descriptions of the human body do, however, appear in Greco-Roman physiognomics too. Following the theoretical discussion, Tractate A of the

[193] It is possible to regard the construction ואיש אשר יהיה[in 4Q186 1 i 7 as the introduction of a protasis ("And if someone [whose] ... will be), but it is impossible to determine where the apodosis begins. See the section on the start of an entry in *4QZodiacal Physiognomy* in Chapter One.

[194] Böck, *Die babylonisch-assyrische Morphoskopie*, 16, 27-28.

[195] *Šumma sinništu qaqqada rabât* 250–255.

[196] Cf. *Alamdimmû* II:44–62; VIII:74–90.

[197] See e.g. the first part of the catalogue of Tractate B (structured from bottom to top!) in *Physiognomonica*, 810a 14–812a 11. For a detailed discussion, see Vogt, *Physiognomonica*, 417-45.

Physiognomonica has appended a catalogue of twenty-two character types.[198] The text does not give definitions of the different character types. It presumes that they are familiar to the reader. For each lemma the catalogue simply strings together bodily characteristics that are the signs of a certain character.[199]

The bodily descriptions in the catalogue of Tractate A do not follow a specific order of parts of the body such as in the Babylonian examples of composite omens for the female where the *a capite ad calcem* principle is applied. Sometimes a description goes back and forth between the area above and below the waist. And many of the character types do not have their bodies fully portrayed. Sometimes just one or two features are described. An example of the first two types demonstrates the nature of the catalogue:

> The characteristic signs of the brave man (ἀνδρείου σημεῖα) are stiff hair, an erect carriage of body, bones, sides and extremities of the body strong and large, broad and flat belly; shoulder-blades broad and far apart, neither very tightly knit nor altogether slack; a strong neck but not very fleshy; a chest fleshy and broad, thigh flat, calves of the legs broad below; a bright eye, neither too wide opened nor half closed; the skin on the body is inclined to be dry; the forehead is sharp, straight, not large, and lean, neither very smooth nor very wrinkled.
>
> The characteristic signs of the coward (δειλοῦ σημεῖα) are soft hair, a body of sedentary habit, not energetic; calves of the legs broad above; pallor about the face; eyes weak and blinking, the extremities of the body weak, small legs and long thin hands; thigh small and weak; the figure is constrained in movement; he is not eager but supine and nervous; the expression on his face is liable to rapid change and is cowed.[200]

[198] *Physiognomonica*, 807a 31–808b 10. The list of character types begins with eight types that form four pairs of opposites (807a 31–808a 11). The brave character is opposed to the cowardly one, the clever character is paired with the stupid one, the shameless with the decent one, and the cheerful character is opposed to the sad one. Another pair appears later in the text: the fierce type is opposed to the gentle one (808a 19–27). Vogt, *Physiognomonica*, 163, indicates that the character types listed in this catalogue are of a predominantly negative kind. Of the remaining twelve character types, seven are negative while the others are ambivalent or neutral. Negative types include the kinaedic (808a 12–16), the malignant (808a 17–19), the mean-spirited (808a 29–31), the abusive (808a 32–33), the lustful (808b 4–6), the one that loves sleep (808b 6–8), and the talkative one (808b 8). The ambivalent or neutral types are the mock modest (808a 27–28), the gamblers and dancers (808a 31), the compassionate (808a 33–37), the gluttonous (808b 2–3), and the one with a good memory (808b 9–10). For a detailed discussion of each character type, see Vogt, *Physiognomonica*, 339-93.

[199] See also the catalogues in Polemo 1.268–282F, and Anonymous Latin, *De physiognomonia liber* §§89–115.

[200] *Physiognomonica* 807a 31–807b 12.

Literary Dependency of the Qumran Physiognomic Texts

Regarding the question of how the Qumran physiognomic catalogues are related to their Babylonian and Greco-Roman counterparts, it is evident that similarities in content can be expected, as the object of descriptions is the human body. Moreover, since we are dealing with the genre of catalogues, albeit from diverse literary traditions, it is not surprising to likewise find similarity with regard to the succinct nature of the literary style used. The words describing the body are strung together.[201]

On the basis of the textual evidence it is difficult to argue for a direct dependence of the Qumran texts on either the Babylonian or the Greco-Roman physiognomic traditions. It has been suggested that the Aristotelian ideal of the golden mean (μεσότης) in Greco-Roman physiognomics lies behind some of the bodily descriptions in *4QZodiacal Physiognomy* and *4QPhysiognomy ar*.[202] Similar descriptions, mainly concerning height ("neither tall nor short"), are also found in the Mandean *Book of the Zodiac*,[203] but in this case it is likely that the Mandean tradition was familiar with the Greco-Roman one. The reference to the type of person born in the middle of *Sagittarius* being of a choleric temperament[204] clearly suggests the theory of humors, which eventually distinguished, in the ninth century, four types of people, *melancholici*, *phlegmatici*, *cholerici*, and *sanguinei*, but the notion of these four types is older than the expressions for them.[205] Such familiarity, however, cannot be presumed for the Qumran texts merely on the basis of descriptions of the body that are couched in terms of the mean between two extremes.

It has been argued that the Aramaic text *4QPhysiognomy ar* was influenced by the Babylonian physiognomic tradition.[206] The evidence for predictions in *4QPhysiognomy ar* seems to suggest such an influence. Then again, there are indications that the Greco-Roman tradition was also familiar with the predictive possibilities of physiognomics, as in the case of Po-

[201] A variety of new words that appear in both catalogues of the *Physiognomonica* further demonstrate this succinct nature. See Vogt, *Physiognomonica*, 188-90: "Es handelt sich dabei zumeist um Komposita, deren etymologische Ableitung klar erkennbar ist und die dem Verfasser die verbale Paraphrase eines Sachverhaltes in einem Nebensatz oder einem Partizip ersparen." (189)

[202] See 4Q186 2 i 3–4; 4Q561 1 i 1–4, cf. Wise, "Horoscope Written in Code," 278. For the golden mean in physiognomics, see Vogt, *Physiognomonica*, 163-66; Sassi, *Science of Man*, 45-50.

[203] Drower, *Book of the Zodiac*, 10, 13, 24, 30

[204] Drower, *Book of the Zodiac*, 27.

[205] Cf. Evans, *Physiognomics*, 18-19; Klibansky, Panofsky and Saxl, *Saturn and Melancholy*, 60. The Mandean *Book of the Zodiac* is thus a fine example of the entanglement of both Babylonian, cf. n. 23 above, and Greco-Roman traditions in one and the same text at a certain stage in its development.

[206] See the position of Mark Geller in n. 191 above.

lemo.[207] Moreover, the form of the bodily descriptions in *4QPhysiognomy ar* is, strictly speaking, not the same as that of the physiognomic omens in the Babylonian tradition. *4QPhysiognomy ar* may very well have had a Babylonian origin, but this cannot be proved on the basis of the text's form and content. It is possible to argue a Babylonian origin on the basis of other considerations. One may point out that other Aramaic texts, like, for example, *4QPrayer of Nabonidus ar* (4Q242),[208] or other technical texts, like, for example, calendrical texts from the Dead Sea Scrolls or the Astronomical Book of *1 Enoch*,[209] also have a Babylonian background. It is possible to conceive of a process of transmission of physiognomic lore from Mesopotamia against such a background. But these factors do not necessarily lead to the conclusion that there was Babylonian influence on *4QPhysiognomy ar* and that physiognomic learning came to Palestine from Mesopotamia. Caution is advised. Jewish culture in Palestine during the Hellenistic-Early Roman period was not influenced either from the East or from the West. It was not a matter of either/or, but rather, at times, of both. The important point is to compare the Qumran physiognomic texts with both traditions and not to exclude one beforehand on the basis of a presumed influence from the other.

Astrology and Physiognomics

The relationship between physiognomics and astrology in *4QZodiacal Physiognomy* needs to be addressed in the light of other Babylonian and Greco-Roman texts that also combine astrology and physiognomics.

It has been argued that due to its rising popularity in the second half of the first millennium BCE, astrology subsumed other disciplines of learning, like magic and medicine, but also physiognomics. The combination of physiognomics with astrology was another illustration of the sympathy between macro and microcosmos, the heavens and human beings.[210]

The details of this development are difficult to trace. The grouping of both disciplines of learning into one sort of text probably began before the Hellenistic period.

[207] See n. 13 above.

[208] See F. García Martínez, "The Prayer of Nabonidus: A New Synthesis," in *Qumran and Apocalyptic: Studies on the Aramaic Texts from Qumran* (F. García Martínez; STDJ 9; Leiden: E.J. Brill, 1992), 116-36, at 135-36.

[209] See M. Albani, *Astronomie und Schöpfungsglaube: Untersuchungen zum astronomischen Henochbuch* (WMANT 68; Neukirchen-Vluyn: Neukirchener, 1994); J. Ben-Dov and W. Horowitz, "The Babylonian Lunar Three in Calendrical Scrolls from Qumran," *ZA* 95 (2005): 104-20.

[210] Cf. Schmidt, "Physiognomik," 1066, 1070; Reiner, *Astral Magic*, 77-79; Böck, "'Esoteric Babylonian Commentary' Revisited," 617, 619-20.

The Late Babylonian text *Esoteric Babylonian Commentary*, perhaps dating to the Persian period, juxtaposes both astrology and physiognomics.[211] The text can be divided into three sections. The first section lists three omen series (*Šumma izbu*, *Sakikkû*, and *Alamdimmû*),[212] as well as three celestial constellations (*Aries*, *Taurus*, and *Orion*). Both the reading and the understanding of this section are disputed. Robert Biggs understands the enumeration of the three omen series as a kind of heading. The culmination of the three constellations refers somehow to the physiognomic series *Alamdimmû*. Böck, on the other hand, takes both lists as a pair that serves to predict the bodily appearance. She interprets the culmination of the constellations as a reference to physical characteristics in general, not to the omen series *Alamdimmû* specifically.[213] Böck adduces the Neo-Assyrian *Diviner's Manual*'s advice to check omens both in heaven and earth, but argues that in this Late Babylonian text astrology plays the dominant part. She suggests that the interpretation of human physiognomy is dependent on the zodiac.[214] It is possible that this text conveys that astrological knowledge of someone's zodiacal sign is instructive in predicting their appearance, but the exact sense of the *Esoteric Babylonian Commentary* remains elusive.

Another Late Babylonian text is more suggestive with regard to the connection between astrology and physiognomics, though it is full of difficulties. Among other things, *LBAT* 1593 is concerned in its first section with determining the physical characteristics as well as some aspects of the future of those born under a certain zodiacal sign, for example:

1. Region of Libra: [(break)] narrow of forehead; variant: the nose? to? [...]
2. and red; (he will have) a long chin; red hair; he will be widowed. Region of Sagittarius: [...]

[211] See Biggs, "Esoteric Babylonian Commentary"; Böck, "'An Esoteric Babylonian Commentary' Revisited."

[212] Böck, "'Esoteric Babylonian Commentary' Revisited," 616, notes that only these three series, out of the wide range of Babylonian omen collections, are concerned with the human body and its appearance. The series *Šumma izbu* ("If a Malformation") is concerned with monstrous births of humans and animals (teratology), see Maul, "Omina und Orakel," 62-64.

[213] Biggs, "Esoteric Babylonian Commentary," 53, reads l.3 until the beginning of l.4 as *ana e-la-nu ki-i ik-šu-du* ALAM.DÍM-*mu-ú iq-ta-bi*. But Böck, "'Esoteric Babylonian Commentary' Revisited," 615, reads *ana* E *la-nu ki-i ik-šu-du alam-dím-mu-ú iq-ta-bi*. Biggs interprets *ana elānu* ("upwards") as an expression together with the verb *kašādu* ("to reach, arrive, approach," for other examples of the verb *kašādu* in an astronomical context see CAD K 273b s.v.) and translates it as "to culminate." Instead, in Böck's interpretation *kî* seems to introduce a new sentence, and before that she reads *ana* E *la-nu* as *ana qabê lāni* ("serving to predict the appearance"). Furthermore, Böck takes the second mention of *alamdimmû* as a parallel of *lānu* ("form, appearance") and translates it accordingly in its general meaning "physical characteristics."

[214] Böck, "'Esoteric Babylonian Commentary' Revisited," 619.

3. (in the?) morning? a woman will (and) kill her? husband – in the middle of Sagittarius a follower? of the [...] constellation [...]
4. is not present?. Region of Aquarius: (the child will be) male, his eyes (or: face) red; region of Aquarius: (it means?) that the lower part? of the ears will be large, a child? to? [...]
5. female, her eyes (will be) ... and beautiful, her? forehead pinched.[215]

This Late Babylonian text provides two important clues for understanding the combination of physiognomics and astrology. First, it demonstrates the notion that zodiacal signs influence the shape and appearance of the human body. Second, *LBAT* 1593 shows a clear structure. The zodiacal information is provided first and only then is the conclusion regarding bodily form and the person's fate given.

The connection of astrology with physiognomics demonstrates the pervasive influence of the zodiacal signs on the human body.[216] In the first century BCE, Cicero clearly attests to the existence of the idea that the zodiac exerts its influence on the body. When presenting the opinion of those who defend the Babylonian genethlialogical predictions (*Chaldaeorum natalicia praedicta*), Cicero says they argue that there is a certain force in the zodiac influencing everything both in heaven and on earth, and:

> so also children at their birth are influenced in soul and body and by this force their minds, manners, disposition, physical condition, career in life and destinies are determined.[217]

This is especially clear in the notion of *melothesia*, the idea that the planets and the zodiacal signs each govern a specific part of the body.[218] In his poem *Astronomica*, probably written at the end of the reign of Augustus and the beginning of that of Tiberius, Manilius describes the parts of the body that are subject to the different signs:

> Now learn how the parts of the human body are distributed among the constellations, and how the limbs are subject each to a particular authority: over these limbs, out of all the parts of the body, the signs exercise special influence. *Aries* as chieftain of them all is allotted the head, and *Taurus* receives as of his estate the handsome neck; evenly bestowed, the arms to shoulders joined are accounted to *Gemini*; the breast is put down

[215] Reiner, "Early Zodiologia," 423.

[216] Cf. Bouché-Leclercq, *L'astrologie grecque*, 428-33; Hübner, *Eigenschaften der Tierkreiszeichen*, 190-201; Sassi, *Science of Man*, 168-72.

[217] Cicero, *On Divination* 2.89. Translation from W.A. Falconer, *Cicero: On Old Age, On Friendship, On Divination* (LCL 154; Cambridge, Massachusetts: Harvard University Press, 2001), 470-73.

[218] Bouché-Leclercq, *L'astrologie grecque*, 319-25; Gundel and Böker, "Zodiakos," 579-82; W. Hübner, "Manilius als Astrologe und Dichter," *ANRW* II 32.1 (1984), 126-320, at 237-42. For decanal *melothesia* see W. Gundel, *Dekane und Dekansternbilder: Ein Beitrag zur Geschichte der Sternbilder der Kulturvölker* (SBW 19; Glückstadt: J.J. Augustin, 1936), 286-87.

> to *Cancer*, the realm of the sides and the shoulder blades are *Leo*'s, the belly comes down to *Virgo* as her rightful lot; *Libra* governs the loins, and *Scorpio* takes pleasure in the groin; the thighs draw near to *Sagittarius*, *Capricorn* is tyrant of both knees, whilst pouring *Aquarius* has the lordship of the lower legs, and over the feet *Pisces* claims jurisdiction.[219]

In the second century CE, Ptolemy of Alexandria devotes a whole chapter in his book on astrology to the influence of each planet and zodiacal sign on the form and mixture of the human body, and another chapter to their effects on injuries and diseases.[220] The powers ascribed to the planets are rooted in the four elements and the mixture of their qualities.[221] This determines their influence on the human body when they rise and set. For example:

> Saturn, when he is rising, makes his subjects in appearance dark-skinned, robust, black-haired, curly-haired, hairy-chested, with eyes of moderate size, of middling stature, and in temperament having an excess of the moist and cold. [...] Mars, when rising, makes his subjects in appearance red and white of complexion, tall and robust, gray-eyed, with thick hair, somewhat curly, and in temperament showing an excess of the warm and dry.[222]

There was a similar system of the mixture of the four elements and their qualities for the zodiacal signs.[223]

The influence of the zodiacal signs on the human body have been collected and organized in astrological lists called *zodiologia*.[224] The *zodiologia* are arranged according to the order of the signs of the zodiac. They disregard any planetary positions and only take into account the position of the sun, which resides in the zodiacal signs for the duration of one month. The assumption is that the influence of the zodiacal sign on the human soul, body, and fate remains the same during this period.

219 Manilius, *Astronomica* 2.453–465. Translation, slightly adapted, from G.P. Goold, *Manilius: Astronomica* (LCL 469; Cambridge, Massachusetts: Harvard University Press, 1997), 118-19. Cf. also Ptolemy, *Tetrabiblos* 3.13.4–5; Sextus Empiricus, *Against the Professors* 5.21-22; Firmicus Maternus, *Mathesis* 2.24. All further translations of the *Astronomica* are taken, sometimes in adapted form, from Goold.

220 Ptolemy *Tetrabiblos* 3.12; 3.13.

221 See also *Tetrabiblos* 1.4. Cf. T. Barton, *Ancient Astrology* (London: Routledge, 1994), 102-11.

222 *Tetrabiblos* 3.12.3, 5.

223 Cf. Hübner, *Eigenschaften der Tierkreiszeichen*, 238-45.

224 See the various *zodiologia* collected in *Catalogus codicum astrologorum Graecorum* (12 vols.; Brussels: Henri Lamertin, 1898-1953), for example, CCAG 4.158–169; 10.212–243; 12.173–191. For more references and discussion, see W. Gundel, "Individualschicksal, Menschentypen und Berufe in der antiken Astrologie," *JdC* 4 (1927): 135-93, at 157-76; E. Svenberg, *Lunaria et zodiologia Latina* (SGLG 16; Gothenburg: University of Gothenburg Press, 1963), 92-104; W. Gundel and H.G. Gundel, *Astrologumena: Die astrologische Literatur in der Antike und ihre Geschichte* (SA.B 6; Wiesbaden: Franz Steiner, 1966), 269-70.

The *zodiologia* list under each sign the various character traits and bodily features ascribed to those born under that sign, as well as the fate that will befall them in life. The "zodiacal children," e.g. "the boy born in the period of the zodiacal sign of *Aries*" (Ὁ γεννηθεὶς νεώτερος ἐν καιρῷ ζῳδίου τοῦ Κριοῦ), could also be referred to by their zodiacal sign in an abbreviated form, such as Λεοντιανοί, Παρθενιανοί, Σκορπιανοί, and Τοξιανοί.[225] Wilhelm Gundel has given the following example of a *zodiologion* from an unpublished manuscript. Many *zodiologia* are quite long, but this short one illustrates the genre very well:[226]

> Die Widderkinder haben ein schönes Gesicht, breite etwas nach der Seite neigende Nüstern, eine Breite Stirn, die in den kahlen Vorderkopf hineinragt, schmale Lippen; sie sind ziemlich groß und rötlich. Ihre Augen sind groß, die Stimme offen, auch eine schöne Hautfarbe gehört zu ihren Naturgaben. Die Teile der Beine zwischen Knie und Knöchel sind kurz; von Natur sind sie fleischig, dichtbehaart, wollig-kraushaarig. Sie schauen zur Erde, sind feig in ihrem Herzen, scherzhaft, schöngeistig, unternehmend, tapfer, roh, waghalsig und leicht veränderlich.[227]

Although Hippolytus of Rome, ca. 170–235/6, criticizes the idea that the zodiacal signs influence the shape of human bodies, he provides an extensive list of all the physical and psychological characteristics attributed to those born under each zodiacal sign, similar to the *zodiologia*.[228] For example:

> Those born in *Taurus* will be of the following type: round head, thick hair, broad (and) square face, black eyes and large eyebrows – in a white man, thin, blood-red veins – long eyelids, thick, huge ears, round mouths, thick nose, round nostrils, thick lips, <short body,> strong in the upper parts,[229] they are sluggish from the legs. The same are by nature: pleasing, perceptive, of a goodly character, pious, just, rustic, complaisant, from twelve years hard workers, touchy, slothful. The stomach of these is small, quickly filled, they plan many things, they are sensible, sparing towards themselves, generous towards others, beneficent, they are partly sad, negligent, in friendship useful on account of mind, distressed.[230]

In theory, the *zodiologia* may have functioned as a reference tool in ancient astrology, astrology of a perhaps more popular form as some have ar-

[225] F. Boll, C. Bezold, and W. Gundel, *Sternglaube und Sterndeutung: Die Geschichte und das Wesen der Astrologie* (5d ed.; Darmstadt: Wissenschaftliche Buchgesellschaft, 1966), 133-34; Gundel and Gundel, *Astrologumena*, 270. Cf. Manilius, *Astronomica* 2.541-642; Sextus Empiricus, *Against the Professors* 5.95–102.

[226] I have been unable to trace the Greek text, which is why I give Gundel's translation.

[227] Gundel, "Individualschicksal," 159.

[228] Hippolytus, *Refutation of All Heresies* 4.15.4–27.2.

[229] Cf. Ptolemy, *Tetrabiblos* 3.12.13.

[230] Hippolytus, *Refutation of All Heresies* 4.16.1–11.

gued.[231] Having determined the solar birth sign of an individual, such texts would have provided easy access to further information concerning the physical and psychological characteristics listed under each zodiacal entry.[232]

Comparing the combination of physiognomics and astrology in *4QZodiacal Physiognomy* with other texts from Babylonian and Greco-Roman traditions yields two interesting results. On the one hand, it suggests that *4QZodiacal Physiognomy* likewise attests to the existence of the belief that the zodiacal signs have an effect on the appearance of the human body. But, on the other hand, the comparison demonstrates that the interest of *4QZodiacal Physiognomy* has a different direction because of its distinctive structure. The connection between astrology and physiognomics is expressed differently in *4QZodiacal Physiognomy*. The astrological data are listed subsequently to the physiognomic descriptions. As has been concluded in the previous chapter, *4QZodiacal Physiognomy* is a physiognomic catalogue in which the descriptions of the human body point the reader to astrological information concerning the types of people described. *4QZodiacal Physiognomy* is similar to other texts in that it *combines* both physiognomics and astrology, but it differs significantly in that the physiognomic descriptions of the human body *precede* the astrological data. This peculiar feature sets this Qumran text somewhat apart from other known ancient texts in which astrology and physiognomic learning are combined.

If in antiquity the belief was held that astrologers were able to predict a child's physical appearance on the basis of his nativity because of the influence of the stars on the human body, it need not surprise us that it must also have been deemed possible to reason the other way round. That is, to determine a person's horoscope from the other known factors in his life, such as his physical characteristics. For example, in Plutarch's account of the life of Romulus, the first-century BCE Roman philosopher and astronomer Tarutius is credited with determining Romulus' horoscope for the moment of his conception and birth from the given facts of his life.[233] His friend Varro put to him this problem, arguing that this could be done just as the solutions of geometrical problems are derived:

[231] Cf. M.P. Nilsson, *Geschichte der griechischen Religion: Zweiter Band, Die hellenistische und römische Zeit* (2d ed.; vol. 2; HdA 5/2; Munich: C.H. Beck, 1961), 486-89; Gundel and Gundel, *Astrologumena*, 269.

[232] Cf. Gundel, "Individualschicksal," 165.

[233] Plutarch, *Romulus* 12.3-5. See the recent and thorough discussion in K. Frommhold, *Bedeutung und Berechnung der Empfängnis in der Astrologie der Antike* (OrA 38; Münster: Aschendorff, 2004), 226-38.

> for the same science, he said, must be capable not only of foretelling a man's life when the time of his birth is known, but also, from the given facts of his life, of hunting out the time of his birth.[234]

Even more interesting is an intriguing remark by the late fourth-century CE astrologer Hephaestion of Thebes:

> If, at some time, from a triplicity there are two zodiacal signs above the earth to which we assume the horoscope (= the ascendant) applies, then we also pay attention to the shape of the man, which one of the zodiacal signs he resembles more, and accordingly we give our decision.[235]

This observation provides an important clue for understanding the relationship between physiognomics and astrology in *4QZodiacal Physiognomy*. If one cannot establish which of two zodiacal signs above the horizon represents the ascendant sign of a certain individual, it is possible, says Hephaestion, to discern this by looking at the shape of his body. And he appends a short list with the physical characteristics of those born at the moment when each of the zodiacal signs is the horoscope. Hephaestion adds that these are the signs of the zodiacal signs themselves when they happen to be the horoscope sign, but that the effects of the planets should also be taken into account, enumerating them briefly. This physiognomic method is, as he says himself, a rough method to find the horoscope, unlike the precise method of "the divine Ptolemy."[236]

Hephaestion thus provides evidence for the belief that astrological information, such as the ascendant zodiacal sign, could be learned through physiognomic inquiry.[237] By means of physiognomic knowledge a person's ascendant zodiacal sign (i.e. his horoscope sign) could be discerned accordingly. This line of reasoning is the guiding principle behind the textual structure of *4QZodiacal Physiognomy*.

234 Plutarch, *Romulus* 12.4. Translation from B. Perrin, *Plutarch's Lives I: Theseus and Romulus, Lycurgus and Numa, Solon and Publicola* (LCL 46; Cambridge, Massachusetts: Harvard University Press, 1967), 122-23.

235 Hephaestion, *Apotelesmatica* 2.2.27: ἐὰν δέ ποτε δύο ζῴδια ὑπὲρ γῆν ᾖ τοῦ τριγώνου εἰς ἃ ὑπονοοῦμεν εἶναι τὸν ὡροσκόπον, καταστοχαζόμεθα καὶ τοῦ εἴδους τοῦ ἀνθρώπου ποίῳ μᾶλλον τῶν ζῳδίων ἀπείκασται καὶ πρὸς ἐκεῖνο ἀποφαινόμεθα. Text from D. Pingree (ed.), *Hephaestio Thebanus: Apotelesmatica* (vol.1; Teubner; Leipzig: B.G. Teubner, 1973), 92.11-15.

236 Hephaestion, *Apotelesmatica* 2.2.28–42. Cf. also n. 104 above.

237 Cf. A. Bouché-Leclercq, *L'histoire de la divination dans l'antiquité* (vol.1; Paris: Ernest Leroux, 1879), 174-75, 266-69; Bouché-Leclercq, *L'astrologie grecque*, 313 n. 1.

CHAPTER III

"IN THE FOOT OF *TAURUS*": ASTROLOGICAL NOTIONS IN *4QZODIACAL PHYSIOGNOMY* (4Q186)

INTRODUCTION

4QZodiacal Physiognomy is a physiognomic-astrological catalogue listing information on astrological matters that concern individual types of people recognizable by their physical descriptions. Although scholars have been trying to understand the text since it was first published, as yet there is no consensus. Consideration of the astrological framework may help to clarify these matters.

4Q186 1 ii 8–9 states that the horoscope (מולד) under which a person was born is "in the foot of *Taurus*" (ברגל השור). But what do these words actually mean? Are they a reference to the position of the sun or the moon at birth; to the first decan of the sign *Taurus* at the moment of conception; or to the ascendant part of the sign at birth?

An important element in the different entries of *4QZodiacal Physiognomy* is the mention of the "house of light" and the "house of darkness," listed together with certain numbers (4Q186 1 ii 7–8; 1 iii 8–9). What is the meaning of this light and darkness terminology? How are the specific combinations of numbers established and how are they assigned to the "house of light" and the "house of darkness?" Furthermore, are the horoscope, the light and darkness terminology, and the realization of the numbers related to each other, and, if so, how? Some scholars have taken the light and darkness terminology as an indication of the text's sectarian worldview, while others have sought to understand it more specifically in terms of different astrological concepts: planetary houses signifying either the duration of day and night, or zodiacal signs of light and darkness; diurnal and nocturnal decans; cosmological rooms above and below the horizon. Also, the realization of the numbers divided between the "house of light" and the "house of darkness" has been related in different ways to the horoscope being "in the foot of *Taurus*": the position of the moon divides the zodiacal sign between the "house of light" and the "house of darkness," signifying the times of day and night; the decan at the moment of conception determines the allocation of diurnal and nocturnal decans; the ascendant position establishes the division of the zodiacal sign into parts above and below the horizon.

Various interpretations have been suggested for the meaning of enigmatic terms and elements in *4QZodiacal Physiognomy* against an astrological background. It will be clear that the significant elusiveness inherent in the terminology and the fragmentary nature of the manuscript preclude any final interpretation. I do not, therefore, intend to attempt any such thing in the following discussion. The different interpretations represent possibilities. Within the realm of possibility, however, some interpretations are better than others because they can account for more features in a coherent way.

In this chapter astrological hypotheses will be discussed, both in terms of their ability to explain certain elements in *4QZodiacal Physiognomy* as well as being considered against the background of ancient astrological notions from Babylonian and Greek astrology. For this reason the chapter begins by addressing some aspects of ancient Babylonian and Greek astrology. These will reoccur throughout the subsequent discussion of the astrological framework of *4QZodiacal Physiognomy*.

SOME ASPECTS OF ANCIENT ASTROLOGY IN BABYLONIA AND GREECE

Evidence for celestial divination is attested in Mesopotamia from the Old Babylonian period in the first half of the second millennium BCE onwards.[1] Like physiognomics, celestial divination was expressed in the form of omens. These were collected, for example, in the first millennium BCE celestial omen series *Enūma Anu Enlil* ("When Anu, Enlil").[2] The importance of celestial divination in the Neo-Assyrian period is demonstrated by the many reports sent to the king by his celestial diviners across the country.[3]

In addition to these divinatory sources, there are also many astronomical texts that show a primarily calendric and mathematical interest, such as the astronomical compendium MUL.APIN ("Plough Star"), from ca. 1000 BCE.[4] Other examples of astronomical writings are non-mathematical texts

[1] See Koch-Westenholz, *Mesopotamian Astrology*; H. Hunger and D. Pingree, *Astral Sciences in Mesopotamia* (HdO I/44; Leiden: Brill, 1999); Maul, "Omina und Orakel," 51-58; Rochberg, *The Heavenly Writing*.

[2] Cf. Hunger and Pingree, *Astral Sciences*, 12-20; E. Reiner, "Babylonian Celestial Divination," in *Ancient Astronomy and Celestial Divination* (ed. N.M. Swerdlow; Cambridge, Massachusetts: MIT Press, 1999), 21-37; Rochberg, *The Heavenly Writing*, 66-78.

[3] See Chapter Two n. 38.

[4] Hunger and Pingree, *MUL.APIN*. It lists, for example, the names and relative positions of fixed stars, the dates of their first rising just before sunrise (heliacal), the simultaneous rising and setting of certain stars and constellations, the constellations through which the moon courses monthly, the periods of visibility and invisibility of the five planets (Mercury,

such as astronomical diaries, goal-year texts, almanacs, and the mathematical ephemerides.[5] Astronomical records reach back to the reign of the Babylonian king Nabonassar (747–734 BCE), beginning with an eclipse on December 6, 747. Suggestive of the transmission of Babylonian astronomical learning is the reference by Ptolemy of Alexandria (second century CE) to records of ancient observations from the beginning of the reign of Nabonassar that had been preserved down to his own time.[6]

Mathematical Astronomy and Zodiacal Astrology

Two fundamental features characterize the development of astrology from the middle of the first millennium BCE onwards: first, the recognition of the periodicity of certain astronomical phenomena and the ability to formulate mathematical models enabling the prediction of the recurrence of those phenomena;[7] second, the introduction of the zodiac. It is because of the second factor particularly that some scholars reserve the term "astrology" for that form of celestial divination that is based on the signs of the zodiac.

The exact relationship between the mathematical astronomical texts and the horoscope texts is not entirely clear due to the lack of evidence. It seems unlikely that advances in mathematical astronomy were made because of astrological concerns regarding horoscopes. Much of the data from the mathematical texts is not used in the horoscope texts. This is not only the case for Babylonian, but also for Hellenistic astronomy and astrology.[8] But what seems clear is that the mathematical character gave astrology its credibility, regardless of its actual use in all aspects. Although "astronomy" and "astrology" were not clearly distinguished before the sixth century CE, ancient scholars were familiar with the distinction between the two concepts implied by these words in modern parlance, as is clear from Ptolemy's opening words in his *Tetrabiblos*:

> Of the means of foreknowledge through astronomy, O Syrus, two are the most important and authoritative. One, which is first both in order and in effectiveness, is that whereby we understand the aspects of the movements of sun, moon, and the stars in relation to each other and to the earth, as they occur from time to time; the second is that in which, through the

Venus, Mars, Jupiter, and Saturn), the length of the daylight scheme, the intercalation schemes, and the lunar visibility scheme. Cf. Hunger and Pingree, *Astral Sciences*, 57-83; Rochberg, *The Heavenly Writing*, 6-7.

[5] See Hunger and Pingree, *Astral Sciences*, 116-270; Rochberg, *The Heavenly Writing*, 147-60.

[6] Ptolemy, *Almagest* 3.7. Cf. Rochberg, *The Heavenly Writing*, 146.

[7] Cf. D. Pingree, *From Astral Omens to Astrology: From Babylon to Bīkāner* (SOR 78; Rome: Istituto Italiano per l'Africa e l'Oriente, 1997), 18-19.

[8] See Rochberg, *The Heavenly Writing*, 160-63. On the interest in applied mathematical science for its own sake on the part of the Babylonian scribes, see e.g. N.M. Swerdlow, *The Babylonian Theory of the Planets* (Princeton: Princeton University Press, 1998).

natural character distinctive of their aspects, we consider the changes they influence with regard to what they surround.[9]

The development cannot be exactly determined, but, roughly speaking, the zodiacal circle was introduced sometime in the fifth century BCE.[10] It is a schematic, symbolic division of the ecliptic, the apparent path followed by the sun around the earth. The ecliptic lies at an angle of approximately 24° to the equator, and the angle that the ecliptic makes with the horizon is dependent on the geographical latitude of the observer or computer's position. The ecliptic or zodiacal circle of 360° was divided into twelve equal sections or signs of 30°. Before the introduction of the zodiac the Babylonians used certain fixed stars for locating bodies in the sky approximately. The astronomical compendium MUL.APIN locates seventeen constellations through which the moon passes every month.[11] By the fifth century BCE these seventeen constellations had been reduced to twelve, and sometime in the fifth century BCE the twelve equal divisions of the ecliptic into zodiacal signs replaced these twelve constellations of unequal size. The signs of the zodiac served as a means of reference for computing and recording the planetary positions more exactly.

There is a difference between zodiacal *constellations* in the zodiacal belt, which comprise actual stars, and zodiacal *signs* on the ecliptic, which are derived from the constellations but are nonetheless symbolic entities of 30° longitude. The constellations are of varying size and some extend well beyond the zodiacal belt, which is given a width of ±12°, i.e. 6° latitude on either side of the ecliptic.[12] Understanding the difference between the zodiac as denoting *signs* and *constellations* is important for the interpretation of the words ברגל השור ("in the foot of *Taurus*") in *4QZodiacal Physiognomy* with regard to identification and localization.

[9] Ptolemy, *Tetrabiblos* 1.1.1 (books, chapters, and sections cited according to Hübner, *ΑΠΟΤΕΛΕΣΜΑΤΙΚΑ*). Translation, slightly adapted, from Robbins, *Ptolemy*, 3 (all translations of the *Tetrabiblos* are taken, sometimes adapted, from Robbins). See W. Hübner, *Die Begriffe 'Astrologie' und 'Astronomie' in der Antike: Wortgeschichte und Wissenschaftsthematik, mit einer Hypothese zum Terminus 'Quadrivium'* (AWLMAGSK 7; Wiesbaden: Franz Steiner, 1989).

[10] See e.g. B.L. van der Waerden, "History of the Zodiac," *AfO* 16 (1952-1953): 216-30; Koch-Westenholz, *Mesopotamian Astrology*, 163-70; Rochberg, *The Heavenly Writing*, 117-18, 129-33, 239. Cf. also the recent contribution by L. Brack-Bernsen and H. Hunger, "The Babylonian Zodiac: Speculations on its Invention and Significance," *Centaurus* 41 (1999): 280-92; Rochberg, *The Heavenly Writing*, 123-33.

[11] See MUL.APIN Tablet I iv 31–39 (cf. Hunger and Pingree, *MUL.APIN*, 67-69, 144).

[12] See e.g. Geminus, *Introduction to Phenomena* 1.3–5. Neugebauer, *HAMA*, 583: "This concept of a latitudinal expanse of 12° of the zodiac is usually motivated as corresponding to the space traversed by the moon and the planets in their latitudinal motion, although it is admitted that Venus might go 2° beyond the limits of ±6°."

Horoscopic Astrology

More or less simultaneously with these two developments the first Babylonian horoscopes are attested. The two earliest horoscopes extant are dated to 410 BCE and the latest is dated to 69 BCE.[13] These texts represent a change from general to personal celestial divination. Traditional Babylonian celestial divination was solely concerned with king, state, and country. Celestial diviners observed the sky for the political and economic benefit of the ruler.[14] With the rise of horoscopic astrology in Mesopotamia somewhere in the fifth century BCE the focus shifted to include individuals and personal predictions.[15] From then on astrology knows two areas of interest:

> the first and more universal is that which relates to entire peoples, countries, and cities, which is called general, and the second and more specific is that which relates to individual men, which is called genethlialogical.[16]

It is in this form that astrology, together with mathematical astronomy, was transmitted from Babylonia to Greece during the Hellenistic period, probably by scribes from the temples of Babylon and Uruk carrying their knowledge with them through the Hellenistic world.[17] Although "the main structure of the astrological theory is undoubtedly Hellenistic,"[18] several elements of Babylonian origin are basic to Hellenistic astrology.[19] In addition, Greek papyri demonstrate the continuing use of Babylonian mathematical methods. These finds minimize a presumed dichotomy between Babylonian and Greek astronomy.[20] However, one main difference is that at the same time Hellenistic astronomy was based on a spherical-geometrical model and

[13] See Rochberg, *Babylonian Horoscopes*, 51-57, 137-40. The earliest Greek horoscope extant, the one for Antiochus I of Commagene, is dated to 62 BCE, practically contemporary with the latest cuneiform horoscope. Cf. Neugebauer and van Hoesen, *Greek Horoscopes*, 14-16, 161-62. Recently it has been re-dated to 109 BCE by M. Crijns in E.M. Moormann and M.J. Versluys, "The Nemrud Dağ Project: First Interim Report," *BABesch* 77 (2002): 73-111, at 97-99. See, however, S. Heilen, "Zur Datierung des Löwenhoroskops auf dem Nemrud Dağ," *forthcoming*. Cf. W. Hübner, "Sulla's Horoscope? (Firm. math. 6,31,1)," in *Horoscopes and Public Spheres: Essays on the History of Astrology* (eds. G. Oestmann, H.D. Rutkin and K. von Stuckrad; RelSoc 42; Berlin: Walter de Gruyter, 2005), 13-35, at 13 n. 2.

[14] See e.g. Pongratz-Leisten, *Herrschaftswissen in Mesopotamien*, 17-46.

[15] Cf. Rochberg, *The Heavenly Writing*, 98-120.

[16] Ptolemy, *Tetrabiblos* 2.1.2. The second book of the *Tetrabiblos* is devoted to general or mundane astrology, while the third book is concerned with individual or genethlialogical astrology. For an overview of both forms, see Bouché-Leclercq, *L'astrologie grecque*, 327-457 (still a standard work on Hellenistic astrology).

[17] Neugebauer, *HAMA*, 613; A. Jones, "Evidence for Babylonian Arithmetical Schemes in Greek Astronomy," in *Rolle der Astronomie*, 77-94.

[18] Neugebauer, *Exact Sciences*, 170.

[19] See Neugebauer, *HAMA*, 589-614; Rochberg-Halton, "Elements of the Babylonian Contribution"; Koch-Westenholz, *Mesopotamian Astrology*, 177-79.

[20] Rochberg, *The Heavenly Writing*, 34-35.

a theory of kinematics of which Babylonian astronomy was apparently completely devoid.[21]

Although the term "horoscope" is used in modern scholarly writings for both Babylonian and Greek texts, it is important to realize that they do not exhibit the same characteristics and actually reflect two different genethlialogical systems.[22] The main difference is that Babylonian horoscopes do not regard as at all important which zodiacal sign is rising above the eastern horizon at a particular time (the ascendant or ὡροσκόπος), which is precisely the most important feature of Greek horoscopes.[23] As Francesca Rochberg explains:

> Babylonian 'horoscopes' are documents that assemble and record a particular series of astronomical data which have been determined to occur either on or near the date of the birth of an individual. A number of examples 'look at the hour,' noting the time of birth occasionally with respect to a seasonal hour (one-twelfth of the length of daylight), and provide planetary positions in the zodiac for the specified time. Only the moon's position is affected by a change in hour, since it moves so much more rapidly than the sun or the five planets. That planetary data are not greatly affected, may perhaps explain why the hour of birth is not noted with regularity in the Babylonian horoscopes. Even when noted, the Babylonian horoscopes' 'inspection of the hour' is not paralleled by the synonymous Greek counterparts, where ὡροσκόπος refers not to the consideration of the time of day, but of the point of the ecliptic (the ascendant) rising at the moment of birth.[24]

Contrary to Babylonian astrology, in Greek astrology the determination of the degree of the zodiac that is rising in the east at the moment of birth is considered the primary important fact.[25] It takes approximately two hours for a 30° section of the ecliptic to rise above the eastern horizon, but the exact duration of ascension varies. The unequal rising times of the different sections of the ecliptic have to do with the position of the ecliptic in relation to the eastern horizon. As this position is also connected with the length of daylight during the year, the determination of the zodiacal rising times is connected with the measurement of the length of daylight at a certain moment of the year. Both zodiacal rising times and length of daylight vary according to geographical latitude.

[21] Cf. Neugebauer, *HAMA*, 675, 771-72; Rochberg, *The Heavenly Writing*, 127.

[22] See Rochberg, *Babylonian Horoscopes*, 1-2; Rochberg, *The Heavenly Writing*, 207-8.

[23] For the meaning of the term ὡροσκόπος, see Gundel and Kehl, "Horoskop," 599-600; Hübner, "Verwendung und Umschreibung des Terminus ὡροσκόπος," 221. Pingree, *From Astral Omens to Astrology*, 20, therefore, argues that the term horoscopy or genethlialogy should be used solely to refer to the particular genre of Greek horoscope texts.

[24] Rochberg, *Babylonian Horoscopes*, 1-2.

[25] See Ptolemy, *Tetrabiblos* 3.3. Cf. Bouché-Leclercq, *L'astrologie grecque*, 86; Gundel and Gundel, *Astrologumena*, 344.

Although the Babylonians were concerned with determining the length of daylight and their methods formed the basis for the Greek ones, the rising times of the zodiacal signs are of less importance in Babylonian than in Greek astronomy.[26] In Greek astronomy, however, the importance of computing the zodiacal rising times, such as in Hypsicles' *On the Ascendant* (second century BCE), seems to go hand in hand with the significance attributed to the ascendant for the nativity in Greek astrology. The actual computation of zodiacal rising times was not a simple matter. From a comment by Hipparchus (second century BCE) it is clear that some of his contemporary astronomers had difficulties understanding the astronomical concept of zodiacal rising times.[27] But to most astrologers this sort of information would probably be at hand in almanacs and ephemerides.[28]

[26] Bouché-Leclercq, *L'astrologie grecque*, 259-69; O. Neugebauer, "The Rising Times in Babylonian Astronomy," *JCS* 7 (1953): 100-2; V. De Falco, M. Krause and O. Neugebauer, *Hypsikles: Die Aufgangszeiten der Gestirne* (AAWGPHK 3/62; Göttingen: Vandenhoeck & Ruprecht, 1966), 5-17; Neugebauer, *HAMA*, 34-45, 301-3, 366-71, 706-33, 763-67, 864-65; B.L. van der Waerden, *Die Astronomie der Griechen: Eine Einführung* (Darmstadt: Wissenschaftliche Buchgesellschaft, 1988), 125-30; J.-H. Abry, "Les *anaphorai* des signes du zodiaque dans les écrits astrologiques," in *Sciences exactes et sciences appliquées à Alexandrie (IIIe siècle av. J.-C. – Ier siècle ap. J.-C.): Actes du Colloque International de Saint-Étienne (6-8 juin 1996)* (eds. G. Argoud and J.-Y. Guillaumin; Mémoires 16; Saint-Étienne: Université de Saint-Étienne, 1998), 305-24; J. Evans, *The History and Practice of Ancient Astronomy* (New York: Oxford University Press, 1998), 95-125; F. Rochberg, "A Babylonian Rising-Times Scheme in Non-Tabular Astronomical Texts," in *Studies in the History of the Exact Sciences in Honour of David Pingree* (eds. C. Burnett et al.; IPTS 54; Leiden: Brill, 2004), 56-94.

[27] Hipparchus, *Commentary on Aratus and Eudoxus* 2.1.2–14. Cf. Neugebauer, *HAMA*, 715.

[28] Neugebauer and van Hoesen, *Greek Horoscopes*, 170, observe that reference to the zodiacal rising times occurs only once in an actual horoscope from 95 CE, whereas in handbooks, such as the astrological treatise of the second-century CE astrologer Vettius Valens, the rising times are mentioned frequently. Jones, *Astronomical Papyri*, 1:282-83; 2:418-19, has published another horoscope (late second or early third century CE) that says explicitly that the computations were carried out on the basis of the table of ascensions from Hipparchus' compilation. Both horoscopes differ from standard horoscopes in that they are elaborate, "deluxe" horoscopes, cast for a socially high class of clients. Jones, *Astronomical Papyri*, 2:249: "They show the astrologer using more resources and making more precise computations than in the other horoscopes. A basic horoscope like 4242 could have been cast in a few minutes by an astrologer equipped with a sign-entry almanac and a set of tables for the sun and moon; but 4277 was the work of several hours of reckoning and careful writing." In so-called standard horoscopes, the sign in which the ascendant is situated is only mentioned in passing, without any further indication of the exact degrees. See Neugebauer and van Hoesen, *Greek Horoscopes*, 18-20, 47-48, 51-53, 61; Baccani, *Oroscopi Greci*, 97, 112, 124, 140; Jones, *Astronomical Papyri*, 2:374-81, 384-87, 394-99.

Astral Influence and the Classification of the Characteristics of Planets and Zodiacal Signs

There is much debate about whether Babylonian astrology conceived of astral influence in the realm of genethlialogy,[29] but in Hellenistic astrology this is clearly one of the basic premises. The planets and zodiacal signs were not seen as mere signs, by their character, qualities and power they were believed to exert influence on heavenly and earthly matters, especially on people from the moment of their birth. The planets were classified, for example, according to their beneficent or maleficent nature, or their gender. The zodiacal signs were also ordered according to their gender, as well as to numerous other characteristics.[30] The ascendant zodiacal sign in the east – literally the "horoscope" (ὡροσκόπος) – established the specific nature of the nativity, determining the configuration of the planets vis-à-vis the zodiacal signs and each other at the moment of a person's birth. The geometrical relationships between planets and zodiacal signs amongst and between each other were expressed in terms of "aspects." In antiquity five aspects were distinguished: conjunction (0°), opposition (180°), trine (120°), quartile (90°), and sextile (60°). These aspects together with the character and qualities of the planets and the zodiacal signs were believed to determine people's nativity.[31]

Planetary Rulership and Zodiacal Houses

The determination of the influence of planets and zodiacal signs was further refined by ascribing planetary rulership over the zodiacal signs in different ways and by making various subdivisions of the zodiacal signs. Including the sun and the moon, seven planets were known in antiquity: moon, Mercury, Venus, sun, Mars, Jupiter, and Saturn.[32] One of the forms of plane-

[29] Cf. Koch-Westenholz, *Mesopotamian Astrology*, 51-52; Rochberg, *The Heavenly Writing*, 293-94. Reiner, *Astral Magic*, 13, argues that there are other areas in which the Babylonians acknowledged the influence of the stars, such as in catarchic astrology and different forms of magic. Cf. also Chapter Two n. 48.

[30] See Hübner, *Eigenschaften der Tierkreiszeichen*.

[31] Cf. Bouché-Leclercq, *L'astrologie grecque*, 72-179; Gundel and Böker, "Zodiakos," 555-59; D. Pingree, "Astrology," in *Dictionary of the History of Ideas: Studies of Selected Pivotal Ideas* (vol. 1; ed. P.P. Wiener; New York: Charles Scribner's Sons, 1973), 118-26; Tester, *A History of Western Astrology*, 4-10; Barton, *Ancient Astrology*, 86-113. The trine aspect is attested in cuneiform sources, but, according to Rochberg-Halton, "Elements of the Babylonian Contribution," 60, "the Babylonian grouping of three signs seems to be the result simply of the schematic arrangement of twelve elements (here zodiacal signs) into four groups of three elements each, rather than the result of some geometrical or spatial relation."

[32] This order was based on the presumed distance of each planet from earth, beginning with the highest one, Saturn. In Hellenistic astrology the position of the planets was imagined as occupying a sphere with its own height and distance from the earth. For different ancient traditions on the order of the planets, see Bouché-Leclercq, *L'astrologie grecque*, 107-8; Nilsson, *Geschichte der griechischen Religion*, 272-73; Neugebauer, *Exact Sciences*, 168-70;

tary rulership was the notion of a planetary or zodiacal house. The twelve signs of the zodiac were divided between the planets. Each of the five planets rules two zodiacal signs, while sun and moon each rule only one sign:

Planets	Zodiacal signs	Zodiacal signs	Planets
Saturn	*Aquarius*	*Capricorn*	Saturn
Jupiter	*Pisces*	*Sagittarius*	Jupiter
Mars	*Aries*	*Scorpio*	Mars
Venus	*Taurus*	*Libra*	Venus
Mercury	*Gemini*	*Virgo*	Mercury
Moon	*Cancer*	*Leo*	Sun

The zodiacal sign ruled by a planet was called its "house" (οἶκος, *domus*) in Hellenistic astrology. This means that, for example, the signs of *Capricorn* and *Aquarius* are both "houses" of the planet Saturn,[33] while the other ten signs are not, or the sign of *Leo* is the "house" of the sun, while the other eleven zodiacal signs are not. The planetary or zodiacal houses affect the influence exerted by the planets when they are in their "house," i.e. in conjunction, making them more powerful.[34]

Subdivisions of the Zodiacal Signs

The 360° of the ecliptic were not only divided into twelve equal parts of 30° distributed among the signs of the zodiac, further subdivisions were made within the signs. Through various systems the 30° of the twelve zodiacal signs could be divided into different parts of smaller sections.[35] Attested subdivisions are, for example, three parts of 10° (decans),[36] twelve parts of 2;30° (*dodecatemoria*),[37] thirty parts of 1°,[38] and, probably merely

Neugebauer, *HAMA*, 604, 690-93; Tester, *A History of Western Astrology*, 4, 18-19, 166-67; W. Hübner, "Der *descensus* als ordnendes Prinzip in der 'Naturalis historia' des Plinius," in *Die Enzyklopädie im Wandel vom Hochmittelalter bis zur frühen Neuzeit* (ed. C. Meier; MMS 78; Munich: Wilhelm Fink, 2002), 25-41, at 28.

[33] These are, respectively, the diurnal and the nocturnal "house," according to the solar semicircle from *Leo* to *Capricorn* and the lunar semicircle from *Aquarius* to *Cancer*. See Ptolemy, *Tetrabiblos* 1.18. Cf. Bouché-Leclercq, *L'astrologie grecque*, 155-57; Hübner, *Eigenschaften der Tierkreiszeichen*, 287-88.

[34] Cf. Bouché-Leclercq, *L'astrologie grecque*, 182-92.

[35] See Gundel and Böker, "Zodiakos," 559-67.

[36] The system of thirty-six decans was developed in Ptolemaic Egypt. Cf. Bouché-Leclercq, *L'astrologie grecque*, 215-40; Gundel, *Dekane und Dekansternbilder*, 1-36; W. Gundel, "Dekane," *RE*, Supp. 7 (1940), 116-24; Neugebauer and van Hoesen, *Greek Horoscopes*, 5-6.

[37] The *dodecatemoria* is an astrological concept of Mesopotamian origin. Cf. Bouché-Leclercq, *L'astrologie grecque*, 299-304; A. Ungnad, "Besprechungskunst und Astrologie in Babylonien," *AfO* 14 (1941-44): 251-84, at 275-76; Sachs, "Babylonian Horoscopes," 71-73; O. Neugebauer and A. Sachs, "The 'Dodekatemoria' in Babylonian Astrology," *AfO* 16 (1952-53): 65-66; B.L. van der Waerden, "Die Zahlen der Texte VAT 7815 und 7816," in *Gestirn-Darstellungen auf babylonische Tontafeln* (E. Weidner; ÖAWPH.SB 254/2; Vienna:

in theory, even smaller sections of 1800 parts of 0;01° (*myriogenesis*).[39] The terms (ὅρια) are a further subdivision of the zodiacal signs into five unequal parts, each of which is allotted to one of the five planets.[40]

These subdivisions show that astrology was believed capable of infinitely precise calculations and, hence, predictions. The connection of these subdivisions with the planets enabled astrologers to extract various interpretations from the computed position of each planet. On the one hand, the possibility of multiple interpretations provided the astrologer with numerous suggestions to interpret people's horoscopes as the clients would wish.[41] On the other hand, the minute distinctions and elaborations were the tools of the trade and the technical complexity had the rhetorical impact of demonstrating the level of knowledge and skillfulness needed to apprehend the art (τέχνη) of astrology.[42]

General and Personal Astrology in the Dead Sea Scrolls: *4QZodiology and Brontology ar* (4Q318) and *4QZodiacal Physiognomy* (4Q186)

Ancient astrology's two areas of interest, viz. general and individual astrology,[43] are both represented in the Dead Sea Scrolls. The Aramaic text 4Q318 (*4QZodiology and Brontology ar*) covers mundane astrology, while genethlialogical astrology is hinted at in *4QZodiacal Physiognomy*. In both cases there is nothing particularly sectarian, or even Jewish, about these texts. If they had been Greek papyri found in Egypt, nothing would suggest a Jewish context. This is actually what makes them so valuable. They testify to a Jewish interest in astrological matters on a scientific level that matches similar texts from the Hellenistic world.

Hermann Böhlaus, 1967), 50-52; Rochberg-Halton, "Elements of the Babylonian Contribution," 57-60; Brown, *Mesopotamian Planetary Astronomy-Astrology*, 259, 264; W. Hübner, "Δωδεκατημόριον," in *Corona Coronaria: Festschrift für Hans-Otto Kröner zum 75. Geburtstag* (eds. S. Harwardt and J. Schwind; Spudasmata 102; Hildesheim: Georg Holms, 2005), 189-217 (I am most grateful to Professor Hübner for sending me a copy of the article prior to its publication).

[38] Cf. Firmicus Maternus, *Mathesis* 8.18–30.

[39] Firmicus Maternus, *Mathesis* 5.1.36; 8.18.1, refers to a Hermetic work on *myriogenesis* by Asclepius that, according to the latter, was revealed to him by Mercury. Cf. Barton, *Power and Knowledge*, 82-83.

[40] E.g. Ptolemy, *Tetrabiblos*, 1.21. Cf. Bouché-Leclercq, *L'astrologie grecque*, 206-15; Tester, *A History of Western Astrology*, 73-77.

[41] Cf. J.D. North, *Horoscopes and History* (WIST 13; London: The Warburg Institute University of London, 1986), xi.

[42] Cf. Barton, *Power and Knowledge*, 71-94.

[43] See n. 16 above.

4QZodiology and Brontology ar consists of two parts: a *selenodromion* and a *brontologion*.[44] Copied somewhere between the late first century BCE and the early first century CE, this text shows that people in Palestine at that time had knowledge of the zodiacal signs and the synodic movement of the moon through the zodiac. In the first part of *4QZodiology and Brontology ar*, the monthly course of the moon through the signs of the zodiac is noted schematically (*selenodromion*).[45] This part gives the Aramaic names of all twelve signs of the zodiac. In the second part predictions are given for when it thunders at the moment when the moon is positioned in one of the zodiacal signs (*brontologion*). In line with the interests of mundane astrology, these predictions concern general matters occurring in the provinces and at the palace. The genres of both parts are fully understandable against the background of Hellenistic astrology, and have their precursors in Babylonian celestial divination.[46]

As has been argued in Chapters One and Two, *4QZodiacal Physiognomy* is not a horoscope or a collection of horoscopes, but a physiognomic catalogue that points readers to certain astrological matters. Since these astrological matters concern individual types of people, the interest of the text is similar to that of personal astrology. *4QZodiacal Physiognomy* specifically mentions the nativity of people's birth (מולד), referring to a part of the zodiacal sign, and perhaps also to people's zodiacal sign (בהמה). Several elements in the text thus suggest an astrological background rooted in genethlialogy.

HYPOTHESES ON THE ASTROLOGICAL FRAMEWORK OF *4QZODIACAL PHYSIOGNOMY* (4Q186)

Undoubtedly, ancient astrological concepts are significant for understanding the background of *4QZodiacal Physiognomy*. The question is which particular notions are of importance. Any hypothesis about the astrological framework of *4QZodiacal Physiognomy* has to start, of course, with what the text itself makes explicit. But here one is confronted with a key problem. The text gives little astrological information in a clear, straightforward

44 Cf. the literature cited in Chapter One nn. 19, 88. Add Albani, "Horoscopes in the Qumran Scrolls," 296-301.

45 4Q317 (*4Qcrypt Phases of the Moon*) concerns the phases of the moon, but the text shows no evidence that it attempts to relate this to the signs of the zodiac. See recently J. Ben-Dov, "The Initial Stages of Lunar Theory at Qumran," *JJS* 54 (2003): 125-38; J.-C. Dubs, "4Q317 et le rôle de l'observation de la Pleine Lune pour la détermination du temps à Qoumrân," in *Le Temps et les Temps dans les littératures juives et chrétiennes au tournant de notre ère* (eds. C. Grappe and J.-C. Ingelaere; JSJSup 112; Leiden: Brill, 2006), 37-54.

46 Cf. Bouché-Leclercq, *L'astrologie grecque*, 363-64; Gundel and Gundel, *Astrologumena*, 260-69.

manner. Apart from knowledge of the zodiacal signs, *4QZodiacal Physiognomy* shows no evident awareness of other astronomical and astrological principles.

Astrology was not a fixed and unified system of concepts and terminology during the Hellenistic and Early Roman period. There was much terminological inconsistency and conceptual confusion, reflecting the still unsettled state of astrology.[47] One has to bear in mind the possibility of multiple developments and trajectories, not all of which are recognizable anymore.

Against this background, and taking into account the late first-century BCE date for the manuscript, *4QZodiacal Physiognomy* can perhaps be seen as a text representative of the incipient stages of horoscopic astrology in Second Temple period Judaism, attempting to render concepts foreign to Jewish culture into Hebrew. The text may represent a translation effort of astrological terminology and concepts into Hebrew. If this is correct, *4QZodiacal Physiognomy* is of importance for the history of astrology in general since it demonstrates, like *4QZodiology and Brontology ar*, the transmission of certain astrological concepts to first-century BCE Palestine.[48]

As concluded in Chapter Two, *4QZodiacal Physiognomy* adheres to the line of reasoning that later found articulated explication through the late antique astrologer Hephaestion. He advised his readers to pay attention to the shape of people's bodies, to see which zodiacal sign they resembled, and to discern their horoscope accordingly. *4QZodiacal Physiognomy*, however, is not only concerned with the entire zodiacal sign, but also with a specific part of it, viz. in 4Q186 1 ii 9 "the foot of *Taurus*" (רגל השור). Most scholars understand this as a reference to a specific part of the constellation or zodiacal sign *Taurus*.[49] For reasons explained below, it should be taken as a reference to a part of the zodiacal *sign*, not the *constellation*, *Taurus*.[50]

47 Cf. Neugebauer, *HAMA*, 278-79, 332-33, 340, 772; Tester, *A History of Western Astrology*, 72.

48 One cannot, of course, exclude the possibility that astrology was known and practiced in Palestine before the first century BCE, but concrete evidence is lacking for this. Furthermore, taking into account the astrological background of the text (a combination of *melothesia* and *dodecatemoria*, see below) a date before the first century BCE for *4QZodiacal Physiognomy* is difficult to prove.

49 The translation by Allegro, DJD 5.89, "on the Festival of Taurus," has met with little approval. Only García Martínez and Tigchelaar, *DSSSE*, 381, seem to accept Allegro's interpretation: "the period of Taurus." But García Martínez, *Dead Sea Scrolls Translated*, 456, has "in the foot of Taurus."

50 On the cloven hoof of the constellation *Taurus*, see Hübner, *Eigenschaften der Tierkreiszeichen*, 126. And on the crooked knee, see W. Hübner, *Grade und Gradbezirke der Tierkreiszeichen: Der Anonyme Traktat De Stellis Fixis, In Quibus Gradibus Oriuntur Signorum: I Quellenkritische Edition, II Kommentar* (SWC; Stuttgart: B.G. Teubner, 1995), 1:140; 2:111.

The terminology ברגל השור ("in the foot of *Taurus*") raises the following questions: what is the significance of this specific reference to a part of the zodiacal sign *Taurus*; is it related to the numbers allotted to the "house of light" and the "house of darkness," and, if so, in what way is it connected? Alongside the physiognomic descriptions, the most significant element in *4QZodiacal Physiognomy* are the numbers linked to the "house of light" (בית האור) and the "house of darkness" (בית החושך). Although references to light and darkness are key concepts in Qumran sectarian texts, the combination with בית in the phrases "house of light" and "house of darkness" does not occur elsewhere and is unique to *4QZodiacal Physiognomy*.[51]

In the extant text of *4QZodiacal Physiognomy*, the element of numbers connected to the "house of light" and "house of darkness" terminology occurs twice, while a third occurrence can be assumed.[52] Despite the uniqueness of these phrases, many scholars relate this to the light/darkness dualism of sectarian texts from the Dead Sea Scrolls, especially the *Two Spirits Treatise* of the *Rule of the Community*.[53] It is possible that the "house of light" and "house of darkness" terminology of *4QZodiacal Physiognomy* was read and understood by members of the Qumran community in light of other texts from Qumran, but it is not necessary, or desirable, to explain this terminology only from that perspective.

The issue is how the numbers allotted to the "house of light" and "house of darkness" came about. Was there some sort of system that regulated the pattern and division of certain numbers between the "house of light" and the "house of darkness?" The proponents of the dualistic interpretation of *4QZodiacal Physiognomy* have not yet provided an answer to this question. Since an astrological framework for this element is denied, such an interpretation has not been able to explain the specific numbers in relation to the "house of light" and the "house of darkness," or to the reference to a specific part of the zodiacal sign, i.e. "the foot of *Taurus*." It is with these considerations in mind that the following discussion on different hypotheses about the astrological notions that are operative in *4QZodiacal Physiognomy* must be approached.

[51] In *T. Jos.* 8:5 the prison is referred to as a "house of darkness" (οἶκος σκότους), but this does not throw much light on the meaning of these words in *4QZodiacal Physiognomy*.

[52] 4Q186 1 ii 7–8; 1 iii 8–9. In both cases the words (ו)רוח לו begin the sentence. Whatever the exact sense of (ו)רוח לו, from its basic connection with the "house of light" and the "house of darkness," one can assume that a third occurrence of these words originally stood in 4Q186 2 i 7, following the words ורוח לון in l.6. But the exact division of numbers in the "house of light" and the "house of darkness" is, unfortunately, lost.

[53] See Chapter Four.

THE WORD "HOUSE" AS A *TERMINUS TECHNICUS* FOR PLANETARY HOUSE

Matthias Delcor and Hermann Lichtenberger have understood the term בית ("house") to be an astrological *terminus technicus* equivalent to Greek οἶκος, Latin *domus*, and Syriac *bēṯ*. Regarding the realization of the numbers in the "house of light" and the "house of darkness," however, their interpretations differ. Delcor understands the division between light and darkness as a reference to the duration of day and night at the moment of birth. Lichtenberger suggests it is dependent on the position of one of the luminaries between two zodiacal signs, the signs being classified either as light or darkness.[54]

Matthias Delcor: Houses of the Sun and the Duration of Day and Night

Delcor is not entirely clear in his understanding of the concept of planetary houses. He seems to equate this concept with the notion of the movement of the sun through the various signs of the zodiac during the year. On the one hand, Delcor describes a planetary house as "la région assignée à chaque planète et plus précisément des 'maisons du cercle du zodiaque.'"[55] But he also states that each zodiacal sign could be called a house of the sun, because the sun seems to travel through each sign of the zodiac circle during a year. Delcor connects this latter idea with a zodiacal interpretation of the six "gates" of the eastern and western horizon where, according to *1 En.* 72, the sun rises and sets. This chapter describes the course of the sun through these gates in relation to the duration of night and day during the year.

Delcor uses the Enochic data on the variable length of day and night to explain the light and darkness terminology in *4QZodiacal Horoscope*, assuming that it provides more or less the same information as *1 En.* 72. According to Delcor, the spirit of each person is conditioned by the duration of day and night at the moment of birth. He suggests that the zodiacal sign *Taurus* corresponds with the fifth gate (*1 En.* 72:11). The duration of day and night during the sun's position in the fifth gate is 11:7. Delcor concludes that this "correspond approximativement aux proportions de notre horoscope, six pour le jour et trois pour la nuit."[56]

There are too many conceptual problems for Delcor's interpretation to be convincing.[57] First, it is correct to say that the sun seems to travel successively through each zodiacal sign during the course of a year, but it is incorrect to apply the notion of planetary houses to this. The "houses" have

[54] Delcor, "Recherches sur un horoscope," 301-4; Lichtenberger, *Studien zum Menschenbild*, 144-46.

[55] Delcor, "Recherches sur un horoscope," 301-2.

[56] Delcor, "Recherches sur un horoscope," 304.

[57] Cf. also Albani, "Horoscopes in the Qumran Scrolls," 304 n. 81.

nothing to do with the movement of the sun as such. This concept simply assigns rulership to the various planets over certain zodiacal signs.

Second, the Enochic gates are not identical with the zodiacal signs. Otto Neugebauer has argued against the interpretation of the "gates" in the *Astronomical Book* as zodiacal signs, suggesting that these gates represent fixed points ("arcs") of the horizon that are related with the rising and setting amplitude of the sun during the course of a year.[58]

Third, Delcor fails to explain the sense of the numbers in *4QZodiacal Physiognomy* and how these come about. Apart from the approximate nature of the numbers, his interpretation can hardly be upheld for the numbers eight and one in 4Q186 1 iii 9.[59] Such a ratio does not occur in *1 En.* 72, nor is it a possible one for the length of day and night. *4QZodiacal Physiognomy*, therefore, does not give data similar to *1 En.* 72.[60]

Hermann Lichtenberger: Zodiacal Signs of Light and Darkness

Lichtenberger asserts that the word בית ("house") represents a technical astrological term expressing the theory of planetary houses, but he does not clarify in what way this concept functions in *4QZodiacal Physiognomy*.

Lichtenberger argues that the relationship between light and darkness is dependent on the position of one of the luminaries or planets between two zodiacal signs, which are classified as either light or darkness. He explains this by referring to symbolism used in the late rabbinic treatise *Pesiqta Rabbati* 20 §5. This text answers the question why God created the world in Nisan, not in Iyyar. The zodiacal sign *Aries* of the month Nisan is associated with light, while the zodiacal sign *Taurus* is linked to darkness. According to the rabbinic tradition in *Pesiq. Rab.* 20 §5, God wished to create the world in light and he therefore told the Prince of Darkness, who looked like a bull, to get out of his way. Lichtenberger concludes that:

> das Tierkreiszeichen des Widders wurde schöpfungstheologisch grundsätzlich mit 'Licht,' das Zeichen des Stiers grundsätzlich mit 'Finsternis' identifiziert. Der jeweilige Stand der Sonne oder des Mondes oder eines

[58] O. Neugebauer, "Notes on Ethiopic Astronomy," *Or* 33 (1964): 50-61. Cf. also Albani, "Horoscopes in the Qumran Scrolls," 295 n. 57. The zodiacal interpretation of the Enochic "gates" is in itself problematic, but Delcor also tries to harmonize the number of twelve zodiacal signs with the description of the sun rising and setting in the same gates opposite each other. If the six gates of the eastern and the six gates of the western horizon are another way of referring to the twelve zodiacal signs, it seems impossible to suggest that the sun rises and sets in the same gate while both represent at the same time one zodiacal sign. Delcor does not clarify how one is to imagine this harmonization.

[59] See also Lichtenberger, *Studien zum Menschenbild*, 145.

[60] Gordis, "Document in Code," 38, simply asserts, without explanation, that the light-darkness terminology refers "to day and night, light and darkness, which were represented in some form as the shrine or dwelling" of the described individuals.

> Planeten zwischen Widder und Stier bei der Geburt ergäbe dann das Verhältnis von Licht- und Finsternisanteilen beim einzelnen Menschen.[61]

Lichtenberger's interpretation is not convincing and the comparative value of the rabbinic text is weak. First, the zodiacal signs *Aries* and *Taurus* are not just "houses" for any luminary, but only for the planets Mars and Venus. The concept of planetary houses stands in the way of Lichtenberger's explanation for the relationship between light and darkness in *4QZodiacal Physiognomy*. A planet is not "in its house" if it is positioned between the zodiacal signs.[62] It seems, however, that the concept of planetary houses is redundant, since Lichtenberger does not refer to it in his explanation.

Second, Lichtenberger cannot explain the specific numbers ascribed to light and darkness in 4Q186 1 ii 7–8. According to *Pesiq. Rab.* 20 §5 *Aries* is completely light and *Taurus* is completely dark. There is no mention of any space between the two signs, or of any gradation between light and darkness. It remains vague how this text helps to understand the concrete division of light and darkness in *4QZodiacal Physiognomy*. The rabbinic text provides no background or system that helps to determine a partition of light and darkness between two zodiacal signs.[63]

Third, *Pesiq. Rab.* 20 §5 qualifies only the first two zodiacal signs *Aries* and *Taurus* as light and darkness, but not the other ten signs.[64] The symbolism behind the zodiacal signs in *Pesiq. Rab.* 20 §5 is determined by a specific theology of creation.[65] The relevance of this text regarding the

[61] Lichtenberger, *Studien zum Menschenbild*, 146.

[62] This does not mean that a planet cannot be positioned between two zodiacal signs. There is the concept of a planet in transit (μετάβασις, *transitus*) between two signs. Serapion, a third-century CE astrologer from Egypt, discusses the power a planet holds when positioned in the last three and the first three degrees of a sign, see CCAG 8/4.230.1–6). Cf. Gundel and Böker, "Zodiakos," 563: "Auf dem Schnitt (τμῆμα, *ultima linea*) zwischen zwei Zeichen ist die Wirkung besonders stark und unermeßlich."

[63] Cf. also Albani, "Horoscopes in the Qumran Scrolls," 304-5 n. 82.

[64] See K.-E. Grözinger, *Ich bin der Herr, dein Gott: Eine rabbinische Homilie zum Ersten Gebot (PesR 20)* (FJS 2; Frankfurt: Peter Lang, 1976), 30-31, 75-104. For the text, see M. Friedman (ed.), *Pesikta Rabbati: Midrash für den Fest-Cyclus und die ausgezeichneten Sabbathe* (Vienna, 1880), 95a–96a. For an English translation, see W.G. Braude, *Pesikta Rabbati: Discourses for Feasts, Fasts, and Special Sabbaths* (2 vols; YJS 18; New Haven: Yale University Press, 1968), 1:400-1. The text does not name the zodiacal sign *Aries* as such, but in one of the manuscripts it does allude to the sign *Taurus* when it mentions "the Prince of Darkness being black as a bull" (ושרו של חושך דומה לשור). Cf. Friedman, *Pesikta Rabbati*, 95a; Grözinger, *Ich bin der Herr*, 30 n. 1.

[65] After light and darkness comes the creation of humanity, symbolized by the sign *Gemini*. Man is to see both light and darkness. *Gemini* represents man. The opposing forces of light and darkness determine the life and fate of every human being. It is up to every individual to choose the light and walk in the path of the Torah. The other nine zodiacal signs can be said to represent phases in the life of man. The next three signs (*Cancer*, *Leo*, and *Virgo*) symbolize the stages in a man's life from infancy to the age suitable for marriage. The signs *Libra* and *Scorpio* stand for the weighing of a man's deeds and the punishment of sins discovered in him with banishment to *gehenna*. After these signs God created *Sagittarius*, because

partition of light and darkness in *4QZodiacal Physiognomy* is therefore limited. It cannot be applied to other signs of the zodiac. Lichtenberger's interpretation, therefore, fails to account satisfactorily for the terminology "house of light" and "house of darkness," or for the concrete numbers connected with it in *4QZodiacal Physiognomy*.

The Evidence for Planetary Houses in Jewish Astrology and 4QZodiacal Physiognomy *(4Q186)*

Later Jewish astrological tradition indeed makes use of the term בית ("house") to express the concept of planetary houses, but this is not the case in *4QZodiacal Physiognomy*. In the medieval astrological text *Baraita de-Mazzalot*, for example, one finds the theory of planetary houses in the ninth section:

> The house of Saturn: *Capricorn* and *Aquarius*; the house of Jupiter: *Sagittarius* and *Pisces*; the house of Mars: *Aries* and *Scorpio*; the house of Venus: *Taurus* and *Libra*; the house of Mercury: *Gemini* and *Virgo*; the house of (the) sun: *Leo*; the house of (the) moon: *Cancer*.[66]

It is evident that this enumeration lists the houses of the seven planets according to Ptolemaic astrology. In *4QZodiacal Physiognomy*, however, this is not the case. In the extant text no mention is made of any planet. Moreover, a planet can only be in one of its houses, not in both at the same time. Even if one were to take recourse to the notion of diurnal and nocturnal houses to somehow explain how a certain planet is in limbo between two of its houses, the fact is that no planet, except Saturn, can be in or near two of its houses at the same time. It is, therefore, extremely unlikely that *4QZodiacal Physiognomy* uses the word בית ("house") as a *terminus technicus* for planetary houses.[67]

when a prayer is made on behalf of the one cast into *gehenna*, he is shot up from there as an arrow from a bow. The sign of *Capricorn* represents the one purified from *gehenna*, while *Aquarius* symbolizes the bucket of water used to purify a man of his sins. The sign of *Pisces*, finally, represents Israel, that is, like the unseen fish in the sea, unaffected by the evil eye or the zodiacal sign, and is destined to inherit the world.

[66] ביתו של שבתי, גדי ודלי, וביתו של צדק, קשת ודגים, ביתו של מאדים, טלה ועקרב, ביתו של נוגה, שור ומאזנים, ביתו של כוכב חמה, תאומים ובתולה, ביתו של חמה, ארי, ביתה של לבנה, סרטן. Wertheimer and Wertheimer (eds.), *Batei Midrashot*, 22. The manuscript used ("New York - Jewish Theological Seminary Rab. 1948 [from the Geniza]") was dated to the tenth century (see Wertheimer and Wertheimer, *Batei Midrashot*, 11), but the text as such is probably older. I owe the reference to the manuscript number to Ronit Nikolsky.

[67] Cf. Bergmeier, *Glaube als Gabe*, 113 n. 447; Alexander, "Physiognonomy," 387 n. 5; Albani, "Horoscopes in the Qumran Scrolls," 308.

"The Foot of *Taurus*" and the Constellation *Taurus*

Several scholars have understood the words "in the foot of *Taurus*" (ברגל השור) as a reference to a part of the constellation in which the sun or the moon was positioned at the moment of birth. Martin Hengel suggested identifying "the foot of *Taurus*" with a certain star on the ecliptic, possibly identical with the "knee" (γόνατος) or "hoof" (ποδὸς ἄκρου) of the constellation *Taurus* as mentioned by Eratosthenes in his third-century BCE star catalogue.[68]

Eratosthenes presents information regarding the *constellation*, not the *sign Taurus*. The "knees" and "hoofs" of the constellation *Taurus* are situated far below the ecliptic. In his star catalogue in the *Almagest*, Ptolemy gives a latitudinal position beneath the ecliptic of -14;50° for the "right hock" (δεξιοῦ σφυροῦ=ν Tau) and -13° for the "left lower leg" (ἀριστεροῦ πήχεως=88[d] Tau).[69]

It is impossible for the sun, the moon, or any of the other five planets to reach this section of the constellation *Taurus* so far below the ecliptic. It lies well beyond the latitudinal width of ±12° of the zodiacal belt, i.e. ±6° above and beneath the ecliptic. The sun's course is identical with the ecliptic (0°). It cannot, therefore, be positioned in different parts of zodiacal constellations above or below the ecliptic.[70] The moon and planets can be positioned within different constellational parts, but only within the width of the zodiacal belt.[71] The orbit of the moon, for example, is inclined to the ecliptic. This causes the moon to cross within an area of 5° latitude to either side of the ecliptic during its periodic movement. The moon, therefore, can only be observed being positioned in those parts of the constellations that are within the latitudinal expanse of 10° of the zodiacal circle.[72]

These considerations argue against understanding the words "in the foot of *Taurus*" as a reference to a part of the *constellation* in which the sun or

[68] Eratosthenes, *Catasterismi* 14. Hengel, *Judentum und Hellenismus*, 435. Cf. Carmignac, "Les Horoscopes," 203; Delcor, "Recherches sur un horoscope," 306-7; Lichtenberger, *Studien zum Menschenbild*; Alexander, "Physiognonomy," 389.

[69] Ptolemy, *Almagest* 7.5[.23]. The constellation *Taurus* is imagined as half of an animal (ἀπότομος). Only the front half is represented, cf. Eratosthenes, *Catasterismi* 14, 23; Hipparchus, *Commentary on Aratus and Eudoxus* 2.6.6; Ptolemy, *Tetrabiblos* 1.9.3. The zodiacal sign *Taurus* is connected with lethal injuries through amputations (ἀποτομαῖς), cf. Ptolemy, *Tetrabiblos*. 4.9.12. See Bouché-Leclercq, *L'astrologie grecque*, 133; Hübner, *Eigenschaften der Tierkreiszeichen*, 113. Ptolemy's catalogue was one of the sources for Albrecht Dürer's famous wood-cut astral map from 1515. See H.G. Gundel, *Zodiakos, Tierkreisbilder im Altertum: Kosmische Bezüge und Jenseitsvorstellungen im antike Alltagsleben* (KAW 54; Mainz: Philipp von Zabern, 1992), 311, 314.

[70] Cf. Albani, "Horoscopes in the Qumran Scrolls," 294.

[71] Cf. n. 12 above.

[72] See e.g. Neugebauer, *Exact Sciences*, 108; Neugebauer, *HAMA*, 68, 80-84, 1107-8, 1111, cf. also 626.

the moon was positioned at the moment of birth. They should rather be taken as a reference to a specific part of the zodiacal *sign Taurus* and indicating the division of that sign.[73]

ROLAND BERGMEIER: DAY AND NIGHT AND THE MOON "IN THE FOOT OF *TAURUS*"

In a short but important excursus, Bergmeier suggested that the realization of the numbers in the "house of light" and the "house of darkness" is the result of the division of the zodiacal sign with the moon positioned in one of its parts. The phrases "house of light" and "house of darkness" represent day and night as times of light and darkness.[74]

The Divided Zodiacal Sign Taurus *in the* Rhetorius-Teucer Text

Bergmeier emphasizes that the astrological background of *4QZodiacal Physiognomy* is determined by understanding the words ברגל השור in 4Q186 1 ii 9 as presupposing a division of the zodiacal sign *Taurus*.[75] As a key text for understanding this division, he uses an excerpt from a text by Teucer "the Babylonian" that has been transmitted by the late antique astrologer Rhetorius. Rhetorius collected and compiled astrological literature. Perhaps originally from Egypt, he was mainly active in Constantinople during the reign of Anastasius I (491–518).[76] Teucer was probably active somewhere during the first century BCE, and most likely of Greek descent from Egypt, rather than from Persian Babylon.[77]

The extant part of the *Rhetorius-Teucer text* is a fascinating example of a short astrological catalogue that deals with the twelve zodiacal signs.[78] It illustrates very neatly the practice of subdivisions, elaborations, classifications, characterizations, and complexity that is typical of astrology as an ancient art (τέχνη).

[73] In addition to the hypotheses discussed below, cf. Wise, "Horoscope Written in Code," 276-77, who correctly inferred that the concept of *dodecatemoria* is involved.

[74] Cf. Bergmeier, *Glaube als Gabe*, 78-81, for the following.

[75] Bergmeier, *Glaube als Gabe*, 79, quotes Boll, Bezold, and Gundel, *Sternglaube und Sterndeutung*, 147: "Weitere praktische Ratschläge für astrologische Konsultationen und für Nativitäten enthalten diejenigen Texte, welche besondere Gutachten abgeben über die Wirkungen der verschiedenen Glieder der Tierkreisbilder."

[76] Gundel and Gundel, *Astrologumena*, 249-51.

[77] Cf. F. Boll, *Sphaera: Neue griechische Texte und Untersuchungen zur Geschichte der Sternbilder* (Leipzig: B.G. Teubner, 1903; repr., Hildesheim: Georg Olms, 1967), 6-11; Gundel and Gundel, *Astrologumena*, 112-13; Hübner, *Grade und Gradbezirke*, 92; W. Hübner, "Teukros [4]," *DNP* 12/1 (2002), 205.

[78] For the text, see CCAG 7.192–213.

The text is arranged into twelve sections according to the number of zodiacal signs, beginning with *Aries*, and each section consists again of twelve subdivisions.[79] Each section opens with a description of some general characteristics regarding the zodiacal sign, such as, for example, whether they are masculine or feminine, diurnal or nocturnal, their seasonal character, and their influence on human matters. Second, the text lists the various connections between the zodiacal sign and the planets, such as planetary houses, exaltations, depressions, and planetary triplicity rulers during day and night. Third, for every three decans of the zodiacal sign the accompanying extra-zodiacal constellations rising simultaneously (the so-called *paranatellonta*) are listed. Fourth, for each decan the so-called planetary "faces" (πρόσωπα) are enumerated.[80] Fifth, bright stars rising simultaneously with a zodiacal sign to the north and south of the ecliptic are listed according to their length, size, and temperament (κρᾶσις). Sixth, the text provides a list of terms (ὅρια).[81] Seventh, an enumeration is given of the different regions of the world (κλίματα) that are under the influence of a particular zodiacal sign.[82] Eighth, the text assigns parts of the body *a capite ad calcem* to the various zodiacal signs, each sign governing a section of the human body and influencing various diseases in that part (*melothesia*). Ninth, the zodiacal sign is allotted two letters from the alphabet.[83] Tenth, the text describes the division of the zodiacal sign into different parts along the 30° longitude, such as, for example, the head, neck, breast, loins, hip joint, hind legs, tail, and feet of *Aries*. Eleventh, the text lists the influences that the zodiacal sign exerts when it is the horoscope (ὡροσκόπος) or ascendant sign. Finally, each section concludes with an account of the various influences attributed to the three decans of the zodiacal sign.

Bergmeier directs attention to the tenth element that the *Rhetorius-Teucer text* lists in each section. This subsection concerns the divisions of

[79] See Boll, *Sphaera*, 5-6.

[80] Cf. Bouché-Leclercq, *L'astrologie grecque*, 216, 224-25; Gundel, *Dekane und Dekansternbilder*, 30-35.

[81] The division into terms corresponds to the Egyptian system presented in Ptolemy, *Tetrabiblos* 1.21. Cf. Bouché-Leclercq, *L'astrologie grecque*, 206-15; Tester, *A History of Western Astrology*, 73-77.

[82] The *Rhetorius-Teucer text* claims the climates to be according to Ptolemy, which is, apart from minor variations, the case, see Ptolemy, *Tetrabiblos* 2.4. Cf. Bouché-Leclercq, *L'astrologie grecque*, 328-47; E. Honigmann, *Die sieben Klimata und die ΠΟΛΕΙΣ ΕΠΙΣΗΜΟΙ: Eine Untersuchung zur Geschichte der Geographie und Astrologie im Altertum und Mittelalter* (Heidelberg: Carl Winter, 1929), 43 n. 1, 47-49; Barton, *Ancient Astrology*, 180-85. See also Chapter Two n. 124.

[83] Cf. Boll, *Sphaera*, 469-72; Gundel and Böker, "Zodiakos," 578; W. Hübner, "Der Tierkreis in der Psalmenmagie," in *Ritual und Poesie: Formen und Orte religiöser Dichtung im Alten Orient, im Judentum und im Christentum* (ed. E. Zenger; HBS 36; Freiburg: Herder, 2003), 293-319, at 303-5.

the zodiacal signs into different parts covering 30° longitude on the ecliptic. The sign of *Taurus* is divided into nine parts:

> From 1° to 3° the head rises, from 4° to 7° the horns, from 8° to 10° the neck, from 11° to 13° the breast, from 14° to 18° the loins, from 19° to 21° the hip joints, from 22° to 24° the feet, from 25° to 27° the tail, from 28° to 30° the hoofs.[84]

In the case of *Taurus* the sign is divided into nine parts, but this is not the case for all zodiacal signs, suggesting that nine is not a set number.[85]

According to Bergmeier the specific division of *Taurus* in *4QZodiacal Physiognomy* is determined by observing the position of the moon.

The Moon in the Feet of Taurus *in Firmicus Maternus,* Mathesis *6.31.88?*

Referring to a passage in the *Mathesis*, an astrological handbook written by Firmicus Maternus around 335 CE, Bergmeier argues that "the foot of *Taurus*" (רגל השור) in 4Q186 1 ii 9 concerns the position of the moon in that part of *Taurus*, though the moon is not mentioned in *4QZodiacal Physiognomy*.

In this particular passage (6.31.88) Firmicus Maternus discusses the position of the moon in certain parts of several zodiacal signs:

> If the moon is found in the feet of *Taurus*, or in the nebula of *Cancer*, or in the mane of *Leo*, or in the front of *Scorpio*, obviously from 8° to 10°, or in *Sagittarius*, or in the spine of *Capricorn*, or in the fishing-line of *Pisces*, or in the head of *Aries*, and if she is without light, that is if all the glare of light disappeared, and if Saturn or Mars cast their light in some way, it will bring forth blind men.[86]

Unfortunately, this passage from Firmicus Maternus cannot be used to support Bergmeier's interpretation that *4QZodiacal Physiognomy* deals with the position of the moon in the partitioned zodiacal signs. There is a textual problem that eliminates it as supportive evidence.

The text does not say "in the feet of *Taurus*" (*in Tauri pedibus*). This reading is a conjecture made by Franz Skutsch without any basis in the manuscript evidence. Instead of *pedibus* the extant manuscripts all have

[84] CCAG 7.197.24–27: 'Ανατέλλει δὲ ἀπὸ μοίρας α´ ἕως γ´ κεφαλή, ἀπὸ δ´ ἕως ζ´ κέρατα, ἀπὸ η´ ἕως ι´ τράχηλος, ἀπὸ ια´ ἕως ιγ´ στῆθος, ἀπὸ ιδ´ ἕως ιη´ ὀσφύς, ἀπὸ ιθ´ ἕως κα´ ἰσχία, ἀπὸ δὲ κβ´ ἕως κδ´ πόδες, ἀπὸ δὲ κε´ ἕως κζ´ οὐρά, ἀπὸ δὲ κη´ ἕως λ´ ὄνυχες.

[85] One zodiacal sign (*Pisces*) has seven parts; one (*Cancer*) has eight parts; eight (*Aries*, *Taurus*, *Gemini*, *Leo*, *Virgo*, *Scorpio*, *Sagittarius*, and *Capricorn*) have nine parts; and two (*Libra* and *Aquarius*) have ten parts.

[86] Firmicus Maternus, *Mathesis* 6.31.88. For the text, see W. Kroll, F. Skutsch and K. Ziegler (eds.), *Firmicus Maternus: Matheseos Libri VIII* (vol. 2; Teubner; Leipzig: B.G. Teubner, 1913), 172-73. For a modern translation, see P. Monat (ed.), *Firmicus Maternus: Mathesis* (vol. 3; Budé; Paris: Les Belles Lettres, 1997), 110.

pl(e)iadibus.[87] This refers to the *Pleiades*, a small group of weak and nebulous stars situated in the front of the *constellation Taurus*.[88] The *Pleiades* were known to cause diseases of the eyes and blindness, a connection probably made because one had to strain the eyes to spot them.[89] As the reading *in Tauri pl(e)iadibus* ("in the *Pleiades* of *Taurus*") makes perfect sense, Skutsch's conjecture has to be rejected.[90] There is, therefore, no mention in Firmicus Maternus, *Mathesis* 6.31.88 of the moon "in the feet of *Taurus*," and thus far there is no other textual evidence for the moon being positioned in this part of the sign.

The Moon "in the Foot of Taurus*" in 4Q186 1 ii 9*

On the basis of these two texts, the *Rhetorius-Teucer text* for the division of the zodiacal sign *Taurus* and the passage from Firmicus Maternus for the position of the moon in the feet of *Taurus*, Bergmeier concludes:

> Steht der Mond 'im Fuß des Stiers' [...], sind die Parzellen 'Füße,' 'Schwanz,' und 'Klauen' die im Qumrantext angesprochenen drei Teile 'im Haus der Finsternis,' während 'Kopf,' 'Hörner,' 'Hals,' 'Brust,' 'Hüfte' und 'Lenden' die sechs Teile 'im Haus des Lichts' darstellen. Die Ausdrücke 'Haus des Lichts/der Finsternis' hängen danach wohl kaum mit dem Licht-Finsternis-Dualismus zusammen, sondern bezeichnen Tag und Nacht als Zeiten des Lichts und der Finsternis.[91]

Thus, the astrological framework in *4QZodiacal Physiognomy* presupposes the division of the zodiacal sign into separate parts. This division is realized by the position of the moon in one of the sign's parts. Bergmeier reads רוח in 4Q186 1 ii 7 as רֶוַח, a suggestion by Robert Gordis, meaning "space, interval" (רוח לו בבית, "it has a space in the house..."').[92] He takes it as a reference to the different areas or parts of the zodiacal sign, and, therefore, draws the conclusion that *4QZodiacal Physiognomy* understands "Aussehen und Schicksal eines Menschen davon abhängig, wie sich der Bereich eines Sternbilds durch den Mondstand in einem seiner 'Glieder' aufteilt."[93]

[87] See the critical apparatus in Kroll, Skutsch and Ziegler, *Firmicus Maternus*, 2:172; Monat, *Mathesis*, 3:88.

[88] See Ptolemy, *Almagest* 7.5[.23].

[89] For further references to *Taurus*, but also to other zodiacal signs, causing deficiencies to the eyes, see Hübner, *Eigenschaften der Tierkreiszeichen*, 193-96; W. Hübner, "Pleiaden," *DNP* 9 (2002), 1127-28.

[90] I am grateful to Professor Hübner for bringing this matter to my attention in a personal communication (e-mail on February 28, 2005).

[91] Bergmeier, *Glaube als Gabe*, 79.

[92] Gordis, "Document in Code," 38.

[93] Bergmeier, *Glaube als Gabe*, 80, refers to Firmicus Maternus, *Mathesis* 8.5.2, saying that the "Tierkreisbilder besitzen 'ihnen anvertraute spatia.'" However, in that passage, introducing the section on the *Sphaera Barbarica*, Firmicus Maternus distinguishes between signs and constellations. He describes the constellations as not erring in their course, but

Furthermore, supporting the interpretation of העמוד השני ("the second column") as a reference to the zodiacal sign *Taurus*,[94] Bergmeier suggests reconstructing 4Q186 2 i 9 as [בפרסות השור וזה ה]ואה בהמתו [ש]ו̊[ר] ("in the hoofs of *Taurus*. And this is his zodiacal sign: *Taurus*"). His reconstruction and interpretation assume the correctness of Allegro's fragments' joins and Strugnell's reconstruction for 4Q186 2 i 6–9.[95] Consistent with the division of the sign *Taurus* in the *Rhetorius-Teucer text*, one would get eight parts in the "house of light" and one part in the "house of darkness" if, according to Bergmeier's interpretation, the moon is positioned in the hoofs of *Taurus*.[96] This reconstruction, however, has lost its textual basis and is therefore no longer feasible.

Bergmeier has made a significant contribution to the elucidation of the astrological framework of *4QZodiacal Physiognomy* by adducing the *Rhetorius-Teucer text*. This text is important evidence for understanding the words ברגל השור ("in the foot of *Taurus*") in 4Q186 1 ii 9 as an indication of the division of the zodiacal sign *Taurus* in *4QZodiacal Physiognomy*, and, presumably, also of the other signs. The passage from Firmicus Maternus, however, fails to support the suggestion that the division of the zodiacal sign depends on the position of the moon in one of its parts.

Even if the reading in *Mathesis* 6.31.88 had been *in Tauri pedibus* ("in the feet of *Taurus*"), this element of Bergmeier's interpretation is not convincing. First, Bergmeier does not explain why those parts from the part in which the moon is positioned onwards until the final part of the zodiacal sign are assigned to the "house of darkness," while the other parts are in the "house of light." This might perhaps be so because the moon is the nocturnal luminary, but this is not clear from the text of *4QZodiacal Physiognomy*. Second, Bergmeier's suggestion that "house of light" stands for day, being the time of light, and "house of darkness" stands for night, being the time of darkness, is problematic. This interpretation fails to explain how day and night are conceptually related to the position of the moon "in the foot of *Taurus*" in a proportion of six to three.[97]

---

possessing the space attributed to them they turn around the world never altering their itinerary. This cannot be taken as a reference to support the notion that the different ecliptical parts of a zodiacal sign are understood to occupy space. But Bergmeier refers to a second passage, Firmicus Maternus, *Mathesis* 8.4.16, stating that one can speak of *unius signi spatium*, "a sign's space." Although this passage is somewhat obscure, it is clear that Firmicus Maternus imagines a zodiacal sign occupying a certain amount of space. Cf. Boll, *Sphaera*, 395 n. 2; Neugebauer, *HAMA*, 953-54.

94 See Chapter One n. 87.

95 See Chapter One and Appendix I.

96 See CCAG 7.197.27.

97 Cf. Albani, "Horoscopes in the Qumran Scrolls," 304.

FRANCIS SCHMIDT: "THE FOOT OF *TAURUS*" AS A DIURNAL DECAN AND THE HOROSCOPE OF CONCEPTION

According to the interpretation of Schmidt, *4QZodiacal Physiognomy* is a text that predicts people's physiognomy, as well as the parts of light and darkness that characterize them, on the basis of their date of conception, not of birth.[98] Schmidt has suggested that the light and darkness terminology of *4QZodiacal Physiognomy* has an astrological background in the concept of diurnal and nocturnal decans. He argues that the realization of the numbers in the "house of light" and the "house of darkness" is based on a calculation of the moment of conception. Accordingly, the text presupposes a set number of nine months for the duration of pregnancy.[99]

Thus, Schmidt proposes, first, that the use of the number nine in *4QZodiacal Physiognomy* can be explained if the two notions of decans and conception are taken into consideration, and, second, that the words "house of light" and "house of darkness" make sense if the astrological classification of the zodiacal signs into diurnal and nocturnal is taken into account.[100]

Diurnal and Nocturnal Zodiacal Signs, Decans, and Quadrants

The subdivision of the entire zodiac into thirty-six decans each of 10° was "one of the means by which Greek astrology could extract multiple interpretations from the computed position of a single heavenly body."[101] According to Manilius' zodiacal decan system:

> no sign has exclusive control over itself: all share their powers with certain signs in equal portions, and in a spirit of hospitality, as it were, they form a heavenly fellowship and surrender the parts of which they are composed to the keeping of other signs. This part the Greeks have termed the system of decans. The name is derived from the numeral, since the signs, which consist of thirty degrees, have a tripartite arrangement and allot ten degrees to each of the signs associating with themselves, the constellations one after the other providing a home for three signs each.[102]

Developed in Ptolemaic Egypt, the system of thirty-six decans originally goes back to the Egyptian calendar of three hundred and sixty days with

98 Schmidt assumes that this is expressed in 4Q186 1 ii 8 by מולד as a reference to conception and by יליד as an indication of birth. See Chapter One.

99 Cf. Wise, "Horoscope Written in Code," 276.

100 Schmidt, "Astrologie juive ancienne"; F. Schmidt, "Astrology and Predestination in Qumran," *Qadmoniot* 30 (1997): 115-18 (in Hebrew); Schmidt, "Ancient Jewish Astrology."

101 Jones, *Astronomical Papyri*, 1:10. Cf. Schmidt, "Astrologie juive ancienne," 130: "Cette division […] leur permettait notamment de diversifier les pronostics."

102 Manilius, *Astronomica* 4.294–302. Cf. also Firmicus Maternus, *Mathesis* 2.4.1–2.

ten-day weeks (not counting the five epagomenal days).[103] As the sun travels more or less 1° per day through the zodiacal circle, decans were originally stars connected with these ten-day week periods. They came to represent divinities ruling these periods and were later called δεκανοί or *decani*, thought to rule ten days or 10° of the ecliptic.

During the Hellenistic period this latter aspect came to the fore in Greek astrology. The decans initially represented individually acting divinities known by names, images and specific effects due to their character. But gradually they lost any such personality traits and were simply seen as a specific 10° part of the ecliptic emanating celestial energy. The thirty-six decans were related to the zodiacal signs or to the planets in various zodiacal and planetary decan systems.[104] The importance of the decans in ancient astrological theory and practice is not clear. Ptolemy, for example, does not discuss them in his *Tetrabiblos*, and they only turn up in a few of the more elaborate Greek horoscopes.[105]

In addition to the concept of thirty-six decans, Schmidt's explanation introduces an astrological classification of the zodiacal signs as diurnal or nocturnal. As with many aspects of ancient astrology, there was no uniform system. Various divisions of the zodiacal signs into diurnal and nocturnal are attested.[106] Manilius, for example, describes three possible divisions. First, he presents a system in which the zodiacal signs are ascribed a diurnal or a nocturnal nature in sections of 60°:

> The zodiacal signs of *Sagittarius* and fierce *Leo*, he who looks round on the golden fleece of his back (sc. *Aries*), then *Pisces* and *Cancer* and *Scorpio* of stinging lash, signs either adjacent or spaced at equal intervals, are all under like estate termed diurnal. The others, identical in number and in the pattern of their spacing, for they are inserted into as many places, are called nocturnal.[107]

Second, Manilius refers to a division in which half of the zodiacal circle is diurnal, while the other half is nocturnal:

[103] For the following, see the literature cited in n. 36 above.

[104] See Manilius, *Astronomica* 4.310–362; Firmicus Maternus, *Mathesis* 2.4.3; Abū Maʿšar, 1.9–77. Cf. Gundel, *Dekane und Dekansternbilder*, 81.246-56; C. Burnett, K. Yamamoto and M. Yano (eds.), *Abū Maʿšar: The Abbreviation of the Introduction to Astrology, Together with the Medieval Latin Translation by Adelard of Bath* (IPTS 15; Leiden: E.J. Brill, 1994), 14-23.

[105] See Neugebauer and van Hoesen, *Greek Horoscopes*, 21-38; Jones, *Astronomical Papyri*, 2:382-83, 420-29, 432-35.

[106] See Bouché-Leclercq, *L'astrologie grecque*, 155-57; Hübner, *Eigenschaften der Tierkreiszeichen*, 286-92.

[107] Manilius, *Astronomica* 2.211–217.

> Some have also asserted that the diurnal belong to the six consecutive stars which begin with *Aries* and that the six from *Libra* count as nocturnal.[108]

Finally, "there are those who fancy that the masculine signs are diurnal and that the feminine class rejoices in the safe cover of darkness."[109] This results in the zodiacal signs being alternately diurnal and nocturnal, beginning with *Aries*. Ptolemy explains that:

> an alternating order was assigned to them because day is always yoked to night and close to it, and female to male. Now as *Aries* is taken as the starting point [...] and as the male likewise rules and holds first place, since also the active is always superior to the passive in power, the signs of *Aries* and *Libra* were thought to be masculine and diurnal [...]. The signs in succession after them correspond, as we have said, in alternating order.[110]

The same binary opposition between male and female, in which the latter is ascribed negative and inferior qualities, plays a significant role in ancient physiognomic theory and, more generally, in ancient "anthropology."[111]

From the various ancient classifications, Schmidt makes use of the second system described by Manilius:[112]

Diurnal	Nocturnal
Aries	*Libra*
Taurus	*Scorpio*
Gemini	*Sagittarius*
Cancer	*Capricorn*
Leo	*Aquarius*
Virgo	*Pisces*

Furthermore, against this background he proposes to divide the zodiacal circle into four quadrants.[113] The first quadrant begins with *Capricorn* (marking the winter solstice), the second with *Aries* (marking the vernal equinox), the third with *Cancer* (marking the summer solstice), and the fourth quadrant begins with *Libra* (marking the autumn equinox). Quadrants two and three correspond with the diurnal half of the zodiac, while quadrants four and one correspond with the nocturnal half. The thirty-six

108 Manilius, *Astronomica* 2.218–220.

109 Manilius, *Astronomica* 2.221–222.

110 Ptolemy, *Tetrabiblos* 1.13.1–2. Ptolemy, *Tetrabiblos* 1.13.3–4, explains another arrangement in which the nature of the zodiacal signs as either diurnal or nocturnal depends on which one is the ascendant.

111 See Chapter Two n. 139.

112 Schmidt, "Astrologie juive ancienne," 129, 133, equates the concept of diurnal and nocturnal zodiacal signs with the notion of solar and lunar "parties" (αἱρέσεις). But the latter concept is based on the planetary houses and the arrangement is different. Cf. Hübner, *Eigenschaften der Tierkreiszeichen*, 287-88. See also n. 33 above.

113 See Chapter One.

decans follow this arrangement. The decans are divided into eighteen diurnal and eighteen nocturnal ones.

The subdivision of the zodiacal signs into thirty-six decans functions as a temporal unit in Schmidt's hypothesis. Each season, between equinox and solstice, corresponds to three zodiacal signs or nine decans. This makes it possible to connect the decans with the determination of the moment of conception. Schmidt argues that people's horoscopes are determined by the decan in which their moment of conception took place. From the moment of conception to that of birth twenty-seven decans are counted for a set period of nine months of embryonic growth.

According to Schmidt the astrological place of conception is established by moving backwards nine signs, starting from the date of birth. On the basis of a text passage by the third-century CE grammarian Censorinus in *On the Day of Birth*, discussing a theory ascribed to the Chaldeans on the duration of pregnancy, he argues that the influence of the sun on the development of the fetus is exerted with differing force due the varying position that the sun has in each decan during the period of pregnancy in relation to its position at the moment of conception. In order to assess this element of Schmidt's hypothesis, it is necessary to pay some attention to the meaning and calculation of the moment of conception in ancient astrology.

The Moment of Conception in Ancient Astrology

Both Babylonian and Greek astrology recognized the importance of the moment of conception for people's horoscopes.[114] Even before the rise of Babylonian horoscopy, the omen series *Šumma ālu* shows the astrological significance ascribed to the moment of conception: "If a man 'approaches' his wife at the rising of the Yoke star (= Bootes), he will get a son with a pleasant spirit."[115] A Babylonian horoscope from 258 BCE provides both the date of conception and of birth showing that the duration of pregnancy amounted to 273 days.[116] In Greek and Roman literary sources it is attested

[114] The following discussion is largely based on the comprehensive study by Frommhold, *Empfängnis in der Astrologie*. Cf. also Bouché-Leclercq, *L'astrologie grecque*, 373-83; Boll, Bezold, and Gundel, *Sternglaube und Sterndeutung*, 153-54; M. Stol, *Birth in Babylonia and the Bible: Its Mediterranean Setting* (CM 14; STYX: Groningen, 2000), 97-98.

[115] Cited from Stol, *Birth in Babylonia*, 97.

[116] See Sachs, "Babylonian Horoscopes," 58-60; Rochberg, *Babylonian Horoscopes*, 73-75. A.J. Sachs, *Late Babylonian Astronomical and Related Texts* (BUS 18; Providence, Rhode Island: Brown University Press, 1955), xxxvii, 256, lists *LBAT* 1588 and 1589 as mentioning "conception of child"; F. Rochberg-Halton, "Babylonian Horoscopes and Their Sources," *Or* 58 (1989): 102-23, at 107 n. 21, mentions an unpublished horoscope in which the moment of conception is accounted for. As an example, she gives lines 8-9: "a child is conceived (when) the sun stands in 5° Aries, Jupiter and Mars stand in 5° Sagittarius. That child (will be) king." Cf. also Reiner, *Astral Magic*, 115, for another possible reference to a gestation period of 277 days.

that astrologers practiced genethlialogy on the basis of the moment of conception.[117] Although in ancient astrological theory the determination of the moment of conception is valued, there are no actual Greek and Roman conception horoscopes extant, only literary examples.[118]

Ptolemy is the only astrologer who has reflected on the importance of the moment of conception for horoscopic astrology. He grants its importance, but in the end favors the time of birth because the moment of conception is usually not known. He circumvents the difficult problem of casting a conception horoscope by assuming that the celestial configuration at birth is similar to that at the moment of conception.[119]

The precise moment of conception was impossible to establish exactly.[120] Some astrologers may have sought the desired information by asking women when according to them conception had taken place,[121] while others may have determined the moment of conception from the given facts of people's lives, as Tarutius is said to have done for Romulus.[122] Most astrologers, however, would probably have assumed a set time for the duration of pregnancy and then counted backwards from the moment of birth.

Knowledge of the exact duration of pregnancy is indispensable for all astrological methods that seek to determine the horoscope for the moment of conception.[123] The gestation period can be expressed either in terms of

[117] Cf. Vitruvius, *On Architecture* 9.6.2; Philo, *On Providence* 1.87; Aulus Gellius, *Attic Nights* 14.1.19; Sextus Empiricus, *Against the Professors* 5.58. Cf. Frommhold, *Empfängnis in der Astrologie*, 3-4, 18-20, 22, 230.

[118] Frommhold, *Empfängnis in der Astrologie*, 226, 241. In a Greek horoscope from 81 CE the time of pregnancy is said to number 276 days, but this not a conception horoscope. Only a set amount of days is given for the duration of pregnancy. The date of conception is not explicitly provided, nor are further astronomical details given for this date. See Neugebauer and van Hoesen, *Greek Horoscopes*, 23-24, 28.

[119] Ptolemy, *Tetrabiblos* 3.2.3–4. Cf. Frommhold, *Empfängnis in der Astrologie*, 10-31.

[120] For criticism of some of the church fathers against astrology because of this, see E. Lesky and J.H. Waszink, "Empfängnis," *RAC* 4 (1959), 1245-55, at 1254-55.

[121] Cf. Bouché-Leclercq, *L'astrologie grecque*, 375. Some claimed that ancient gynecology gave clues for recognizing conception. Cf. Soranus, *Gynecology* 1.44. Cf. Lesky and Waszink, "Empfängnis," 1248-50.

[122] See Chapter Two nn. 233, 234.

[123] On different opinions on conception, duration of pregnancy, and stages of growth of the fetus in Babylonian and Greek cultures, see e.g. E. Lesky and J.H. Waszink, "Embryologie," *RAC* 4 (1959), 1236-40; H. King, "Making a Man: Becoming Human in Early Greek Medicine," in *The Human Embryo: Aristotle and the Arabic and European Traditions* (ed. G.R. Dunstan; Exeter: University of Exeter Press, 1990), 10-19; D.M. Balme, "Ἄνθρωπος ἄνθρωπον γεννᾷ: Human is Generated by Human," in *The Human Embryo*, 20-31; D. Gourevitch, "La gynécologie et l'obstétrique," *ANRW* II 37.3 (1996), 2098-2103; H. King, "Geburt: Medizinisch," *DNP* 4 (1998), 835; Stol, *Birth in Babylonia*, 1-26; Frommhold, *Empfängnis in der Astrologie*, 32-39.

That ancient Jewish authors were probably familiar with certain, widely current, ideas about conception, pregnancy, and gestation is convincingly argued by P.W. van der Horst, "Seven Months' Children in Jewish and Christian Literature from Antiquity," in *Essays on the Jewish World of Early Christianity* (P.W. van der Horst; NTOA 14; Freiburg, Switzerland:

months or days. The ten-month period for pregnancy seems to have been the traditional number of months.[124] This is related to the fact that the period of pregnancy is commonly counted according to the menstruation period, which in turn was seen as related to the period of the moon, since both cover a period of about 28 days. Ten moon months result in an average number of 280 days for pregnancy. This same number of days, however, can also be numbered as nine months and 10 days when one takes schematic months of 30 days as a measuring unit.[125] In some of the ancient astrological treatises the mean values of 273 and 273 1/3 days for pregnancy were taken as the basis for arithmetical and geometrical methods to compute the moment of conception exactly.

Katrin Frommhold distinguishes between four methods for computing the astrological configuration at the moment of conception.[126] Characteristic for all methods is that they count back from the moment of birth. The calculations involve the determination of the position of the sun, the moon, and the ascendant at the moments of conception and birth.[127]

A rule attributed to the Egyptian priest Petosiris, but probably originating in the second century BCE,[128] is concerned with the position of the moon and the ascendant at the moment of conception and birth. The rule is that the ascendant zodiacal sign at birth is the same sign in which the moon

---

Universitätsverlag and Göttingen: Vandenhoeck & Ruprecht, 1990), 233-47; P.W. van der Horst, "Sarah's Seminal Emission: Hebrews 11:11 in the Light of Ancient Embryology," in *Hellenism-Judaism-Christianity: Essays on Their Interaction* (P.W. van der Horst; CBET 8; Kampen: Kok Pharos, 1994), 203-23.

124 Aulus Gellius, *Attic Nights* 3.16.1.

125 Frommhold, *Empfängnis in der Astrologie*, 33-34, 38 (for the confusion between nine schematic and ten sidereal months, see also 56, 195 n. 486). Expressed in weeks, the total is 40 in both cases. Neugebauer distinguishes between three types of "months." First, there is the schematic month of 30 days. Second, one can reckon with sidereal months of about 27 1/3 days. A sidereal month represents the average interval between the moon's consecutive returns to the same fixed star. Third, there is the synodic month of about 29 1/2 days. A synodic month takes the course of the sun and the moon in relation to each other into consideration. It denotes the interval between consecutive conjunctions of sun and moon, i.e. between two new moons. The synodic month provides the basis for calendrical months of 29 ("hollow") or 30 ("full") days. In antiquity the sidereal month was also defined more exactly as a period of 27 1/3 days. This results in a period of 273 1/3 days as the mean value for pregnancy, which is almost identical to the 273 days of the Babylonian horoscope from 258 BCE. See O. Neugebauer, "*Decem Tulerunt Fastidia Menses*," *AJP* 84 (1963): 64-65; Neugebauer, *HAMA*, 1083-84. Cf. Frommhold, *Empfängnis in der Astrologie*, 93-94, 133, 177.

126 Two of these methods were not influential in astrological tradition and they will therefore be further ignored here. See Frommhold, *Empfängnis in der Astrologie*, 173-90.

127 Frommhold, *Empfängnis in der Astrologie*, 240.

128 To the Egyptian pharaoh Nechepso (seventh century BCE) and the priest Petosiris (either contemporary with Nechepso or fourth century BCE) is ascribed an astrological compendium that shows Babylonian, Greek, and Egyptian influences. It is a pseudepigraphon that probably originated in second-century BCE Alexandria. Cf. Gundel and Gundel, *Astrologumena*, 27-36; D. Pingree, *The Yavanajātaka of Sphujidhvaja* (vol. 2; HOS 48; Cambridge, Massachusetts: Harvard University Press, 1978), 436-37.

is positioned at conception, and, vice versa, that the zodiacal sign in which the moon stands at birth is the same sign that ascends at conception.[129]

Censorinus has transmitted a method in which the course of the sun through the zodiac is combined with the astrological theory of aspects to discern both the moment of conception and of birth.[130] Since Censorinus' text is important for Schmidt's interpretation, it is necessary to examine what Censorinus has to say on the "Chaldean" method and how ancient astrologers have appropriated it.[131]

In his work *On the Day of Birth* Censorinus is, among other things, concerned with theories regarding the duration of pregnancy. In this context he discusses a theory, ascribed to the Chaldeans,[132] according to which births are possible in the seventh, the ninth, and the tenth months. The reason why births are possible in these months is because of the aspect of the sun with regard to its position at the moment of conception.[133] The sun causes the moment of birth by means of three aspects (opposition, trine, and quartile) that are regarded as powerful and beneficial. On its course through the zodiac, the sun enters each month in a different relationship or aspect with regard to its starting position from the moment of conception.[134] The aspects influence the gestation of the embryo and the birth of

[129] Frommhold, *Empfängnis in der Astrologie*, 70-172. Cf. J.-F. Bara (ed.), *Vettius Valens d'Antioche: Anthologies, Livre I* (EPRO 111; Leiden: E.J. Brill, 1989), 214-28. Frommhold points out that there are two assumptions for the computability of the conception moon sign on the basis of the ascendant birth sign. First, the duration of pregnancy is expressed as a ten-month period counted on the basis of the sidereal moon period. Second, the pregnancy period is exactly ten months if the moon is positioned in the ascendant at the moment of birth. If the moon stands elsewhere, its elongation from the ascendant is used to add or subtract from the mean number of days for pregnancy, thereby determining the exact number of days for a particular pregnancy. Together with an estimation of the moon's velocity through the zodiacal circle, these two assumptions provide the basis for calculating the moment of conception in terms of a calendar date at which the moon always occupies the same zodiacal sign that is ascending at the moment of birth.

[130] See n. 31 above.

[131] For the following, see Frommhold, *Empfängnis in der Astrologie*, 40-69.

[132] Censorinus, *On the Day of Birth* 8.1: *Chaldaeorum ratio* (K. Sallmann [ed.], *Censorinus: De die natalie liber* [Teubner; Leipzig: B.G. Teubner, 1983], 12.18). Cf. also 10.48.

[133] Censorinus, *On the Day of Birth* 8.13: "according to this theory seven months' children are born in opposition, whereas nine months' children in trine, ten months' children indeed in quartile." Cf. G. Rocca-Serra, *Censorinus: Le jour natal* (HDAC 5; Paris: Librairie Philosophique J. Vrin, 1980), 12.

[134] C.S.F. Burnett, "The Planets and the Development of the Embryo," in *The Human Embryo*, 95-112, points out that the exact correlation between consecutive months of gestation and specific planets, familiar from medieval literature, does not seem to appear in sources from antiquity, although several elements in ancient astrology might be regarded as having been conducive to the development of such a concept. One of these elements is the connection between the sun's course through the zodiac and the different months of embryonic growth in relation to the theory of aspects in Censorinus' account. According to Burnett, "Planets and the Development of the Embryo," 96 n. 8, no clear ancient examples are brought forward by Bouché-Leclercq, *L'astrologie grecque*, 508-11.

the child in different ways. When the sun enters the fourth and fifth zodiacal signs, it enters respectively the aspects of quartile and trine with regard to place of conception. These are regarded as the first effective aspects that the sun enters during the gestation period of the fetus. In the seventh zodiacal sign the sun stands in opposition to the place of conception. This is the fullest and strongest aspect, which already brings forth mature infants called *septemmestres*.[135] A birth in the eighth month is not possible because the sun does not stand in any aspect with regard to the place of conception.[136] The eighth zodiacal sign, like the sixth, is a powerless aspect. Birth was deemed possible again in the ninth and tenth months because the sun regards the place of conception again in respectively trine and quartile, two very powerful aspects.

Astrologers used Censorinus' "Chaldean" theory in reversed manner. The zodiacal position of the sun at the moment of birth was the basis to determine the sun's position at conception. The similarity, however, between number of pregnancy months and number of zodiacal signs remained the same. Astrologers distinguished between seven and ten months' children. Regarding ten months' children the assumption was that at the time of conception the sun was situated in the left quartile with regard to its position at the moment of birth. Frommhold explains that "left" refers to the direction of the annual course of the sun through the zodiac, which runs anticlockwise. The "left quartile," therefore, refers to the fourth zodiacal sign to the left of the birth sign, again, counting the latter as one of the four signs. If one assumes, for example, a ten months' child to have been born when the sun was positioned in *Aries*, the fourth zodiacal sign to the left is *Cancer*. The sun was, therefore, positioned in the zodiacal sign *Cancer* ("left quartile") at the moment of conception according to this rule.[137]

[135] Frommhold, *Empfängnis in der Astrologie*, 43, points out that in Censorinus' account a zodiacal sign equals one month, and that the sun's zodiacal position at conception is already counted as one month. That way after the sun has traveled through six more signs after the moment of conception seven months are counted. Furthermore, the aspects do not represent schematic months of 30 days that have passed. Rather, the seventh month begins at 180°, the ninth month at 240°, and the tenth month at 270°. However, if one counts sidereal moon months of about 28 days, one is in the middle of the seventh month at 180°, in the middle of the ninth month at 240°, and at the middle of the tenth month at 270°. See Frommhold, *Empfängnis in der Astrologie*, 44 n. 149.

[136] It was the general opinion in antiquity that an eighth months' birth was not viable. Cf. the references cited in n. 123 above.

[137] Schmidt's interpretation assumes a mean period of nine months of pregnancy. As this is equal to twenty-seven decans or 270° the result is almost the same. However, to be exactly the same, and for the sun to be positioned at the place of birth in the left quartile, someone conceived in the zodiacal sign of *Taurus* has to be born in *Aquarius*, not in the third decan of *Capricorn* (see n. 157 below).

Notwithstanding the literary testimonies, conception horoscopy seems not to have had as large a following among astrologers as birth horoscopy. Frommhold concludes:

> Bis auf das fiktive, bei Plutarch überlieferte Empfängnishoroskop des Romulus und die beiden Beispielhoroskope, welche die Astrologen Vettius Valens und Hephaistion zur Veranschaulichung der Petosiris-Regel auf ihre eigen Konzeption gestellt haben, ist in der griechischen und römischen Astrologie kein einziges originales Empfängnishoroskop überliefert. Diese Form der Horoskopie als konkurrierende Methode zur Geburtshoroskopie hat sich praktisch offensichtlich nicht durchsetzen können.[138]

The Division of Light and Darkness in 4QZodiacal Physiognomy *(4Q186)*

The point of departure for understanding the astrological framework is to take the number of nine zodiacal signs or twenty-seven decans as representative for the mean period of pregnancy. Accordingly, one zodiacal sign or three decans equal one month.

Adding the notion of diurnal and nocturnal zodiacal signs according to Manilius' second arrangement, according to which half from *Aries* to *Virgo* is diurnal, while the other half from *Libra* to *Pisces* is nocturnal, Schmidt reasons that any fetus can have a maximum of six diurnal zodiacal signs and a minimum of three nocturnal signs during the gestation period. Any person having a lesser share of diurnal signs has at least three nocturnal zodiacal signs, or nine nocturnal decans. This number of three nocturnal signs or nine nocturnal decans Schmidt calls the "common fund," which cannot be altered.[139] Regardless of the date of conception, therefore, every individual has a common fund of nine diurnal and nine nocturnal decans. This number of eighteen invariable decans is common to every mean period of pregnancy of nine months, no matter during what period of the year the fetus develops. This means that nine variable decans remain to be designated as diurnal or nocturnal.[140]

According to Schmidt, *4QZodiacal Physiognomy* addresses the question of how to divide the nine variable decans into diurnal and nocturnal decans. He suggests that the text locates the conception of those with a maximum of nine diurnal decans in the first quadrant, between the first decan of *Cap-*

[138] Frommhold, *Empfängnis in der Astrologie*, 241.

[139] Equally, an embryo having the maximum of six nocturnal zodiacal signs has a minimum of three diurnal signs. Any individual with fewer nocturnal signs has at least three diurnal zodiacal signs, or nine diurnal decans. Again, this represents what Schmidt calls the "common fund."

[140] Schmidt, "Astrologie juive ancienne," 133-34.

ricorn and the third decan of *Pisces*.[141] Those people with a maximum of nine nocturnal decans were conceived in the opposite, third quadrant, between the first decan of *Cancer* and the third decan of *Virgo*.[142] The second quadrant contains the intermediate positions from the most diurnal to the most nocturnal categories, while the third quadrant includes the transitional positions between the most nocturnal and the most diurnal categories.[143]

Interpreting "the second column" (העמודהשני) as a reference to the second quadrant, Schmidt argues that the remaining fragments of *4QZodiacal Physiognomy* successively listed the physiognomic and spiritual descriptions of those types of people conceived in the three decans of *Aries*, *Taurus*, and *Gemini*, respectively.[144] He suggests that the words "in the foot of *Taurus*" (ברגל השור) are a reference to "la première partie de la constellation du Taureau apparaissant à l'Orient du ciel, ou le premier décan du Taureau."[145] He understands it together with מולד as a reference to the place of conception. Schmidt suggests that the moment of birth of this individual is to be found in the third decan of *Capricorn*. This means that 4Q186 1 ii 7–8 provides the numbers of diurnal and nocturnal decans at the moment of conception in terms of six in the "house of light" and three in the "house of darkness." Thus, the "house of light" corresponds with the diurnal decans, while the "house of darkness" refers to the nocturnal decans.[146]

141 All those conceived in the first quadrant share the same division of eighteen diurnal and nine nocturnal decans during their periods of gestation (covering twenty-seven decans). The nine-month gestation period of every person conceived during the zodiacal signs from *Capricorn* to *Pisces* necessarily passes through the diurnal half of the zodiacal circle from *Aries* to *Virgo*, thereby benefiting maximally from the diurnal decans.

142 All those conceived in the third quadrant, therefore, possess the same number of eighteen nocturnal and nine diurnal decans, because their gestation period necessarily covers the nocturnal half of the zodiac from *Libra* to *Pisces*.

143 Schmidt, "Astrologie juive ancienne," 134-36.

144 Schmidt, "Astrologie juive ancienne," 136-38. The second quadrant contains the intermediate positions from the most diurnal to the most nocturnal category of individuals. The nine variable decans are divided accordingly, beginning with a division into nine diurnal and zero nocturnal decans in the first decan of *Aries* and ending with a partition of one diurnal and eight nocturnal decans in the third decan of *Gemini*. That the first decan of *Aries* has the maximum number of diurnal decans, like those in the first quadrant, is necessarily so. The gestation period of someone conceived at this time benefits from the maximum amount of eighteen diurnal decans until the third decan of *Virgo*.

145 Schmidt, "Astrologie juive ancienne," 136.

146 Schmidt, "Astrologie juive ancienne," 136-37. The total division of diurnal and nocturnal decans is respectively fifteen and twelve. For those conceived in the first decan of *Taurus*, the influence of the sun on the development of the fetus is exerted for the period of fifteen diurnal decans until the third decan of *Virgo*. Then the sun exerts its influence for another period of twelve nocturnal decans until the time of birth in the third decan of *Capricorn*. Subtracting the minimum amounts of nine diurnal and nine nocturnal decans leaves the variable decans to be numbered as six diurnal and three nocturnal decans. Although no zodiacal information or reference to "the second column" are provided for the type listed in 4Q186 1 iii, Schmidt situates this entry in the second quadrant because it is listed next to 4Q186 1 ii. Having more parts of darkness, eight in the "house of darkness" and one in the

On the basis of the date of conception predictions could be made regarding the physiognomic and spiritual state of categories of people. It is important to notice that Schmidt connects the division of diurnal and nocturnal decans to a certain spiritual state of each type. The ratio of diurnal and nocturnal decans at the time of conception apparently reflects people's spiritual share in light and darkness. But Schmidt rejects the idea that this is related to the dualism envisaged in the *Two Spirits Treatise* in 1QS 3:13–4:26, which represents an absolute dualism. Because of the common fund of nine diurnal and nine nocturnal decans, it is impossible for people's spiritual share to be completely within the "house of light" or the "house of darkness." *4QZodiacal Physiognomy* represents only a relative dualism.[147]

Conception, "the Foot of Taurus, *" and Schmidt's Decanal Interpretation*

The ingenuity of Schmidt's interpretation lies in its combination of different elements from ancient astrology in a comprehensive manner. But under closer scrutiny several issues appear problematic.

First, according to Schmidt's model, most moments of conception result in a similar physiognomic and spiritual state. There are only ten different divisions of diurnal and nocturnal decans.[148] One would, however, anticipate the use of the concept of thirty-six decans to result in more diversification of the prognostications regarding the shape and appearance of the human body and the character of people's spirit. The expectation would be that every decan governs its own type of people in terms of their physique and spirit; the effect of Schmidt's hypothesis is merely ten types, implying that this is not the case.[149] Related to this issue is the function of the decans. They have no actual influence on the development of the embryo; the

---

"house of light," this type is further towards the third quadrant of the most nocturnal entries. Thus, the place of conception is positioned in the third decan of *Gemini* and the moment of birth is expected to occur in the second decan of *Pisces*. Finally, Schmidt situates the conception of the type listed in 4Q186 2 i in the second decan of *Aries* and its place of birth in the first decan of *Capricorn*.

[147] Schmidt, "Astrologie juive ancienne," 126-27, 138-39.

[148] Those conceived in the zodiacal signs of *Capricorn*, *Aquarius*, *Pisces* and the first decan of *Aries* apparently share the same spiritual make-up of eighteen diurnal and nine nocturnal decans. Schmidt does not comment on whether their physiognomies might differ, but this does not seem to be the case. The second type concerns those conceived in the zodiacal signs of *Cancer*, *Leo*, *Virgo*, and the first decan of *Libra*, who are credited with an identical physiognomic and spiritual portrait of nine diurnal and eighteen nocturnal decans. Finally, those conceived in one of the eight remaining decans from the second of *Aries* until the third of *Gemini* have identical physiognomic and spiritual characters to those conceived in one of those from the second decan of *Libra* until the third of *Sagittarius*. Thus, one arrives at a total of ten physiognomic and spiritual character types. Cf. Schmidt, "Astrologie juive ancienne," figure 3.

[149] Schmidt, "Astrologie juive ancienne," 125: "A chaque division du zodiaque correspond un portrait-type, physique et spirituel." Cf. Albani, "Horoscopes in the Qumran Scrolls," 328. See also n. 101 above.

decans are redundant in this sense. They merely signal a moment in time and serve as a simple arithmetical device. The key element is the division of the zodiacal circle in a diurnal and a nocturnal half. According to their diurnal and nocturnal nature the decans are equally indicative of people's spiritual character and have no further importance.

Second, Schmidt suggests that the place of conception "in the foot of *Taurus*" (ברגל השור) refers to the first decan of *Taurus* or the first part of the constellation *Taurus* to appear in the eastern sky, but this is incorrect for several reasons. First, the decans are a subdivision of the zodiacal *signs* schematically dividing each sign into three parts of 10°. As such they have nothing to do with the actual *constellations*. Secondly, there is no evidence from ancient astrology that the first decan of *Taurus* is referred to as the "foot of *Taurus*."[150] Finally, even if the words ברגל השור are taken as a reference to the forefeet[151] of the zodiacal *constellation Taurus* rising above the eastern horizon, it is impossible to understand it as the first part of the constellation to appear, because the constellation *Taurus* rises backwards and not head first.[152]

Third, like all astrological theories concerning the establishment of the date of conception, Schmidt assumes that such a determination takes its starting point from the date of birth. However, Albani rightly remarks that "one should also expect, therefore, a statement concerning the place of birth within the decans of zodiacal signs" in *4QZodiacal Physiognomy*.[153] The text, however, does not provide this. The date of birth is pivotal for calculating the moment of conception. It not being mentioned seriously hampers Schmidt's interpretation. This means that an ancient astrologer had to make the calculations before use could be made of the text. Taking the date of birth he would count back twenty-seven decans and only then probably turn to *4QZodiacal Physiognomy* to find the information needed. But how would an ancient astrologer know where to look in the text?

Fourth, Schmidt argues that *4QZodiacal Physiognomy* provides the predictions regarding the physiognomic and spiritual characters of categories of individuals on the basis of the date of conception, but he does not explain how the text is structured to facilitate access to this kind of information. Taking Schmidt's hypothesis as our point of reference, only two elements seem to present themselves as possible markers for retrieving the data

150 Cf. Gundel, *Dekane und Dekansternbilder*, 119-21, 160-61. See Albani, "Horoscopes in the Qumran Scrolls," 326-28.

151 Schmidt, "Astrologie juive ancienne," 136 n. 22: "les pattes antérieures," wrongly translated in Schmidt, "Ancient Jewish Astrology," 199 n. 28, as "rear hooves." This is impossible because the constellation *Taurus* is imagined as a halved animal, cut from the middle, cf. n. 69 above.

152 Cf. Hübner, *Eigenschaften der Tierkreiszeichen*, 102.

153 Albani, "Horoscopes in the Qumran Scrolls," 326.

needed: (1) the signs of the zodiac and their decans; (2) the numbers allotted to the "house of light" and the "house of darkness." Regarding the former element, moving backwards nine zodiacal signs or twenty-seven decans from the moment of birth, an ancient astrologer would arrive at a certain zodiacal sign and its decan. These would function as his indicators in the auxiliary text before him. But the zodiacal information is provided at the end of the account in 4Q186 1 ii. A similar problem arises regarding the latter element of the numbers in the "house of light" and the "house of darkness". Having knowledge of the system according to which the zodiacal signs from *Aries* to *Virgo* are diurnal and those from *Libra* to *Pisces* nocturnal, an ancient astrologer could do the math and arrive at a certain division of variable diurnal and nocturnal decans, which he would then look up. But, again, the data concerning the "house of light" and the "house of darkness" seem to stand somewhere in the middle of an account, which does not help to find it easily. A more important objection is that one would have to assume knowledge on the part of an ancient user of half of the information he wants to retrieve, which does not make much sense.[154]

Fifth, the argument that the astrological framework of *4QZodiacal Physiognomy* is based on the idea that a person's horoscope is determined by the moment of his conception is not convincing. A more likely interpretation for מולד is that it refers to the horoscope, not in the sense of the ascendant (ὡροσκόπος), but, equivalent to the Greek γένεσις, in the sense of the nativity, i.e. the configuration of heavenly bodies in relation to the zodiacal circle at the moment of birth.[155] If, therefore, no distinction is made in 4Q186 1 ii 8 between the moment of conception and the moment of birth by means of the words מולד and ילוד, there is no need to assume that *4QZodiacal Physiognomy* is based on the astrological notion to determine a person's horoscope from the moment of conception.[156] This means that *4QZodiacal Physiognomy* does not provide evidence for an adapted applica-

[154] Schmidt does not explain this sufficiently, but he seems to make two assumptions. First, he suggests that a system dividing the zodiacal circle in a diurnal half from *Aries* to *Virgo* and a nocturnal half from *Libra* to *Pisces* is the astrological background for the numbers in the "house of light" and the "house of darkness." Second, although such a system forms the background, one must assume that it was not known to the users of *4QZodiacal Physiognomy* who interpreted the references to the "house of light" and the "house of darkness" as a division of the spiritual character of a type of person. After all, if the ancient users had been familiar with this system, there would not have been much need to look for the diurnal and nocturnal division after they had counted backward from the moment of birth. The only new type of information would be the physiognomic description of the newborn on the basis of his moment of conception. Another option is that the ancient reader was not familiar with Manilius' second system. But then the division of light and darkness could not have functioned as a marker in the text. One thus returns to the question of how the information in the text was to be found if the date of birth is not mentioned.

[155] See Chapter One. For "nativity," see Gundel and Kehl, "Horoskop," 600.

[156] See n. 98 above and the section on מולד ("horoscope") in Chapter One.

tion of the method used by astrologers to establish the moment of conception on the basis of the sun's movement during pregnancy through the zodiacal circle, as described by Censorinus.[157]

If, however, the key to understanding the astrological framework in *4QZodiacal Physiognomy* is not conception horoscopy, the interpretation of the words "house of light" and "house of darkness" as a reference to the variable diurnal and nocturnal decans becomes difficult to maintain. Schmidt's explanation for the number nine is entirely based on the combined assumptions that the horoscope is determined by the moment of conception and that twenty-seven decans equal a mean period of pregnancy. If these assumptions do not hold water, the reference to a typology of diurnal and nocturnal zodiacal signs as a third assumption loses its explanatory function.[158]

MATTHIAS ALBANI: "THE FOOT OF *TAURUS*" ASCENDING ABOVE THE HORIZON INTO LIGHT

Following Bergmeier, Albani has argued that the key to understanding the astrological character of *4QZodiacal Physiognomy* is the phrase "in the foot of *Taurus*" (ברגל השור). These words presuppose a partition of the zodiacal sign *Taurus*, and they indicate an exact localization in that sign. Instead of the moon, however, it is the horizon that functions as the dividing line between the different parts of the sign. The "house of light" and the "house of darkness" are related to cosmological rooms above and below the horizon, not to day and night. The "house of light" contains the parts of the zodiacal sign that have risen above the horizon, while the "house of darkness" refers to those parts that are still below the horizon.[159]

Albani characterizes *4QZodiacal Physiognomy* as "a list or compilation of options for astrological interpretations systematically arranged according to certain astrological criteria." It is "an auxiliary astrological resource for creating horoscopic prognostications." More specifically, *4QZodiacal*

[157] The difference being that in *4QZodiacal Physiognomy* the number of twenty-seven decans guides the computations, whereas in the astrological method based on Censorinus' account the idea is to find the sun's place at the moment of conception in the left quartile with regard to its position on the date of birth (see also n. 137 above).

[158] In addition, Schmidt's choice for the second system described by Manilius is not arbitrary but necessary if one assumes that twenty-seven decans equal the mean period of pregnancy for explaining the actual numbers used in 4Q186 1 ii 7 and 4Q186 1 iii 9. If, for example, one assumes another arrangement according to which the zodiacal signs are alternately diurnal and nocturnal beginning from *Aries*, the result for someone conceived in the first decan of *Taurus* is three diurnal and six nocturnal decans and not the numbers given.

[159] Albani, "Horoscopes in the Qumran Scrolls." Cf. also Albani, "Horoscopes."

Physiognomy provides "the astrological possibilities of interpretation for the observed ascendant."[160]

The Ascendant Part of the Divided Zodiacal Sign

Albani calls his hypothesis for the astrological framework of *4QZodiacal Physiognomy* an ascendant interpretation. This one is preferable to solar and lunar interpretations, since neither of them "can offer a satisfactory explanation for the statements about the light-darkness ratio in connection with the astronomical position 'in the foot of Taurus.'"[161]

A solar interpretation of *4QZodiacal Physiognomy* has difficulty accounting for "the foot of *Taurus*" (רגל השור) as a part of the constellation *Taurus*, because the sun cannot reach this area.[162] One could propose instead that it refers to an ecliptical part of the zodiacal sign *Taurus*, but Albani makes two further objections against such a solar interpretation. First, he objects that in solar *zodiologia* the position of the sun is not further specified as being in a specific part of the zodiacal sign, and that only complete zodiacal signs are relevant. Second, a solar interpretation does not explain the relationship between the different light-darkness ratios and the positions of the sun in the zodiacal sign. This also counts for the lunar interpretation.[163] A solar interpretation seems unable to account for the realization of the numbers in the "house of light" and the "house of darkness," but it is not true, as Albani states, that solar *zodiologia* disregard distinctions within a sign. There is clear evidence that a precise localization of the sun in a zodiacal sign was of significance.[164] In some Babylonian as well as Greek horoscopes the solar longitude in degrees of a zodiacal sign is provided.[165] As these data obviously cannot be derived from direct observation, one has to assume that calculations were made, or, more probably, that ephemeris tables or almanacs were available from which to retrieve the required data. There is abundant evidence for the existence of these kinds of texts for the Babylonian and Greco-Roman astronomical traditions,[166] but none are known from ancient Palestine.

160 Albani, "Horoscopes in the Qumran Scrolls," 309-10.

161 Albani, "Horoscopes in the Qumran Scrolls," 305.

162 See n. 70 above.

163 See n. 97 above. But Albani, "Horoscopes in the Qumran Scrolls," 303 n. 79, 307 n. 87, admits the possibility that sun or moon are in conjunction with the ascendant.

164 See Hübner, *Grade und Gradbezirke*.

165 See Rochberg, "Babylonian Horoscopes," 65-67, 97-81, 83-85; Jones, *Astronomical Papyri*, 1:47; 2:374-77, 382-83, 416-17.

166 See A. Jones, "Studies in the Astronomy of the Roman Period II: Tables for Solar Longitude," *Centaurus* 39 (1997): 211-29; Rochberg, *Babylonian Horoscopes*, 8-9; Jones, *Astronomical Papyri*, 1:281; 2:191, 199; Rochberg, *The Heavenly Writing*, 145-63.

Like Bergmeier, Albani uses the *Rhetorius-Teucer text* as an important key text for his interpretation. In this text the enumeration of the nine parts of the zodiacal sign *Taurus* indicates the successive rising of the ecliptical parts of that sign,[167] imagined as the limbs of the sign. The aspect of rising is an important clue for understanding the astrological framework of *4QZodiacal Physiognomy*. Albani proposes that the words ברגל השור ("in the foot of *Taurus*") are a reference to the ascendant, i.e. that ecliptical part of the zodiacal sign *Taurus* rising above the eastern horizon at the time of birth. It takes approximately two hours before the entire 30° section has entirely risen above the horizon. This means that during the time of ascension an ever-greater part appears above the horizon, leaving an ever-smaller part below the horizon.

Using a text such as that of Rhetorius-Teucer, which divides the sign *Taurus* into nine parts, an ancient astrologer could, in theory, count which parts of the sign had risen above the horizon and which parts still remained below the horizon. According to Albani, if the ascendant is "in the foot of *Taurus*" that means that this part is rising from below the eastern horizon. In the ascendant interpretation it is the ascendant, not the sun or the moon, which divides the rising zodiacal sign into parts of light and darkness. This explains the numbers assigned to the "house of light" and the "house of darkness" in 4Q186 1 ii 7–8. Six ecliptical parts of the sign *Taurus* have risen above the horizon, while three parts are still below the horizon. At the moment of birth six parts of *Taurus* were "in the house of light," while three parts were still "in the house of darkness." Consequently, Albani takes the zodiacal sign as the object of reference of רֶוַח לֹו: "it has a space (in the house of light of six [parts], and three in the house of darkness)." He prefers this reading because, contrary to the reading רוּחַ ("spirit"), it allows an astrological interpretation of the division of light and darkness in *4QZodiacal Physiognomy* without necessarily invoking the theological anthropological background of the *Two Spirits Treatise* in 1QS 3:13–4:26.[168]

"House of Light" and "House of Darkness" as Cosmological Rooms above and below the Horizon

According to the ascendant interpretation, the "house of light" denotes the part above the horizon while the "house of darkness" refers to the area below

[167] This is expressed by the use of the verb ἀνατέλλω ("to rise"). For *Taurus*, see CCAG 7.197.24 (cf. n. 84 above). For the importance of establishing the rising degree of the zodiacal sign at the moment of birth, see n. 25 above.

[168] Albani, "Horoscopes in the Qumran Scrolls," 285, 308-9, 312. Cf. n. 92 above. For Albani's previous position on this matter and on *4QZodiacal Physiognomy* in general, see Albani, "Der Zodiakos in 4Q318," 7-8, 40; Albani, *Astronomie und Schöpfungsglaube*, 343 n. 262.

the horizon. Such an understanding is supported by Greco-Roman astrological texts that show that the area above the horizon was associated with light and the area below the horizon with darkness.

Albani seeks confirmation for his interpretation of the "house of light" and the "house of darkness" in a commentary by the Greek astronomer Hipparchus on the third-century BCE poem *Phenomena* by Aratus. Hipparchus comments on a section in which Aratus describes how some zodiacal constellations rise in the east while others set in the west. The relevant comment is in a passage in the section describing the rising of *Cancer* where Aratus says:

> No more will *Bootes* bulk large above and below the horizon, the lesser part being 'day,' and the greater already in darkness.[169]

Hipparchus explains:

> 'day' signifies the part of the cosmos above the earth, and 'night' the part below the earth.[170]

Albani concludes, therefore, that "the phrases 'house of light/darkness' [...] are related to the cosmological rooms above and below the horizon."[171]

In addition to these texts adduced by Albani, one can also point to the section in the *Tetrabiblos* on determining the length of a person's life. Here Ptolemy discusses which places are vital for a planet to be positioned in with regard to obtaining lordship. In this context he states that:

> the whole region below the earth must, as is reasonable, be disregarded when a domination of such importance is concerned, except only those parts which in the ascendant sign itself are coming into the light.[172]

I suggest that this passage supports the ascendant hypothesis in two important ways. First, it corroborates the idea that the separate ecliptical parts of a zodiacal sign are important to take account of in an astrological procedure. Second, as the area "below the earth" (ὑπὸ γῆν) is disregarded except for those parts of the zodiacal sign that are rising above the horizon into the area of light, Ptolemy, by extension, evidently designates the part "above the earth" (ὑπὲρ γῆν) as an area of "light" (φῶς). In another passage

[169] Aratus, *Phenomena* 579–580. Translation, slightly adapted, from D. Kidd (ed.), *Aratus: Phaenomena* (CCTC 34; Cambridge: Cambridge University Press, 1997), 114-15, 382. Cf. also 575–576 where the upper parts of a setting constellation are described as moving into the night, and 581–582 where Aratus says of the setting *Bootes* that it is "satiated with light" (φάεος κορέσηται).

[170] Hipparchus, *Commentary on Aratus and Eudoxus* 2.2.15.

[171] Albani, "Horoscopes in the Qumran Scrolls," 308.

[172] Ptolemy, *Tetrabiblos* 3.11.4. Cf. Hephaestion, *Apotelesmatica* 2.11.19.

Ptolemy states that a zodiacal sign is diurnal if it is above the earth, and nocturnal if it is below the earth.[173]

In addition, Albani puts forward that the interpretation of cosmological areas is supported by the spatial notion suggested by the use of "house" (בית) in *4QZodiacal Physiognomy*. Here it is interesting, I suggest, to refer to the use of *bītu* ("house") in cuneiform texts in the phrase *bīt niṣirti* ("house of the secret"), which is also found in a number of Babylonian horoscopes. This concept is identified as a forerunner of the Greek astrological theory of exaltations (ὑψώματα), according to which the planets hold a special influence in certain parts of the zodiacal signs. The difference is that in cuneiform texts the place of influence is referred to as the entire zodiacal constellation or sign, not a specific part of it. Also, the *bīt niṣirti* refers to the place or region in which a planet's position causes auspicious omens, but this is not exactly the same as the exaltation referring to a planet's greater planetary influence. The word *bītu* ("house") was originally used in the term *bīt niṣirti* as a reference to the general region of a constellation in the sky, but in horoscope texts it seems to be a reference to a zodiacal sign. Its spatial notion is, nonetheless, clearly attested.[174]

With regard to 4Q186 1 ii, the ascendant interpretation succeeds in explaining the realization of the specific numbers in the "house of light" and the "house of darkness" in connection with an astrological explanation for the words "in the foot of *Taurus*." Taking the words "the second column" (העמוד השני) to refer to *Taurus* as the second zodiacal sign, the idea of a division of the sign makes it possible that different physiognomic types were classified under the zodiacal sign *Taurus*, and other signs. These types of people corresponded to the various divisions of the signs.

The implication of the ascendant interpretation for *4QZodiacal Physiognomy* is that it presupposes that different parts of the zodiacal sign influence the shape and appearance of the human body. Albani, therefore, suggests that "the idea seems to be that one obtains a more differentiated physiognomic classification by dividing the zodiacal sign – in other words, not only twelve but 9 x 12 physiognomic types."[175]

This notion finds support, I suggest, in a remark by the Skeptic philosopher Sextus Empiricus (second century CE). An important argument brought forward against astrology in antiquity was that people born under the same sign had different fates (the case of twins is a classic example).[176]

[173] Ptolemy, *Tetrabiblos* 3.11.20. This variable system is not to be confused with Ptolemy's account of diurnal and nocturnal zodiacal signs in *Tetrabiblos* 1.13.

[174] See Rochberg-Halton, "Elements of the Babylonian Contribution," 53-57; Koch-Westenholz, *Mesopotamian Astrology*, 134-35; Rochberg, *Babylonian Horoscopes*, 46-50.

[175] Albani, "Horoscopes in the Qumran Scrolls," 312.

[176] For this and other arguments against astrology in antiquity, see Chapter Five n. 55.

Similarly, Sextus argued that those born in the same sign of the zodiac are not similar in shape or character, unless, he added, the proponents of astrology say that the degrees and minutes into which each sign is divided are capable of causing these differences.[177] Despite Sextus' criticism, his remark shows that the different parts and subdivisions of the zodiacal signs could be taken into account to explain physical and psychological differences between people born under the same zodiacal sign. That the separate degrees of the signs were indeed thought to be decisive for people's fates is demonstrated by different astrological systems.[178] With regard to *4QZodiacal Physiognomy* this means that different physiognomic types of people could belong to one zodiacal sign, viz. each corresponding to one of its different divisions.[179]

4QZodiacal Physiognomy *(4Q186) and Casting Horoscopes at Qumran*

Albani classifies *4QZodiacal Physiognomy* as a text that is structured according to astrological criteria. As such it was an auxiliary text for creating horoscopic predictions concerning people's fates and physical appearance. He provides some suggestions regarding practical requirements for the use of a text like *4QZodiacal Physiognomy*, and he addresses the question of whether the text was actually used for casting horoscopes.

Practical usage of *4QZodiacal Physiognomy* would have required knowledge of the rising times of the zodiacal signs, either by computation or observation. If the text lists data for each ascending part of the zodiacal signs, one needs to know which parts are ascending at what time. According to Albani, this required computational competence by someone at Qumran to determine the rising times.

The rising times are an indication of the number of degrees, or arcs, of the equator that cross the horizon of a given geographical latitude simultaneously with the consecutive signs of the zodiac. The rising times of the zodiacal signs are connected with the length of daylight and vary accordingly during the year due to the angle of the ecliptic in relation to the eastern horizon. The computations must take the vernal equinox, the point on the ecliptic at which the length of both day and night is equal, as their start-

[177] Sextus Empiricus, *Against the Professors* 5.99. As a critic of astrology, Sextus Empiricus concluded that astrologers could never determine exactly the ascendant or degree of the zodiacal sign rising at the moment of people's birth. He ridiculed the impossibility of exact observations, ignoring the largely non-observational character of astrology. See *Against the Professors* 5.27–28, 68–72, 74, 80–85. Cf. E. Spinelli, "Sesto Empirico e l'astrologia," in *Traditions of Theology: Studies in Hellenistic Theology, Its Background and Aftermath* (eds. D. Frede and A. Laks; PhA 89; Leiden: Brill, 2002), 239-73.

[178] Cf. Gundel and Böker, "Zodiakos," 563-66.

[179] Cf. Chapter One n. 94.

ing point.[180] From sunrise to sunset the sun travels half a circle, i.e. 180°. If at a certain moment of the year the sun is known to be at the beginning of a zodiacal sign, this means that between sunrise and sunset six signs, or 180°, have crossed the eastern horizon. Depending on the length of daylight, which depends on the sun's position in the ecliptic in relation to the equator, the signs of the zodiac rise quickly or slowly. Importantly, computations apply only to particular geographical latitudes.[181]

But Albani notes that "records of such relatively sophisticated computations are not attested in the Qumran texts."[182] He suggests that a possible clue to the practice of actual observation of the heavens might be found in a stone disc excavated from the settlement of Khirbet Qumran.[183] Albani has interpreted this object as a sundial, but this identification, as well as how it works, is, at the moment, far from clear.[184]

Albani draws the conclusion that, thus far, "there is no clear proof [...] for observing or computing the rising times of the zodiacal signs at Qumran," although "if the ascendant-interpretation is correct, this would be the

180 In antiquity opinions differed as to the exact point of the ecliptic at which the vernal equinox was thought to occur. In Babylonian astronomy the vernal equinox was placed either at 10° *Aries* (System A) or 8° *Aries* (System B). In Greek astronomy the vernal equinox was placed either at 0° *Aries*, 8° *Aries*, or 15° *Aries* (10° *Aries* and 12° *Aries* being isolated occurrences). The position of the vernal equinox at 0° *Aries* is related to the discovery of precession, while the placement of the vernal point at 15° *Aries* has to do with calendrical considerations to have the equinox on the fifteenth day of a given month (this is also attested in Babylonian tradition). See Neugebauer, *HAMA*, 368-69, 593-600.

181 See Geminus, *Introduction to Phenomena* 7.9–11. Cf. n. 26 above.

182 Albani, "Horoscopes in the Qumran Scrolls," 310.

183 Locus 45, object 1229: "disque de pierre," see J.-B. Humbert and A. Chambon (eds.), *Fouilles de Khirbet Qumrân et de Aïn Feshkha I: Album de photographies, Répertoire du fonds photographique, Synthèse des notes de chantier du Père Roland de Vaux OP* (NTOA.SA 1, Fribourg, Switzerland: Éditions Universitaires and Göttingen: Vandenhoeck & Ruprecht, 1994), 308.

184 Albani, "Horoscopes in the Qumran Scrolls," 310 n. 97; U. Gleßmer and M. Albani, "An Astronomical Measuring Instrument from Qumran," in *The Provo International Conference on the Dead Sea Scrolls: Technological Innovations, New Texts, and Reformulated Issues* (eds. D.W. Parry and E. Ulrich; STDJ 30; Leiden: Brill, 1999), 407-42; M. Albani and U. Gleßmer, "Un instrument de mesures astronomiques à Qumrân," *RB* 104 (1997): 88-115. See also Pfann, "Writings in Esoteric Script," 186, 188-90; G.M. Hollenback, "The Qumran Roundel: An Equatorial Sundial?," *DSD* 7 (2000): 123-29; B. Thiering, "The Qumran Sundial as an Odometer Using Fixed Lengths of Hours," *DSD* 9 (2002): 347-63; Ben-Dov, "Initial Stages of Lunar Theory," 134 n. 28; Hirschfeld, *Qumran in Context*, 151-152; G.M. Hollenback, "More on the Qumran Roundel as an Equatorial Sundial," *DSD* 11 (2004): 289-92.

For more general discussions of sundials, see S.L. Gibbs, *Greek and Roman Sundials* (YSHSM 11; New Haven: Yale University Press, 1976); R.R.J. Rohr, *Sundials: History, Theory, and Practice* (trans. G. Godin; New York: Dover Publications, 1996). For Vitruvius' first-century BCE book on sundials and clocks, see J. Soubiran (ed.), *Vitruve: De l'architecture, Livre 9* (Budé; Paris: Les Belles Lettres, 1969), 2-36. Cf. I.D. Rowland and T.N. Howe (eds.), *Vitruvius: Ten Books on Architecture* (Cambridge: Cambridge University Press, 1999), 107-18, 281-91; I.K. McEwen, *Vitruvius: Writing the Body of Architecture* (Cambridge, Massachusetts: MIT Press, 2003), 229-50.

prerequisite for the practical use of a zodiacal text like 4Q186."[185] From this lack of evidence it seems that *4QZodiacal Physiognomy* did not have a practical function in a horoscopic practice in the Qumran community.

In Albani's argumentation the issues of actual observation and computational competence figure prominently. However, I think it is important to point out that lack of evidence for either of them does not necessarily mean that a horoscopic practice could not have taken place. Rather than actually observing, most astrologers would have depended on computations. Neugebauer pointed out that ancient astronomy is overwhelmingly mathematical:

> Both Babylonian and Greek astronomy are based on a set of relatively few data, like period relations, orbital inclinations, nodes and apogees, etc. The selection of these data undoubtedly required a great number of observations and much experience to know what to look for. Nevertheless, a mathematical system constructed at the earliest possible stage of the game was generally no longer systematically tested under modified conditions.[186]

The results of these computations were at hand in almanacs and ephemerides.[187] No individual astrologer had to calculate for himself the positions of the luminaries, the planets, and the ascendant; quite the contrary. Unfortunately, no evidence of such records has been found at Qumran. But in this case Albani rightly remarks that this silence on the part of the sources does not mean that Jewish users of texts like *4QZodiacal Physiognomy* had no access to such resources from their Hellenistic environment.[188] So the possibility that the rising times of zodiacal signs were used in a horoscopic practice at Qumran cannot be ruled out.

Besides the possibility of almanacs and ephemerides with the zodiacal rising times, one need not even assume a real understanding of the matter. Reference has already been made to Hipparchus' comment that some of his contemporary astronomers did not fully grasp this complicated concept.[189] Moreover, many astrological texts from antiquity have been transmitted that are worthless and useless from an astronomical perspective.[190] Astrological

[185] Albani, "Horoscopes in the Qumran Scrolls," 311. Although there is no evidence for the observation or computation of the zodiacal rising times, it is perhaps possible that observation of the phases of the moon is indicated by 4Q317 (*4Qcrypt Phases of the Moon*), see Dubs, "4Q317 et le rôle de l'observation de la Pleine Lune."

[186] Neugebauer, *HAMA*, 14.

[187] Rochberg, *Babylonian Horoscopes*, 7-11; Rochberg, *The Heavenly Writing*, 105.

[188] Albani, "Horoscopes in the Qumran Scrolls," 295.

[189] Cf. n. 27 above.

[190] Regarding, for example, the issue of planetary visibility, Neugebauer, *HAMA*, 830-31, states that "most of the data found in the popular literature or in the astrological treatises have no theoretical background whatsoever and probably often enough not even an observational basis." In a more condescending tone, *HAMA*, 943: "Astrology is a dogmatic discipline, following a strict ritual in combining certain data without worrying how reliable these data were. This attitude is reflected in the fact that astrologers for centuries used arithmetical

handbooks such as the *Anthology* of Vettius Valens (second century CE) or the *Mathesis* of Firmicus Maternus have been transmitted for centuries. Although Vettius Valens refers explicitly to Hypsicles' *On the Ascendant*,[191] it is doubtful whether he really used the mathematical principles. He seems unaware of inconsistencies in the different doctrines that he had inherited from his predecessors and confuses data regarding the rising times for different *climata*. Firmicus Maternus exhibits the same unawareness and confusion when he lists the rising times of the zodiacal signs for the different *climata* in *Mathesis* 2.11 and assigns the same values to Babylon and Alexandria.[192] Despite these apparent astronomical shortcomings, these texts were, nonetheless, transmitted, so astronomical validity is perhaps not a good criterion for deciding on their usefulness.

With regard to *4QZodiacal Physiognomy*, this means that the text's astronomical background of zodiacal rising times need not have been understood or even be astronomically correct for people to have somehow made use of it. An illustrative example may be found in astrological practices in present-day India. According to traditional schools of astrologers, which have a limited membership, the ascendant is decided by divination.[193] A girl throws stones on a diagram to determine the ascendant. For the traditional schools it seems to be entirely beside the point whether in actuality this was people's birth ascendant. Of course, they believe that this is so. But they are rivaled by new schools of astrologers, open to everybody, that make use of modern resources like computers to determine the ascendant exactly in terms of modern astronomy and to point out the mistakes of the traditional way. Regardless of these modern developments, which have much to do with the democratization of astrology as a profession in the villages, this example vividly demonstrates how horoscopy could be practiced without a "real" anchor in astronomy.[194]

---

models, e.g. for planetary positions or for determining the length of daylight, which were long superseded by more accurate procedures. No astrologer cared about the reliability of the basic parameters of his planetary tables."

191 Vettius Valens, *Anthology* 3.13.8. For the text, see D. Pingree (ed.), *Vettius Valens: Anthologiae* (Teubner; Leipzig: B.G. Teubner, 1986), 149.8.

192 Cf. Neugebauer, *HAMA*, 719, 729, see also 823-24, 953-54. As to the practical astrological value, Tester, *A History of Western Astrology*, 142, emphasizes that "although Firmicus' work is long and very detailed, if confusing, nobody could actually have practiced astrology with only the *Mathesis* to hand. To use the book at all one would have needed to be expert, to sort out his muddles; or unprincipled or stupid enough to ignore them."

193 A fragmentary Greek magical papyrus gives a hint of astrological divination from antiquity, see *PGM* CX.1–12. Cf. H.D. Betz, *The Greek Magical Papyri in Translation, Including the Demotic Spells* (Chicago: University of Chicago Press, 1996), 312.

194 See G. Tarabout, "Pratiques astrologiques, double autorité textuelle (monde indien)," paper read at the CNRS Symposium *Divination et écriture, écriture de la divination*, Paris, December 10, 2005. The proceedings are to be published by J.-J. Glassner.

MODIFICATIONS OF AND FURTHER ADDITIONS TO ALBANI'S ASCENDANT INTERPRETATION

From the foregoing discussion it has become apparent that Albani's ascendant interpretation comes closest to coherently explaining the different elements in *4QZodiacal Physiognomy* against an astrological background and in accordance with notions from Greek astrology.

The words "in the foot of *Taurus*" (ברגל השור) refer to a certain part of the sign *Taurus*, and reflect the notion of a division of the zodiacal signs into different parts. The terminology "house of light" (בית האור) and "house of darkness" (בית החושך) has a spatial sense, referring to the hemispheres above and below the horizon. The realization of the numbers assigned to them is a result of the ascendant being in a part of the zodiacal sign, dividing it into parts above and below the earth.

Despite the fact that the ascendant interpretation is the most convincing possibility, it is evident, from the understanding of the textual structure of *4QZodiacal Physiognomy* and the relationship between physiognomics and astrology in Chapters One and Two, that I cannot agree with Albani's characterization of the text as an astrological one structured according to astrological criteria. *4QZodiacal Physiognomy* does not provide physiognomic descriptions on the basis of astrological possibilities. It is the other way around. The text gives astrological information on the basis of the physiognomic descriptions. The fact that astrological data have not governed the way in which the information in *4QZodiacal Physiognomy* has been processed and catalogued makes it unlikely that the text functioned in a horoscopic practice at Qumran. That is, it could not be used to look up the physiognomic possibilities for an individual after the astrologer knew the exact section of a rising zodiacal sign, as with the Greek *zodiologia*.

This different assessment of the character of *4QZodiacal Physiognomy*, however, does not weigh against the plausibility of the ascendant interpretation with regard to the sense of certain terminology in the text and the astrological concepts it reflects. On the contrary, the ascendant interpretation helps to understand the nature of the astrological matters that are signified by the human body according to *4QZodiacal Physiognomy*. But a few modifications and additions must be made.

"The Foot of Taurus*" in the "House of Light"*

The ascending part of the zodiacal sign is counted as still belonging to the "house of darkness," below the earth. As Albani says: "the 'feet of Taurus' are in the act of leaving the 'house/pit of darkness.'"[195] This interpretation

[195] Albani, "Horoscopes in the Qumran Scrolls," 307.

results in an exact match between the numbers in *4QZodiacal Physiognomy* and the division of *Taurus* in the *Rhetorius-Teucer text*. There the "feet" (πόδες) of *Taurus* are listed as the seventh element of a list that enumerates nine sections for *Taurus*.[196] If one assumes that the "foot" (רגל) in *4QZodiacal Physiognomy* is equivalent to the "feet" (πόδες) in Rhetorius-Teucer, the 6:3 division in *4QZodiacal Physiognomy* implies that the "foot" (רגל), being the seventh element, belongs in the "house of darkness."

However, from an astrological perspective such a classification does not make sense. Ptolemy, in the statement regarding the importance of those parts of the ascendant sign that are coming into the light, makes clear that the parts of the sign below the earth are to be ignored.[197] It makes more sense to regard the "foot of *Taurus*" (רגל השור) part as belonging to the area above the earth (ὑπὲρ γῆν) in Ptolemy's terminology and in the "house of light" (בית האור) in the terminology of *4QZodiacal Physiognomy*. It is almost a tautology to state that the ascendant, as that point of the ecliptic or part of the sign that is rising above the eastern horizon, belongs to the area above the earth, in the "house of light." This means that there is not an exact match between the data in *4QZodiacal Physiognomy* and Rhetorius-Teucer. Assuming the division in the latter text and counting the position "in the foot of *Taurus*" as belonging to the "house of light" results in a different division of light and darkness; seven in the "house of light" and two in the "house of darkness," numbers not given in 4Q186 1 ii 7–8. This suggests a different division of *Taurus* in *4QZodiacal Physiognomy*.

Albani suggests that the entire text of *4QZodiacal Physiognomy* could have contained one hundred and eight physiognomic types. According to his interpretation this means that it is perfectly possible that the complete text of *4QZodiacal Physiognomy* contained entries with a division of nine parts in the "house of darkness" and zero parts in the "house of light," but not vice versa.[198] This is problematic. If one part of a zodiacal sign must at least be visible on the eastern horizon for it to radiate its influence, this excludes the possibility of all zodiacal parts being assigned to the "house of darkness." The other way around, however, viz. the last part being the ascendant and therefore all parts of the sign being in the "house of light," makes sense from an astrological perspective. But this would then concern

[196] CCAG 7.197.26. See also n. 84 above.

[197] See n. 172 above.

[198] On the one hand, the first part of the ascending zodiacal sign, for example the head of *Taurus*, is regarded as belonging to the "house of darkness" because it is in the act of leaving the area below the horizon. On the other hand, it is not possible for all the parts of the sign to be above the earth in the "house of light", because the last ascending part must be regarded as being below the earth and leaving the "house of darkness." If all signs have nine parts this means that there are nine possible divisions between the "house of light" and the "house of darkness", resulting in Albani's hundred and eight physiognomic types.

types of people connected with entire zodiacal signs, which seems unlikely in a text dealing with divisions of the signs. So if the last parts were ignored – this is pure speculation – then there would be ninety-six physiognomic types in the text at an average of nine parts per sign; if not, then there would be one hundred and eight.

Melothesia *and* Dodecatemoria *in* 4QZodiacal Physiognomy *(4Q186)*

It is worthwhile to have a closer look at the sort of text the tenth section of the *Rhetorius-Teucer text* represents. This section concerns the partition of the zodiacal sign into several ecliptical parts of longitudinal degrees referred to as the sign's body parts. In the case of *Taurus*, however, Albani correctly states that the schematic distribution of the limbs of Taurus contradicts the representation of the zodiacal constellation, which is imagined as a halved animal.[199] In addition, the constellation *Taurus* rises backwards and not head first.[200]

This raises the question as to the exact sense of the division of the zodiacal signs in the tenth section of the *Rhetorius-Teucer text* where *Taurus* is a whole animal, rising head first. One possible explanation is this text, or its *Vorlage*, followed an older iconographic example, perhaps Egyptian, in which *Taurus* was represented as a whole animal.[201] But Neugebauer suggests another explanation.

In a short article Neugebauer discusses an astrological tradition he found in two Vatican codices in which the concept of *dodecatemoria* has been mixed with another astrological concept, namely that of *melothesia*.[202] He illustrates the general scheme of this particular astrological tradition with the sign of *Cancer*. Twelve places in this sign are associated with different parts of the body, and the numbers in degrees for each part show alternating differences of 2 and 3:

2°	head	17°	left claw
5	face	20	back

[199] Albani, "Horoscopes in the Qumran Scrolls," 303 n. 78. See nn. 69 and 151 above.

[200] Cf. n. 152 above. This is also acknowledged in the *Rhetorius-Teucer text*: "it rises from its hinder parts, setting straight, belonging for the most part in the invisible cosmos and looking south" (ἀνατέλλον ἐκ τῶν ὀπισθίων μερῶν, δῦνον ὀρθῶς, οὗ τὸ πλεῖον μέρος ἐν τῷ ἀφανεῖ κόσμῳ κεῖται, ἀποβλέπον εἰς τὸν νότον, CCAG 7.196.21–23).

[201] See Boll, *Sphaera*, 395 n. 3; Gundel and Böker, "Zodiakos," 560. In ancient iconography *Taurus* is primarily represented as a whole animal. For images, see Gundel, *Zodiakos*, 69-70, 161-63, 174, 205, 211, 231. Cf. also the fourth- to sixth-century CE synagogue zodiac mosaics from Beth 'Alpha, Ḥammath Tiberias, and Na'aran in R. Hachlili, *Ancient Jewish Art and Archaeology in the Land of Israel* (HdO: Siebente Abteilung: Kunst und Archäologie 1/2B4; Leiden: E.J. Brill, 1988), 304.

[202] O. Neugebauer, "Melothesia and Dodecatemoria," AnBib 12 (SBO 3, *OrAnt*; Rome: Pontificio Istituto Biblico, 1959), 270-75. I owe this reference to Professor Hübner. For *dodecatemoria* see n. 37 above, and for *melothesia* see Chapter Two n. 218.

7	mouth	22	abdomen
10	shoulder blades	25	hands
12	throat	27	feet
15	right claw	30	haunches[203]

Neugebauer suggests that this is the result of leaving out in every second case the fraction 1/2 in an original sequence of 2 1/2, 5, 7 1/2, 10, 12 1/2, and so on up to 30. This original sequence represents the concept of *dodecatemoria* dividing each sign of the zodiac into twelve equal parts of 2;30°.

As for the parts of the body that are associated with each twelfth section, Neugebauer argues that this connection forms "a curious mixture of the well-known zodiacal 'melothesia' and specific features of the constellation in question," representing however "purely astrological speculation without any contact with astronomical reality."[204] In the twelfth divisions of the zodiacal signs, an attempt is made to integrate certain features of the zodiacal constellations, thereby modifying the partition somewhat. For example, the two so-called bi-corporeal signs, *Gemini* and *Pisces*, are divided into two sequences each. Also, constellational parts like *Heart* of *Leo*, *Ear of Corn* of *Virgo*, and *Bow* of *Sagittarius* are given longitudinal degrees that do not correspond astronomically with the constellation. The *Heart* of *Leo* is at 10° longitude, whereas according to Ptolemy's star it (α Leo) has a longitude of 2;30°.[205] The list also assigns *Virgo*'s *Ear of Corn* to 10° longitude, while Ptolemy gives a longitude of 26;40° for *Spica* (α Vir).[206] In the case of *Sagittarius* the list assigns the *Bow* to 15° longitude, but in the *Almagest* this part of the constellation has a longitude between 6;40° and 9° (δ, ε, λ, and μ Sgr).[207]

Neugebauer suggests that this peculiar astrological tradition that combines *dodecatemoria* and *melothesia* forms the ultimate source of the tenth section in the *Rhetorius-Teucer text* on the division of the zodiacal signs. He takes the division of *Taurus* as an example and concludes:

> Obviously the compiler of this list (Rhetorius? or Teucros?) no longer understood that he was dealing with rounded dodecatemoria. Thus he changed the numbers, adopting the scheme 'from *a* to *b*, from *b* + 1 to *c*, from *c* + 1 to *d*' etc., thus reducing their number to nine. Furthermore, he

203 Neugebauer, "Melothesia and Dodecatemoria," 270. This scheme is only correctly preserved for the signs *Cancer*, *Capricorn*, and *Aquarius*. In other instances scribal errors have altered this underlying format. Cf. the tables on 271, 273.

204 Neugebauer, "Melothesia and Dodecatemoria," 272. Cf. Boll, *Sphaera*, 395: "eine von aller astronomischen Richtigkeit gänzlich entfernte Phantasieschöpfung, in der die Sternbilder des Tierkreises in höchst willkürlicher Weise nach dem Prokrustesbett der je 30 Grade umfassenden Zeichen verzerrt werden."

205 Cf. Neugebauer, "Melothesia and Dodecatemoria," 271; Ptolemy, *Almagest* 7.5[.26].

206 Cf. Neugebauer, "Melothesia and Dodecatemoria," 271; Ptolemy, *Almagest* 7.5[.27].

207 Cf. Neugebauer, "Melothesia and Dodecatemoria," 273; Ptolemy, *Almagest* 8.1[.29].

> did not understand the purpose of this list and simply took it as a list of consecutively rising parts.[208]

The original purpose of the text would presumably have been to inform which (rounded) dodecatemorial sections were deemed to hold authority over each part of the human body. In the *Rhetorius-Teucer text* this was apparently no longer understood. The described body parts were transferred from referring to the human body to the members of the zodiacal signs. The list was understood to give an account of the consecutively rising limbs of these signs. This also explains why *Taurus* is imagined as a whole animal, rising head first.

The astrological tradition that combines *melothesia* and *dodecatemoria* suggests that the difference noted above between the division of *Taurus* in Rhetorius-Teucer and in *4QZodiacal Physiognomy* may be due to different trajectories of transmission. Although it is evident that there are differences between textual traditions with regard to the exact division, the basis is the same.

There is another example of this astrological tradition. In *Mathesis* 8.4.1–13, Firmicus Maternus provides a division of the twelve zodiacal signs. He also regards the different (longitudinal) sections as the body parts of the signs:

> Now, I will endeavor to teach you which degrees you must seek in which parts of the zodiacal signs. For all 30° are divided across all the bodies of the zodiacal signs.[209]

Neugebauer apparently did not know of this passage, but it fits the same pattern as that in the Vatican codices and in the *Rhetorius-Teucer text*.[210] It is evident that the system of rounded *dodecatemoria* originally must have been the basis for the partition. The exact scheme has not been preserved in any of the divisions, probably because of scribal errors due to the many numbers used. In the case of *Sagittarius*, for example, the alternating differences of 2 and 3 are maintained until the eleventh part of 26°-27°, resulting in a total of thirteen parts.[211] In the cases of *Aries*, *Gemini*, *Libra*, *Capricorn*, and probably *Aquarius* the original number of twelve parts has been preserved, but the alternating scheme of 2 and 3 has been disturbed.[212] Furthermore, as in the Vatican codices presented by Neugebauer, Firmicus Maternus divides the bi-corporeal signs *Gemini* and *Pisces* into two se-

[208] Neugebauer, "Melothesia and Dodecatemoria," 274.

[209] Firmicus Maternus, *Mathesis* 8.4.1.

[210] Although Albani, "Horoscopes in the Qumran Scrolls," 327, mentions this passage in Firmicus Maternus, he does not elaborate the relationship with Rhetorius-Teucer in the way set out here. Cf. also Hübner, "Δωδεκατημόριον," 198 n. 54.

[211] Firmicus Maternus, *Mathesis* 8.4.9.

[212] Firmicus Maternus, *Mathesis* 8.4.1, 3, 7, 10, 11.

quences.[213] But this is not the case in Rhetorius-Teucer,[214] which demonstrates that there were different traditions in which the concepts of *dodecatemoria* and *melothesia* were mixed. The following example for *Taurus* from Firmicus Maternus is illustrative:

> [Division of *Taurus*.] 1° and 2° are in the horns of *Taurus*, 3°, 4°, and 5° in the whole face, 6° and 7° in the back part of the neck, 8°, 9°, and 10° in the forehead, 11° and 12° in the heart, 13°, 14°, and 15° in the shoulders, 16° and 17° are assigned to the forefeet of *Taurus*, 18°, 19°, and 20° in the belly, 21° in the knees, 22°, 23°, 24°, and 25° in the hinder feet, 26° and 27° are in the genital parts of *Taurus*. In the haunch are 28° and 29°, in the tail is found 30°. Such is the number of parts divided across the whole body of *Taurus*.[215]

It is evident that the division of *Taurus* in Firmicus Maternus is not the same as in the *Rhetorius-Teucer text*.[216] Common to both, however, is the image of *Taurus* as a whole animal. In both texts it was no longer understood that the dodecatemorial sections held influence over various parts of the human body (*melothesia*). The enumerated body parts were transferred from referring to the human body to referring to the imagined body of the zodiacal signs. It is evident that in Firmicus Maternus the two concepts of *dodecatemoria* and *melothesia* have been merged, just as in Rhetorius-Teucer and the two Vatican codices.

I suggest that *4QZodiacal Physiognomy* belongs to a similar astrological tradition in which the concepts of *dodecatemoria* and *melothesia* were merged together. According to the division of *Taurus* in this text, "the foot of *Taurus*" (רגל השור) in 4Q186 1 ii was the sixth section of nine that ascended above the horizon, into the "house of light." The words "foot of *Taurus*" (רגל השור) seemingly indicate one of the limbs of the zodiacal sign. Whereas the dodecatemorial part behind it originally controlled both or one of the feet of the human body, it is now understood as that body part of the zodiacal sign and influencing the shape and appearance of the entire human body. If this interpretation is correct, *4QZodiacal Physiognomy* provides important, although very implicit, evidence for the antiquity of this astrological tradition.[217] This background makes clear that not all signs were divided into nine sections and also that there were different divisions for the same sign so that there was no fixed set of divisions for the zodiacal signs in ancient astrology.

213 Firmicus Maternus, *Mathesis* 8.4.3, 12–13.

214 See CCAG 7.199.8–12; 212.3–5.

215 Firmicus Maternus, *Mathesis* 8.4.2.

216 But note that Rhetorius-Teucer locates the feet in 22°-24°, while Firmicus Maternus gives 22°-25° as the position.

217 The manuscript is dated to the end of the first century BCE and the beginning of the first century CE. For Teucer's date, see n. 77 above.

The different texts discussed are rooted in the same astrological tradition that merged *melothesia* and *dodecatemoria*. It is clear that there were variants according to the way in which the zodiacal sign could be divided into its imagined body parts, but because the notion is so specific it seems probable that they all go back to one source. It is difficult to trace the origin of this tradition. Although the concept of *dodecatemoria* is Babylonian, the specific idea of *melothesia* lacks any decisive evidence for a Babylonian origin.[218] One should allow for the possibility that it originated in Egypt with Teucer as an element of Greek astrology.[219]

The Human Body Signifying Greek Astrology in 4QZodiacal Physiognomy *(4Q186)*

An important corollary of the ascendant interpretation is that it points decidedly to a Greek background as against a Babylonian one for the astrological tradition in *4QZodiacal Physiognomy*.

The genre of *4QZodiacal Physiognomy* is clearly not that of a horoscope or a collection of horoscopes, but its interest is horoscopic. The physiognomic descriptions of the human body invite the reader to learn about someone's zodiacal sign, its ascendant position, and its division. If the words "in the foot of *Taurus*" (ברגל השור) refer to that part of the sign ascending above the eastern horizon at the moment of birth, then the interest of the text is clearly horoscopic. The main difference between Babylonian and Greek horoscopy is that the former was not at all concerned with determining the ascendant at the time of birth, whereas this is the latter's primary concern. If the ascendant interpretation is correct, this means that *4QZodiacal Physiognomy* shows evidence of the Hellenistic concern with determining the ascendant.

THE ASTROLOGICAL BACKGROUND OF *4QZODIACAL PHYSIOGNOMY* (4Q186)

The astrological background of *4QZodiacal Physiognomy* consists of two elements. First, that the position of the ascendant determines the number of parts of the zodiacal signs above and below the horizon, understood in the text in terms of "house of light" and "house of darkness." Second, the divi-

[218] The passage in Manilius seems the oldest attestation (see Chapter Two n. 219). There is some Babylonian evidence that certain parts of the human body were correlated with the planets (the spleen with Jupiter, and the kidney with Mars), but it is not clear if this is directly related to the Hellenistic concept of *melothesia*. E. Reiner, "Two Babylonian Precursors of Astrology," *NABU* (1993): 21-22; Koch-Westenholz, *Mesopotamian Astrology*, 178; Reiner, *Astral Magic*, 59-60.

[219] On Teucer's Egyptian background, see n. 77 above.

sion of the zodiacal signs is according to their imagined bodies ("in the foot of *Taurus*"), which combines the notions of *melothesia* and *dodecatemoria*.

Ancient readers of *4QZodiacal Physiognomy* need not have been aware, of course, that the astrological background consisted of these elements. It has been shown that those transmitting this tradition did not understand the merging of *melothesia* and *dodecatemoria* anymore, because the rising ecliptical parts were taken to be the sign's body parts. It has also been shown that ancient astrologers did not always properly understand certain astronomical and astrological notions, such as some specific calculations and concepts.

The original, complete text of *4QZodiacal Physiognomy* was an elaborate physiognomic catalogue (see Chapter One on columns and measurements) that listed separate entries for every division of the twelve zodiacal signs in order to determine someone's ascendant sign and its division between the "house of light" and the "house of darkness." The ancient users need not have been aware that this division was a result of the zodiacal position. Or even if they did know, they need not have known anything about the complicated methods for calculating zodiacal rising times. They were content to know the division of the zodiacal sign between light and darkness. That was enough. How this came about was probably of no interest to them because the text listed what they wanted to know. This astrological information could be attained through physiognomic observation, so it is irrelevant with regard to the use and understanding of the text of *4QZodiacal Physiognomy* how one could also arrive at this astrological information from an astronomical perspective, either observational or mathematical.

CHAPTER IV

"THERE IS A SPIRIT FOR HIM…": HUMAN OR ZODIACAL SPIRITS IN *4QZODIACAL PHYSIOGNOMY* (4Q186)

INTRODUCTION

In the previous chapter I concluded that the goal of physiognomic inquiry according to the list in *4QZodiacal Physiognomy* is to discern a person's ascendant zodiacal sign and its division between the "house of light" and the "house of darkness." However, an important element has not been considered yet. The text explicitly connects the numbers in the "house of light" and the "house of darkness" with the word רוח (in the construction רוח לו), for example:

> There is a spirit for him in the house of light (of) six (parts), and three (parts) in the house of darkness.[1]

It may be, as was argued in Chapter Three, that this sentence refers to the ascendant sign that is divided between the area above and below the horizon, but the text speaks of רוח ("spirit"). What is the meaning of the word and the phrase in this text and its context? Is it possible to understand its sense within the framework of the modified ascendant interpretation?

Scholars are divided over the reading and meaning of this word. Contrary to the general understanding, Gordis proposed to read רֶוַח לו ("it has a space…"). Bergmeier and Albani accepted this reading and understand the zodiacal sign as the object of reference.[2] However, the suffix is used throughout *4QZodiacal Physiognomy*, as far as the extant text is concerned, for referring to the described individual.[3] Furthermore, the entry in 4Q186 1 ii does not show evidence of a change of subject between the physiognomic part and the one dealing with רוחלו. What does this mean for this interpretation of רוחלו?

[1] 4Q186 1 ii 7–8. Cf. also 1 iii 8–9.

[2] 4Q186 1 ii 7–8: "It (i.e. the zodiacal sign) has a space in the house of light of six [parts], and three in the house of darkness." Bergmeier and Albani preferred this reading because it allows for an astrological interpretation of the division of light and darkness in *4QZodiacal Physiognomy* without necessarily invoking the theological anthropological background of the *Two Spirits Treatise*.

[3] See Chapter One n. 84 and Appendix I n. 35. This seems also to be the case in *4QPhysiognomy ar*, see Chapter One n. 192.

Most scholars, however, read רוח as רוּחַ ("spirit") and assume that רוח לו ("his spirit") refers to the human spirit, i.e. the spirit of the described person. This interpretation understands *4QZodiacal Physiognomy* as a text that visualizes the human spirit as divided between light and darkness, for example: "His spirit has six (parts) in the house of light and three in the house of darkness" (4Q186 1 ii 7–8). This division of the human spirit between light and darkness is taken as a dualistic feature of the text and it has, therefore, been related to the so-called *Two Spirits Treatise* in the *Rule of the Community* (1QS 3:13–4:26). In order to understand this connection, it is necessary to take a closer look at the *Two Spirits Treatise*. Does it attest a division of the human spirit between light and darkness, and, if so, in what way? The text of *4QZodiacal Physiognomy* must first of all be understood in its own right before it is related to other Qumran texts. Therefore, how does the understanding of רוח as the human spirit relate to other elements in the text, such as the physiognomic descriptions and the astrological information concerning the zodiacal sign? Moreover, can this traditional interpretation account for the realization of the numbers divided between the "house of light" and the "house of darkness?" What explanation is given for the terminology used, is it adequate, and if not, what other sense can רוח have in *4QZodiacal Physiognomy*?

In addition to a discussion of these previous interpretations, in this chapter I propose a new interpretation of the words רוח לו ("there is a spirit for him"); one that makes sense within the astrological framework of the modified ascendant interpretation.[4] If the allocation of numbers between the "house of light" and the "house of darkness" is astrologically the result of the ascendant zodiacal sign that is divided between the areas above and below the horizon, what then is the meaning of רוח ("spirit")? Below I shall argue that רוח is used to refer to spirits that are related to the zodiacal signs; each of the twelve signs has a spirit.[5] These zodiacal spirits have a close relationship with human beings from the moment of their birth. The text, therefore, introduces them with רוחלו ("there is a spirit for him"). This in-

[4] On the basis of the modified ascendant interpretation proposed in Chapter Three, an astrological explanation for the words רוח לו that retains the sense of "spirit" is possible without necessarily being dualistic.

[5] This notion not only has ancient roots, it also appears in much later times. The Venetian scholar Vincenzo Maria Coronelli (1650–1718) included in his cosmographic conception "relations between planets, zodiacal spirits, and the metals," thus D. Cosgrove, "Global Illumination and Enlightenment in the Geographies of Vincenzo Coronelli and Athanasius Kirchner," in *Geography and Enlightenment* (eds. D.N. Livingstone and C.W.J. Withers; Chicago: University of Chicago Press, 1999), 33-66, at 39. This notion of relations between different cosmic elements is also a suggestive analogy for the relations made between the human body, zodiacal spirits, ascendant zodiacal signs, and stones in *4QZodiacal Physiognomy*. See also D. Cosgrove, *Apollo's Eye: A Cartographic Genealogy of the Earth in the Western Imagination* (Baltimore: Johns Hopkins University, 2001), 171.

terpretation has the advantage of explaining the different elements in the entries of the text of *4QZodiacal Physiognomy* in a comprehensive manner.

SPACE OR SPIRIT IN THE "HOUSE OF LIGHT" AND THE "HOUSE OF DARKNESS"

If one assumes that *4QZodiacal Physiognomy* provides information regarding the division of people's ascendant zodiacal sign between the area above the horizon ("house of light") and the area below the horizon ("house of darkness"), the question is what can רוח mean in such a context.

Albani understands רֶוַח ("space") as a reference to the space occupied by the different parts of the zodiacal sign in the areas above and below the horizon.[6] The suffix in לו ("for him") then refers to the sign itself. In other words, it is the zodiacal sign that is the subject of רֶוַח לו: the sign has thus many parts of space in "the house of light" and thus many parts in "the house of darkness." This seems plausible in the context of Albani's ascendant interpretation, but in the sequence of the text it is unlikely. It is improbable that the suffix in לו refers to the zodiacal sign itself.

This understanding presupposes a change of subject in the entries that is unwarranted on the basis of the text. In the physiognomic descriptions, the suffixes evidently refer to the types of people whose bodies are described. In 4Q186 1 ii 6 the physiognomic section is followed by the phrase "and he is from the second column" (והואה מן העמוד השני). In this case the word והואה ("and he") continues the object of reference of the physiognomic description, viz. the type of person. A new subject has not been introduced in the text. Furthermore, subsequently to the division of light and darkness, the text continues with the described individual as the subject when it mentions his birth in 4Q186 1 ii 8: הואה ילוד ("he was born").[7]

As in the case of והואה ("and he") in 4Q186 1 ii 6 and הואה ילוד ("he was born") in 4Q186 1 ii 8, the subject of רוח לו in 4Q186 1 ii 7 is the individual type of human being with which this entry of the catalogue is concerned, not the zodiacal sign. A change of subject for רוח לו in between is not indicated by the text in any way and is, therefore, doubtful. This is supported by the other two occurrences of the phrase ורוח לו in 4Q186 1 iii 8 and 4Q186 2 i 6. These follow immediately after the physiognomic descriptions. Again, to understand the suffix here as having the same object of reference as in the descriptions of the human body seems most likely.

[6] Albani, "Horoscopes in the Qumran Scrolls," 285, 308-9.

[7] A change of subject occurs in between in 4Q186 1 ii 8 – וזה הואה ("And this is") –, but its object of reference is immediately explained by the word מולד ("horoscope"). As this is not the case with רוח לו ("There is a spirit for him"), there is no reason to assume a change of subject here.

If the suffix in לו ("for him") refers to the types of people described in the entries of *4QZodiacal Physiognomy*, it raises the question what the reading רֶוַח ("space, room, interval") contributes to the understanding of the text. Gordis merely comments that if the terms "house of light" and "house of darkness" refer to day and night, which were represented in some form as the shrine or dwelling of the described people, then the vocalization רֶוַח ("space, interval") seems plausible.[8] But he does not explain what he means by the representation in the form of a shrine or dwelling. Nor does he make clear what the sense of רֶוַח ("space") is here. It is not evident what improvement this interpretation and reading offer for the understanding of the text. The most plausible reading, therefore, remains רוּחַ ("spirit").

THE HUMAN SPIRIT IN *4QZODIACAL PHYSIOGNOMY* (4Q186) AND THE *TWO SPIRITS TREATISE* (1QS 3:13–4:26)

Allegro regarded *4QZodiacal Physiognomy* as an astrological text dealing with the influence of the stars on the human body and spirit, the latter in terms of a division between light and darkness. Because he thought it dealt with the human spirit, Allegro suggested that *4QZodiacal Physiognomy* must be read along with the *Two Spirits Treatise* in the sectarian *Rule of the Community* (1QS 3:13–4:26).[9]

Many scholars have accepted this understanding of the text. They assume that *4QZodiacal Physiognomy* determines the division of parts of light and darkness within the spirit of each human being. This partition in a person's spirit is expressed mathematically. The idea that people are torn between two principles is clarified by referring to the *Two Spirits Treatise*. *4QZodiacal Physiognomy* is seen as an element of the dualistic worldview of the Qumran Community. The proponents of this interpretation, however, do not explain how the specific numbers in the text might have been established. It is simply assumed that in *4QZodiacal Physiognomy* the human spirit is divided according to an apparent set of nine parts between light and darkness.[10]

[8] Gordis, "Document in Code," 38.

[9] Allegro, "Astrological Cryptic Document," 291.

[10] See Dupont-Sommer, "Deux documents horoscopiques," 243-46; Delcor, "Recherches sur un horoscope," 303, 310-11, 319; J.H. Charlesworth, "A Critical Comparison of the Dualism in 1QS 3:13-4:26 and the 'Dualism' Contained in the Gospel of John," in *John and the Dead Sea Scrolls* (ed. J.H. Charlesworth; New York: Crossroads, 1991), 76-106, at 86 (see also 80 n. 14, 83 n. 29) (originally published in *NTS* 15 [1968-69]: 389-418); P. von der Osten-Sacken, *Gott und Belial: Traditionsgeschichtliche Untersuchungen zum Dualismus in den Texten aus Qumran* (SUNT 6; Göttingen: Vandenhoeck & Ruprecht, 1969), 185-89; G. Maier, *Mensch und freier Wille: Nach den jüdischen Religionsparteien zwischen Ben Sira und Paulus* (WUNT 12; Tübingen: J.C.B. Mohr [Paul Siebeck], 1971), 252-53; M. Hengel,

Other scholars, however, have denied a close relationship between the two texts for different reasons. One reason is that the *Two Spirits Treatise* is understood in terms of an absolute dualism according to which light and darkness, symbolizing righteousness and wickedness, are irreconcilable principles in conflict with each other. People are divided between the two, i.e. individual persons belong to either one. A mixture of both elements within an individual would not be possible. This latter feature, however, is a characteristic of *4QZodiacal Physiognomy*, which allows a coexistence of light and darkness within individual types of people, being an example of relative dualism.[11]

Another reason put forward against a close connection between the two texts, one based on a diachronic reading of the *Two Spirits Treatise*, is that the light and darkness dualism of 1QS 3:13–4:14 is not the same as the opposition in 1QS 4:15–26 between the spirits of truth and iniquity. The spirits of truth and iniquity feud in man's heart, as a result of which each individual has a greater or lesser share in either one. But this latter concept is different from the notion that light and darkness mix in various proportions in people's spirits, allegedly demonstrated by *4QZodiacal Physiognomy*. The *Two Spirits Treatise* allows for a mixture of two elements in a human being, but only regarding the spirits of truth and iniquity, not with regard to light and darkness.[12]

Before turning to the issue of the division of the human spirit in *4QZodiacal Physiognomy*, it is necessary to deal briefly with the *Two Spir-*

"Qumrān und der Hellenismus," in *Qumrân: Sa piété, sa théologie et son milieu* (ed. M. Delcor; BETL 46; Paris-Gembloux: Duculot and Leuven: University Press, 1978), 333-72, at 367, 371; Lichtenberger, *Studien zum Menschenbild*, 147; H. Stegemann, "Zu Textbestand und Grundgedanken von 1QS III, 13-IV, 26," *RevQ* 13/49-52 (1988): 95-131, at 117-18, 126-27; A.E. Sekki, *The Meaning of* Ruaḥ *at Qumran* (SBLDS 110; Atlanta, Georgia: Scholars Press, 1989), 123; M.J. Davidson, *Angels at Qumran: A Comparative Study of 1 Enoch 1-36, 72-108 and Sectarian Writings from Qumran* (JSPSup 11; Sheffield: Sheffield Academic Press, 1992), 160; Lange, *Weisheit und Prädestination*, 159-60; Alexander, "Physiognonomy," 387, 390; Alexander, "Wrestling Against Wickedness," 331-33; J. Frey, "Different Patterns of Dualistic Thought in the Qumran Library," in *Legal Texts and Legal Issues*, 275-335, at 293-94; J.J. Collins, *Apocalypticism in the Dead Sea Scrolls* (London: Routledge, 1997), 47; R.V. Kvalvaag, "The Spirit in Human-Beings in Some Qumran Non-Biblical Texts," in *Qumran Between the Old and New Testaments* (eds. F.H. Cryer and T.L. Thompson; JSOTSup 290; Sheffield: Sheffield Academic Press, 1998), 159-80, at 169-71; Dimant, "Dualism at Qumran," 62-64; M. Broshi, "Predestination in the Bible and the Dead Sea Scrolls," in *Bread, Wine, Walls, and Scrolls* (M. Broshi; JSPSup 36; Sheffield; Sheffield Academic Press, 2001), 238-51, at 240-41; F. García Martínez, "Magic in the Dead Sea Scrolls," in *The Metamorphosis of Magic from Late Antiquity to the Early Modern Period* (eds. J.N. Bremmer and J.R. Veenstra; GSCC 1, Leuven: Peeters, 2002), 13-33, at 30-32; VanderKam and Flint, *The Meaning of the Dead Sea Scrolls*, 234.

[11] Licht, "Legs as Signs"; Gordis, "Document in Code from Qumran," 37; Schmidt, "Astrologie juive ancienne," 139.

[12] Bergmeier, *Glaube als Gabe*, 80-81; Albani, "Horoscopes in the Qumran Scrolls," 313-14.

its Treatise, how this text has been interpreted, and how *4QZodiacal Physiognomy* has been related to it.

The Two Spirits Treatise *in the* Rule of the Community

The *Two Spirits Treatise* has long been recognized as a separate composition that was incorporated into the sectarian *Rule of the Community* (*Serekh ha-Yaḥad*) at a certain stage of its development.[13] It is only completely preserved in the *Serekh* copy from Cave 1 (1QS 3:13–4:26), dated to 100–75 BCE. Other *Serekh* manuscripts show that textual developments occurred during the transmission of the *Two Spirits Treatise* within the *Serekh*, but they do not provide enough evidence to assess this growth in detail.[14]

13 The different manuscripts of the *Serekh ha-Yaḥad* from Caves 1, 4, and 5 are evidence of its composite character and attest to its growth and redaction. For a possible fragment from Cave 11, see E.J.C. Tigchelaar, "A Newly Identified 11QSerekh ha-Yaḥad Fragment (11Q29)?," in *Proceedings of the Jerusalem Congress*, 285-92.

Various proposals have been put forward for the literary development of the *Serekh*, but there is no agreement on the details of the reasons for and chronology of these developments. See e.g. J. Murphy-O'Connor, "La genèse littéraire de la *Règle de la Communauté*," *RB* 76 (1969): 528-49; J. Pouilly, *La Règle de la Communauté de Qumrân: Son évolution littéraire* (CahRB 17; Paris: Gabalda, 1976); Stegemann, "Zu Textbestand," 96-100; R.A.J. Gagnon, "How Did the Rule of the Community Obtain its Final Shape? A Review of Scholarly Research," *JSP* 10 (1992): 61-72; P.S. Alexander, "The Redaction-History of *Serekh ha-Yaḥad*: A Proposal," *RevQ* 17/65-68 (1996): 437-56; G. Vermes, "The Leadership of the Qumran Community: Sons of Zadok-Priests-Congregation," in *Festschrift für Martin Hengel*, 375-84, at 379-81; P. Garnet, "Cave 4 ms Parallels to 1QS 5.1-7: Towards a *Serek* Text History," *JSP* 15 (1997): 67-78; S. Metso, *The Textual Development of the Qumran Community Rule* (STDJ 21; Leiden: Brill, 1997); P.S. Alexander and G. Vermes (eds.), *Qumran Cave 4.XIX: Serekh ha-Yaḥad and Two Related Texts* (DJD 26; Oxford: Clarendon, 1998), 9-12; M. Blockmuehl, "Redaction and Ideology in the *Rule of the Community*," *RevQ* 18/72 (1998): 541-60.

For scribal marks and layout as evidence for the *Two Spirits Treatise* as a separate composition, see Lange, *Weisheit und Prädestination*, 165-68; Tov, *Scribal Practices*, 179-84, 206-8.

14 Two *Serekh* manuscripts, 4Q255 (4QpapS^{a}) and 4Q257 (4QpapS^{c}), preserve fragmentary remnants of the *Two Spirits Treatise*. Tigchelaar, "Names of the Spirits," 538-47, enlarged the text of 4QpapS^{c} through the reattribution of several fragments. He also identified another manuscript as a possible copy of the *Two Spirits Treatise* (1Q29a). The manuscripts 4QpapS^{c} and 1Q29a primarily contain text that overlaps with the part on the Two Ways in the *Two Spirits Treatise* (1QS 4:2–14). The differences between these manuscripts show that the exact wording of this list changed over time. In the case of the very fragmentary text 4QpapS^{a} A, there is a possible witness to an alternative version of the first part of the *Two Spirits Treatise*. There is no exact correspondence, but many words and phrases are highly suggestive. Reference is made to the "ways of man" (l.2: דׄרכי איש). Someone is said "to instruct" (l.3: להשכיל), and "the spirits of the sons of man" (l.4: רׄוחוׄתׄ בנׄי איש) are mentioned. Finally, the words "light" and "darkness" occur (l.5: אור, חושכ) in this context. Cf. Alexander and Vermes, DJD 26.31, 36-37; Metso, *Textual Development*, 18-21, 68, 90-91, 106, 113-14.

The *Two Spirits Treatise* is not extant in the remaining fragments of 4Q256 (4QS^{b}), although text corresponding to 1QS columns 1–2 and 5 has survived. It is not clear whether it was missing in the actual text of 4Q256 (4QS^{b}), thus Lange, *Weisheit und Prädestination*, 126, or that this is simply due to chance of survival, see Metso, *Textual Development*, 25; Alexander and Vermes, DJD 26.42. With regard to 4Q262 (4QS^{h}) Metso, *Textual Development*, 91-

While the final form of the *Two Spirits Treatise* in 1QS has structural unity and coherency of content,[15] explanations from a literary-critical perspective have been put forward for its redaction history. Peter von der Osten-Sacken took his lead from the heading in 1QS 3:13–15 and argued on terminological, syntactical, and thematic grounds that 1QS 4:15–23a is a secondary addition to 1QS 3:13–4:14, while the final part of 1QS 4:23b–26 represents a third phase.[16] Eibert Tigchelaar distinguished basically two hypothetical groups in the text of the *Two Spirits Treatise*: (1) 1QS 3:18–4:14, and (2) 1QS 3:13–18 and 4:15–26. A first redactional layer is represented by 1QS 3:18–4:1, to which the list of virtues and vices in 1QS 4:2–14 was added at some stage. In a second phase this first layer was reworked in a new framework, adding 1QS 4:15–23 together with the introduction (1QS 3:13–18) and résumé (1QS 4:23–26).[17]

A salient consequence of these literary-critical explanations for the redaction history of the *Two Spirits Treatise* is that the light and darkness terminology, which is deemed to be an outstanding feature of the text, is lacking completely in what are assumed to be later additions to the text, except for the heading "to teach all the sons of light" in 1QS 3:13 according to Tigchelaar's scheme.

Many scholars consider that the *Two Spirits Treatise* was a core theological document of the Qumran community, expressing the group's dualistic and deterministic views. Parallels can be found in other Qumran texts and it is believed to have influenced various other writings of the sect.[18] Others, however, doubt the central importance of the *Two Spirits Treatise* for the Qumran community.[19] The *Two Spirits Treatise* was most likely inserted as a later addition to the *Serekh*, but some scholars recognize it to be a pre-sectarian composition antedating the establishment of the Qumran

---

92, suggests that it is possibly part of the *Two Spirits Treatise*, but according to Alexander and Vermes, DJD 26.190, 194, the text is unparalleled and probably belongs to another scroll.

15 See J. Licht, "An Analysis of the Treatise on the Two Spirits in DSD," in *Aspects of the Dead Sea Scrolls* (eds. C. Rabin and Y. Yadin; ScrHier 4; Jerusalem: Magnes, 1958), 88-100; Lange, *Weisheit und Prädestination*, 130-32, 140-43, 165-68; J. Duhaime, "Les voies des deux esprits (*1QS* iv 2-14): Une analyse structurelle," *RevQ* 19/75 (2000): 349-67; J. Duhaime, "Cohérence structurelle et tensions internes dans l'Instruction sur les Deux Esprits (1QS III 13 – IV 26)," in *Wisdom and Apocalypticism in the Dead Sea Scrolls and in the Biblical Tradition* (ed. F. García Martínez; BETL 168; Leuven: Peeters, 2003), 103-31 (Duhaime argues both for coherency of the final text and a preceding redactional history).

16 Von der Osten-Sacken, *Gott und Belial*, 17-27, 116-89. Cf., however, Lange, *Weisheit und Prädestination*, 131.

17 Tigchelaar, *To Increase Learning*, 201-3.

18 Parallels to the *Two Spirit Treatise* can be found in 4Q525 (*4QBeatitudes*) 11–12 1–4 and CD 2:2–13 (*Damascus Document*), see É. Puech (ed.), *Qumrân Grotte 4.XVIII: Textes Hébreux (4Q521-4Q528, 4Q576-4Q579)* (DJD 25; Oxford: Clarendon, 1998), 141-42; Alexander and Vermes, DJD 26.3. For the relationship between the *Two Spirits Treatise*, the *Instruction* text, and the *Hodayot* text 1QH^a 5, see Tigchelaar, *To Increase Learning*, 194-207.

19 Cf. Stegemann, "Zu Textbestand," 125-30.

community, which is set around 150 BCE.[20] Whether the text is seen as representing earlier or later developments of dualistic thought, one should allow for the possibility that the origin of some of its views lies in Persian influence on Second Temple period Judaism.[21]

As a religio-historical phenomenon, dualism can be defined as a concept according to which two fundamentally opposed, causal principles underlie the existence of the world and its constitutive elements. This concept of dualism can be further refined according to a typology of forms in which it is expressed.[22] It is acknowledged that in the *Two Spirits Treatise* dualistic notions are expressed on different levels, such as cosmic, ethical, and psychological.[23] Furthermore, dualism as an idea of two opposing principles that constitute all existence does not appear in such a radical form in the *Two Spirits Treatise* because the two spirits and their ways are presented as subordinate to God who is the one determining everything.[24]

[20] Stegemann, "Zu Textbestand," 128; Lange, *Weisheit und Prädestination*, 127-30; Frey, "Patterns of Dualistic Thought," 295-300.

[21] See e.g. K.G. Kuhn, "Die Sektenschrift und die iranische Religion," *ZThK* 49 (1952): 296-316; A. Dupont-Sommer, "L'instruction sur les deux Esprits dans le 'Manuel de Discipline,'" *RHR* 142 (1952): 5-35; A. Dupont-Sommer, "Le Problème des influences étrangères sur la secte juive de Qoumrân," *RHPR* 35 (1955): 75-94; H. Michaud, "Un mythe zervanite dans un des manuscrits de Qumrân," *VT* 5 (1955): 137-47; E. Kamlah, *Die Form der katalogischen Paränese im Neuen Testament* (WUNT 7; Tübingen: J.C.B. Mohr [Paul Siebeck], 1964), 49-71, 163-68; S. Shaked, "Qumran and Iran: Further Considerations," *IOS* 2 (1972): 433-46; M. Philonenko, "La doctrine qoumrânienne des deux Esprits: Ses origines iraniennes et ses prolongements dans le judaïsme essénien et le christianisme antique," in *Apocalyptique iranienne et dualisme qoumrânien* (eds. G. Widengren, A. Hultgård and M. Philonenko; RI 2; Paris: Adrien Maisonneuve, 1995), 163-211; F. García Martínez, "Iranian Influences in Qumran?," in *Apocalyptic and Eschatological Heritage: The Middle East and Celtic Realms* (ed. M. McNamara; Dublin: Four Courts, 2003), 37-49. Cf. also e.g. O.J.F. Seitz, "Two Spirits in Man: An Essay in Biblical Exegesis," *NTS* 6 (1959): 82-95, who argues for an Old Testament background concerning the notion of two opposing spirits; P. Wernberg-Møller, "A Reconsideration of the Two Spirits in the Rule of the Community (1QSerek III,13-IV,26)," *RevQ* 3/11 (1961): 413-41, who understands the *Two Spirits Treatise* as a purely Jewish document, although he earlier admitted Persian influence, see P. Wernberg-Møller, *The Manual of Discipline: Translated and Annotated with an Introduction* (STDJ 1; Leiden: E.J. Brill, 1957), 70.

[22] See Charlesworth, "Critical Comparison," 76 n. 1; U. Bianchi and Y. Stoyanov, "Dualism," *ER* 4 (2005), 2504-17.

[23] Charlesworth, "Critical Comparison," 77-89; Frey, "Patterns of Dualistic Thought," 289-95.

[24] See Collins, *Apocalypticism in the Dead Sea Scrolls*, 43-44; García Martínez, "Iranian Influences," 44.

The Two Spirits: Angels, Demons, and Dispositions

The *Two Spirits Treatise* can be divided into five sections.[25] First, it begins with a grammatically well-structured heading that states the subject matter of the composition.[26] It entrusts the *Maskil*:[27]

> to instruct and to teach all the sons of light about the nature of all the sons of man, concerning all the types of their spirits with their signs, concerning their deeds in their generations, and concerning the visitation of their punishments as well as the times of their reward.[28]

If "nature" is the correct sense here for the Hebrew word תולדות,[29] then this heading makes clear that the text is not in the first place about the history or origin of mankind, but, in a more specific sense, about mankind's na-

[25] I follow the structural division presented by Lange, *Weisheit und Prädestination*, 141-43. Cf. Licht, "Analysis of the Treatise"; Duhaime, "Cohérence structurelle."

[26] Von der Osten-Sacken, *Gott und Belial*, 18-21, argues that the introduction only covers 1QS 3:17b–4:14, while Lange, *Weisheit und Prädestination*, 141-43, 148-50, assumes that it covers the entire composition.

[27] For the leadership figure of the *Maskil*, see e.g. Lange, *Weisheit und Prädestination*, 144-48; C. Hempel, "The Qumran Sapiential Texts and the Rule Books," in *The Wisdom Texts from Qumran and the Development of Sapiential Thought* (eds. C. Hempel, A. Lange and H. Lichtenberger; BETL 159; Leuven: Peeters/Leuven University Press, 2002), 277-95, at 286-94; C.A. Newsom, *The Self as Symbolic Space: Constructing Identity and Community at Qumran* (STDJ 52; Leiden: Brill, 2004), 165-74.

[28] 1QS 3:13–15.

[29] Most scholars translate תולדות with "genealogies," "generations," "origins," or "history," in accordance with its use in the Hebrew Bible. See e.g. W.H. Brownlee, *The Dead Sea Manual of Discipline: Translation and Notes* (BASORSup 10-12; New Haven: American Schools of Oriental Research, 1951), 12-13; Dupont-Sommer, "L'instruction sur les deux Esprits," 7, 12-13; Wernberg-Møller, *Manual of Discipline*, 25, 66-67; H.G. May, "Cosmological Reference in the Qumran Doctrine of the Two Spirits and in Old Testament Imagery," *JBL* 82 (1963): 1-14, at 2; Lange, *Weisheit und Prädestination*, 148-49; J.R. Levison, "The Angelic Spirit in Early Judaism," in *Society of Biblical Literature 1995 Seminar Papers* (ed. E.H. Lovering, Jr.; *SBLSP* 34; Atlanta, Georgia: Scholars Press, 1995), 464-93, at 483. However, Scholem, "Physiognomy," 477-79; Scholem, "Fragment," 180-81, argues that תולדות is used, analogous to Greek φύσις and Latin *natura*, as a *terminus technicus* in early medieval Jewish physiognomic texts when they quote Gen 5:1 (זה ספר תולדת אדם, "This is the book of the generations of man"). He suggests that this sense is also implied in the *Two Spirits Treatise*. Schäfer, "Metoposkopie und Chiromantik," 91 n. 31, remarks that it is somewhat speculative to suggest that תולדות represents a conscious effort to translate Greek terminology. That may be, but the suggested sense remains possible.

4Q418 77 2 (*4QInstruction*[d]) is too fragmentary to decide whether תולדות is a reference to the predestined history of mankind, thus Lange, *Weisheit und Prädestination*, 149 n. 115, or to the nature/characteristics of mankind, which cannot be ruled out, see Strugnell, Harrington and Elgvin, DJD 34.297-98. The occurrence of ותולדות in 4Q230 9 1 is suggestive but too fragmentary to be conclusive, see Tigchelaar, "Names of the Spirits," 535. For תולדות as "nature," see Wernberg-Møller, "Reconsideration of the Two Spirits," 419; Shaked, "Qumran and Iran," 434 n. 4; Schiffman, *Reclaiming the Dead Sea Scrolls*, 362; Alexander, "Physiognonomy," 390-91; J. Duhaime, "Dualism," in *Encyclopedia of the Dead Sea Scrolls*, 215-20, at 216; Tigchelaar, *To Increase Learning*, 196; Wold, *Women, Men and Angels*, 105-6; J. Vázquez Allegue, *La "Regla de la comunidad" de Qumrán* (BEBM 8; Salamanca: Sígueme, 2006), 79.

ture, concerning people's types of spirits, their deeds, and their punishments and rewards. Following the heading there is an introductory key passage expressing the deterministic worldview of the *Two Spirits Treatise*:

> From the God of knowledge comes all there is and there shall be. Before they existed he determined all their plans and when they come into existence at their ordained time they will fulfill all their work in accordance with his glorious plan and without alteration. In his hand are the laws of all things and he supports them in all their affairs. He created man to rule the world and placed before him two spirits to walk with them until the moment of his visitation.[30]

The God of knowledge is presented as the ontological basis of everything and everyone. The dualism in the text is moderate in form because the two spirits come from God and are not on the same ontological level.

The second section (1QS 3:18–4:1) first characterizes the two spirits ethically by calling them the spirits of truth and deceit. The dualistic opposition between the spirits is emphasized by light and darkness imagery:

> From the spring of light comes the nature of truth, and from the source of darkness comes the nature of deceit.[31]

Next, a cosmic perspective strengthens the dualism of the categories of truth and deceit. People are divided into two groups under two angelic leaders:

> And in the hand of the Prince of Lights is dominion over all the sons of justice who walk on paths of light. And in the hand of the Angel of Darkness is total dominion over the sons of deceit who walk on paths of darkness.[32]

Further, when the text explains why the sons of justice sin, it says that all the spirits from the lot of the Angel of Darkness cause the sons of light to fall (1QS 3:24). The division of mankind into two different groups is extended to the angelic world with groups of spirits belonging to light and darkness. The ethical and cosmic dualistic categories are interlocked with each other, which is reinforced by the ending of the second section:

> He created the spirits of light and darkness, and established on them every deed, [o]n their [path]s every labor.[33] God has loved one of them for all

30 1QS 3:15–18. For וישם לו ("he placed before him"), see Charlesworth, "Critical Comparison," 83-84.

31 1QS 3:19.

32 1QS 3:20–21.

33 With Brownlee, *Dead Sea Manual of Discipline*, 15 n. 41, I assume a dittography here. Cf., however, Stegemann, "Zu Textbestand," 101-3, for another suggestion that is based on his understanding that the theme of divine judgment is central to the entire treatise. But in the following sentence the deeds and paths of the two spirits are referred to again, suggesting that no other element figures prominently in the ending of the second section.

> eternal [a]ges and with all his deeds he is pleased forever; the other he has abhorred very much[34] and all his paths he has hated forever.[35]

The spirits of light and darkness created by God are identical with the spirits of truth and deceit placed before man, demonstrating the relative form of dualism in the *Two Spirits Treatise*.

The third section (1QS 4:2–14) is a list consisting of two parts in which the text enumerates the paths of the two spirits in the world with their characteristics: on the one hand, humility, compassion, goodness, and understanding; on the other hand, greed, wickedness, falsehood, and cruelty. If people follow the first path there will be everlasting rewards, but if they walk in the other there will be eternal punishments. This section translates the heavenly opposition of the second section into an earthly parallel because it lists the ethical realization in human conduct of the dualism between the two spirits as well as its eschatological consequences for mankind.[36]

In the fourth section (1QS 4:15–23) the eschatological perspective dominates. Humanity is divided into armies and, according to the path they walk and the deeds they do, individual people fall into either one of the divisions belonging to the two spirits. However, people's paths and deeds and consequently their belonging to one of the divisions are dependent on whether people's allotment is great or small. God established in equal measure the two spirits and their divisions until the final age and between them he put eternal enmity and violent conflict (1QS 4:15–18), but God also determined an appointed time for judgment when he will cleanse and purify the upright people whom he has chosen for an everlasting covenant and to them shall belong all the glory of Adam (1QS 4:18–23).

The fifth section (1QS 4:23–26) is a résumé that reiterates some notions from the previous sections, clarifying, or modifying, meanings that were implicit before:

> Until now the spirits of truth and injustice strive in the heart of man.[37]

The text has made clear before that humanity is divided into two groups according to the two spirits, but it did not present the inner person as the battleground of the two spirits.[38] This adds a psychological dimension to the notion of dualism in the *Two Spirits Treatise*. The result of this fight within people's hearts is that they either walk in wisdom or in folly. The implication seems to be that both spirits exist within human beings, but

[34] Cf. Stegemann, "Zu Textbestand," 104.
[35] 1QS 3:25–4:1.
[36] Cf. Lange, *Weisheit und Prädestination*, 157-58.
[37] 1QS 4:23.
[38] In 1QS 3:18 they are placed before man.

that people act according to either one depending on the outcome of their fight. The idea of allotment is taken up again to explain that some people will be righteous and hate injustice in accordance with their allotment in truth, while others will be wicked and hate the truth in accordance with their share in the lot of injustice. It is repeated that God has established the two spirits in equal measure, but now the text explicitly states the reason for the existence of the two spirits:

> He has given them as an allotment to the sons of man so that they know good [and evil, for Go]d casts the lot of all living beings according to his spirit in him [until the time of] the visitation.[39]

This statement seems to make clear that God determines people's belonging to either spirit according to their own spirit.

An important issue over which scholars are in disagreement is whether the two spirits should be interpreted primarily as cosmic, angelic spirits that influence people externally or as psychological dispositions within people.[40] It is not, however, necessary to interpret the two spirits in the entire text as references to either angelic beings or psychological dispositions. The *Two Spirits Treatise* conveys notions of dualism on different levels, including angelic, psychological, and ethical modes. In the text as we have it the psychological realm of the inner person is interconnected with the cosmic realm of supernatural angels and demons, being expressed in and recognizable by human conduct in terms of ethical dualism.[41]

Whether or not one assumes a literary growth in different phases of the *Two Spirits Treatise*, certain tensions remain in the final text. For example, on the one hand, the two spirits are placed before human beings who are rigorously assigned to the division of one of the two spirits, but, on the other hand, the spirits are presented as battling within people's hearts, suggesting that people partake of both spirits. Even before the publication of *4QZodiacal Physiognomy*, some scholars assumed that the *Two Spirits Treatise* demonstrates a notion according to which both of the two spirits exist within human beings in different proportions. The share of the two spirits in the world, i.e. presumably the numerical strength of their divisions, is equal (1QS 4:16.25), but each individual has a greater or smaller share in either one of the two spirits that are fighting within his heart (1QS 4:16.23).[42] A passage from Philo's commentary on Exodus was first ad-

[39] 1QS 4:26. Cf. Stegemann, "Zu Textbestand," 105-10.

[40] Cf. Wernberg-Møller, "Reconsideration of the Two Spirits"; Lichtenberger, *Studien zum Menschenbild*, 123-42; Sekki, *Meaning of* Ruaḥ *at Qumran*, 193-219; Levison, "Angelic Spirit," 480-86.

[41] Cf. Collins, *Apocalypticism in the Dead Sea Scrolls*, 40-41.

[42] Dupont-Sommer, "L'instruction sur les deux Esprits," 28-29; J. Daniélou, "Un source de la spiritualité chrétienne dans les manuscrits de la Mer Morte: la doctrine des deux es-

duced as a parallel for the idea that a mixture of two elements inhabits every human being.[43] And when *4QZodiacal Physiognomy* was published it was understood to confirm this understanding of the *Two Spirits Treatise*.

The Two Spirits and Light and Darkness in Human Beings

The fierce opposition between the divisions of the two spirits of light and darkness as well as the notion that human beings belong to either one or the other seems to conflict with the idea that both spirits fight their battle within people's hearts. However, the *Two Spirits Treatise* also mentions that the belonging of human beings to either group is dependant on their allotment being great or small (1QS 4:16), and that the sons of light can sin because their wrong conduct falls under the dominion of the Angel of Darkness and the spirits of his lot cause them to fall, which is in compliance with the mysteries of God (1QS 3:21–24).

Scholars have interpreted this to mean that people partake of both spirits, but that their shares differ and that the balance is tipped in favor of either one of the two spirits. This would result in the end in the allotment of people to either the division of the spirit of light or that of the spirit of darkness.[44] In this context *4QZodiacal Physiognomy* is invoked to demonstrate how the Qumran community understood this to have worked.

Hartmut Stegemann, for example, takes the statement that God established the two spirits in equal measure (1QS 4:16.25) to mean that there is quantitatively an equal amount of good and evil in the world. This balance, however, applies to the world, not to individual people. In the case of individuals, Stegemann assumes that it must be possible to determine everyone as either predominantly good or predominantly evil. The statement that each person has a greater or smaller share in either one of the two spirits suggests the solution to this matter. This implies that no one can be com-

---

prits," *DV* 25 (1953): 127-36, at 128; Michaud, "Un mythe zervanite," 146. For other explanations, Wernberg-Møller, *Manual of Discipline*, 84; Licht, "Analysis of the Treatise," 91 n. 13.

[43] Philo, *Questions and Answers on Exodus* 1.23. See Dupont-Sommer, "Le Problème des influences étrangères," 85-86; J. Daniélou, *Philon d'Alexandrie* (Paris: Arthème Fayard, 1958), 53-57; A. Wlosok, *Laktanz und die philosophische Gnosis: Untersuchungen zu Geschichte und Terminologie der gnostischen Erlösungsvorstellung* (AHAWPHK 2; Heidelberg: Carl Winter, 1960), 107-11; Kamlah, *Form der katalogischen Paränese*, 39-56; Philonenko, "La doctrine qoumrânienne des deux Esprits," 186-88; Hengel, *Judentum und Hellenismus*, 418-19.

[44] See Charlesworth, "Critical Comparison," 83; Von der Osten-Sacken, *Gott und Belial*, 24-25, 172-74; Maier, *Mensch und freier Wille*, 252-53; Hengel, "Qumrān und der Hellenismus," 371; Lichtenberger, *Studien zum Menschenbild*, 147-48; Stegemann, "Zu Textbestand," 117-20, 128-29; Schiffman, *Reclaiming the Dead Sea Scrolls*, 362-64; Lange, *Weisheit und Prädestination*, 159-60; A. Lange, "Wisdom and Predestination in the Dead Sea Scrolls," *DSD* 2 (1995): 340-54, at 347-48; Alexander, "Physiognonomy," 388; Kvalvaag, "Spirit in Human-Beings," 168-71.

pletely righteous or completely wicked. There is always a mixture of both spirits within individual people, but in such a way that they belong predominantly to either one or the other.[45]

This mixture of both spirits in human beings is best illustrated, according to Stegemann and others, by *4QZodiacal Physiognomy*. According to the traditional scholarly understanding of this text the human spirit is divided between light and darkness on the basis of a nine-point scale:

> Bei diesem System gewährleistet die *ungerade* Zahl 9, daß eine Parität nie zustandekommen kann, sondern immer eine der beiden Seiten das Übergewicht hat und somit stets die Zuordnung des einzelnen Menschen zu derjenigen Grundkategorie, die bei ihm überwiegt, möglich ist. In *1QS* III, 13-IV, 26 werden keine derartige Zahlen genannt; doch lösen sich alle entsprechenden Vorstellungsprobleme zufriedenstellend, wenn man auch hier ein rechnerisches Grundmodell entsprechend dem von *4Q186* voraussetzt mit irgendeiner ungeraden Zahl als 'Schlüssel'; diese sollte nicht zu niedrig angenommen werden, damit sich hinreichend viele verschiedene 'Menschenklasse' oder 'Menschentypen' ergeben, wie sie offenbar mit באותותם כול בני איש לכול מיני רוחותם III, 13f in den Blick genommen worden sind. Ungerade Zahlen in der Größenordnung von mindestens 9 dürften aber für derartige Klassifizierungszwecke völlig hinreichen.[46]

The purpose of *4QZodiacal Physiognomy* is understood to be the determination of the division of parts of light and darkness within the spirit of each human being. This partition would have been expressed arithmetically on a nine-point scale, which makes an equal division of light and darkness impossible. The uneven number nine is taken to imply the allocation of people to either predominantly the "house of light," and by extension the division of the spirit of light, or predominantly the "house of darkness," and the spirit of darkness, but this is not made explicit by the extant text of *4QZodiacal Physiognomy*.

The argument that the relative form of dualism attested in *4QZodiacal Physiognomy* militates against a close connection with the *Two Spirits Treatise* is questionable.[47] First of all because the *Two Spirits Treatise* demonstrates various levels of dualism in a moderate form as God created the two spirits. Second, because it also allows for a mixture of the two spirits within individual people. The *Two Spirits Treatise* expresses the idea that every person partakes of the two spirits, belonging to the realms of light and darkness, in varying proportions.

The argument against connecting the two texts that is based on a diachronic reading is not strong either.[48] Later additions may lack light and

[45] Stegemann, "Zu Textbestand," 117.
[46] Stegemann, "Zu Textbestand," 118. Cf. the references in n. 10 above.
[47] Cf. n. 11 above.
[48] See n. 12 above.

darkness terminology, but they were probably understood on the same level. It is evident that in the second section (1QS 3:18–4:1) the different dualistic categories of truth and deceit, justice and injustice, light and darkness are related to each other.[49] The fourth section (1QS 4:15–23), and by extension the fifth (1QS 4:23–26) too, continues the text and can, therefore, be assumed to know that these different categories are interrelated. This means that *4QZodiacal Physiognomy*, a manuscript dated to the turn of the era, could very well have been read in comparison with the entire *Two Spirits Treatise* in 1QS, a manuscript dated to ca. 100–75 BCE.

These two issues, therefore, cannot decide the possible connection between *4QZodiacal Physiognomy* and the *Two Spirits Treatise*. A more important, and decisive, matter, however, is whether the text of *4QZodiacal Physiognomy* is really concerned with the human spirit being divided between light and darkness.

THE HUMAN SPIRIT IN *4QZODIACAL PHYSIOGNOMY* (4Q186)

From the outset scholars have interpreted *4QZodiacal Physiognomy* within the context of the *Two Spirits Treatise*. This seems sensible given the light and darkness imagery in both texts as well as the prominence of the word רוח ("spirit"). It is, however, questionable whether this presumed context has done justice to the meaning of the different elements in the text of *4QZodiacal Physiognomy*, if the text is taken on its own merits first.

In the previous chapters I have been concerned with the interpretation of the physiognomic structure and sense of the text and the astrological framework that it presupposes. To understand these elements, other physiognomic and astrological literature has been adduced, but the interpretation of *4QZodiacal Physiognomy* was not placed beforehand in a Qumran context. The understanding of other writings from the Qumran community did not, and should not, predetermine the horizon of interpretation for *4QZodiacal Physiognomy*. The text should first be taken on its own merits. In a subsequent stage of interpretation it should be brought into context with other Qumran writings in order to inquire if and how it possibly relates to them, but a "mismatch" does not necessarily alter the understanding of *4QZodiacal Physiognomy*. One should allow for the possibility that, although the non-biblical Qumran writings show a large degree of coherence as a collection,[50] not all texts found there need fit within a unified frame-

[49] Cf. Charlesworth, "Critical Comparison," 78.

[50] See e.g. F. García Martínez, "Qumran Origins and Early History: A Groningen Hypothesis," *FO* 25 (1988): 113-36, at 116; F. García Martínez and A.S. van der Woude, "A 'Groningen' Hypothesis of Qumran Origins and Early History," *RevQ* 14/56 (1990): 521-41; Dimant, "Qumran Manuscripts."

work of reference and ideology. This does not imply a rejection of such texts by the community. It merely signals the possibility of a varied collection of writings and suggests that "scientific" writings such as *4QZodiacal Physiognomy*, *4QPhysiognomy ar*, 4Q318 (*4QZodiology and Brontology ar*), and also 4Q317 (*4Qcrypt Phases of the Moon*) should not necessarily be interpreted primarily against the religio-ideological background of the Qumran community.

Turning to the issue of whether the division of the human spirit between light and darkness is the subject matter of *4QZodiacal Physiognomy*, I intend to take the interpretation of Alexander, whose views on the matter can be taken as representative for the general understanding, as the point of departure for my discussion.[51]

Philip Alexander: Physiognomy, Human Spirit, Astrology, and Initiation

Alexander acknowledges the connection between physiognomics and astrology and the primacy of physiognomics in *4QZodiacal Physiognomy*. However, he argues that a third element is involved, viz., in line with the more general interpretation, the spiritual character of the described individuals. He suggests that the purpose of *4QZodiacal Physiognomy* is to provide a means of determining a subject's true nature: his spirit. These three elements are interrelated. Observing people's bodily form and appearance makes it possible to predict the nature of their spirits, which in turn points to their time of birth and sign of the zodiac. The goal of the physiognomic inquiry is to discern people's spiritual share in light and darkness.

Alexander argues that *4QZodiacal Physiognomy* is related to the *Two Spirits Treatise* because both texts share the same dualistic worldview and use more or less similar light and darkness terminology.

The point of *4QZodiacal Physiognomy* is understood to be the division of people's spirit on a nine-point scale between the "house of light" and the "house of darkness." Alexander rejects any physiognomic or astrological sense for these phrases. He assumes that the term "house of light" is set in clear contrast with the "pit of darkness" (בור החושך) in 4Q186 1 ii 7–8, and that both phrases have a strongly sectarian meaning.[52] Instead of attributing physiognomic or astrological meaning to this phraseology, Alexander re-

[51] See Alexander, "Incantations and Books of Magic," 364-65; Alexander, "Physiognonomy"; P.S. Alexander, "Predestination and Free Will in the Theology of the Dead Sea Scrolls," paper read at a seminar on *Divine and Human Agency in Paul and his Cultural Environment* at the University of Aberdeen, August 18-21, 2004 (http://www.abdn.ac.uk/divinity/Gathercole/paper-alexander.htm).

[52] Alexander, "Physiognonomy," 387-88, reads "in the pit of darkness" (בבור החושך) in 4Q186 1 ii 7–8. The double use, however, of בית ("house") is clearly attested and the reading in 4Q186 1 ii 7 should be בבית החושך ("in the house of darkness"), see Appendix I. Alexander, however, ignores the parallel in 4Q186 1 iii 8–9.

lates it to the light and darkness terminology in the *Two Spirits Treatise* because it recalls the distinctive theology of the Qumran sect that is expressed in that core document. The terminology used to express the dualism of light and darkness in both texts is synonymous. The *Two Spirits Treatise* states:

> From the spring of light comes the nature of truth, and from the source of darkness comes the nature of deceit.[53]

According to Alexander and others the "spring of light" (מעין אור) equals the "house of light" (בית האור) of *4QZodiacal Physiognomy* and the "source of darkness" (מקור חושך) equals the "pit of darkness" (בור החושך).[54]

On the basis of the context of the *Two Spirits Treatise*, Alexander interprets the light and darkness imagery in *4QZodiacal Physiognomy* in ethical terms. The point of the text is to measure the goodness and badness of people on a nine-point scale according to the degree in which their spirits participate in both the "house of light" and the "house of darkness."[55] The nine-point scale excludes the possibility that people might be in spiritual balance. Because of this inevitable imbalance Alexander assumes that people must be either predominantly good or predominantly bad. Those whose spiritual balance is predominantly in the "house of light" belong with the "sons of light" and those whose spiritual balance is predominantly in the "house of darkness" belong with the "sons of darkness."[56]

Alexander suggests that the function of *4QZodiacal Physiognomy* was to help the leadership figure of the *Maskil* in discovering who had been elected a "son of light," and, hence, was eligible to join the community, and who a "son of darkness." A physiognomic inquiry was a means of controlling the admission of candidate members into the community.[57] This may well have been done in secret by the leadership of the group, because the people observed need not have been aware that they were being physiognomized.[58]

Many scholars, however, have noted that the *Two Spirits Treatise* does not deal at all with physical characteristics by means of which people could

[53] 1QS 3:19.

[54] Alexander, "Physiognonomy," 390. Cf. Charlesworth, "Critical Comparison," 86.

[55] Alexander, "Physiognonomy," 387, excludes the possibility that the scale included zero and that people were either totally good or totally bad. Cf. Dupont-Sommer, "Deux documents horoscopiques," 245-46; Schmidt, "Astrologie juive ancienne," 138-39.

[56] Cf. n. 44 above.

[57] Alexander, "Physiognonomy," 390-93. Cf. Licht, "Legs as Signs," 25-26; Hengel, "Qumrān und der Hellenismus," 371; I. Gruenwald, *Apocalyptic and Merkavah Mysticism* (AGJU 14; Leiden: E.J. Brill, 1980), 79 n. 21, 218; E. Regev, "Comparing Sectarian Practice and Organization: The Qumran Sects in Light of the Regulations of the Shakers, Hutterites, Mennonites and Amish," *Numen* 51 (2004): 146-81, at 152.

[58] Polemo stressed the importance of physiognomizing without the knowledge of the subject that is being observed so that the latter cannot attempt to conceal his true character from the physiognomist. Cf. Gleason, *Making Men*, 40-41.

be recognized as belonging to either the realm of light or that of darkness. There are no evident references that imply a physiognomic test of the human spirit. The heading of the *Two Spirits Treatise* makes mention of "signs" (אותות) to understand the nature of all the sons of man (1QS 3:13–14), but these seem to consist of the different ways of human conduct as listed in the virtues and vices discussed in the third section (1QS 4:2–14). According to the fourth section of the *Two Spirits Treatise* (1QS 4:15–23), people are judged to belong to either the division of the spirit of light or that of the spirit of darkness according to the path they walk and the deeds they do. The text does not state that these are registered on and discernible from the human body.

The *Two Spirits Treatise* may not contain concrete references to physiognomics, but Alexander argues that its language is nonetheless highly suggestive. He deems it possible that the "signs" (אותות) mentioned in 1QS 3:14 may at one time have been extended to include physiognomic criteria. On an intertextual level Alexander considers *4QZodiacal Physiognomy* to be a secondary reading of the *Two Spirits Treatise* that attempts to apply the teaching of the latter in a concrete and practical way. Furthermore, he understands the phrase "the nature of all the sons of man" (כול בני איש תולדות) from the *Two Spirits Treatise* (1QS 3:13) to be a clear echo of Gen 5:1, "This is the book of the generations of man" (זה ספר תולדת אדם), and notes that this verse is used in medieval Jewish physiognomic literature.[59]

Alexander, therefore, suggests that a physiognomic test of admission was applied at Qumran. He reasons that the use of physiognomics as a divinatory art must be understood against the dualistic and eschatological background of the Qumran sect. Magic and divination were used in order to fight off the sons of darkness and to keep them out of the world of the elect ones of the sect.[60] However, Alexander also notes that a physiognomic test of people's spirit upon entry into the community conflicts with the reference to an annual examination of people's spirits and deeds in 1QS 5:23–24. The fixed division of light and darkness in people's spirits is predetermined astrologically in *4QZodiacal Physiognomy*, while 1QS 5:23–24 envisages the possibility of upgrade and downgrade mobility within the sectarian hierarchy on the basis of a person's spirit and deeds.[61] Alexander suggests that different views prevailed at different times at Qumran, or that 1QS 5:23–24 is the official position of the leadership towards the commu-

[59] Cf. n. 29 above.

[60] Alexander, "Wrestling Against Wickedness." See also García Martínez, "Magic in the Dead Sea Scrolls."

[61] Cf. Sekki, *Meaning of* Ruaḥ *at Qumran*, 98-99.

nity, while the *Maskil* used *4QZodiacal Physiognomy* secretly; a possibility suggested by the way that the text is written.[62]

Some Problems with Alexander's Interpretation

Prima facie, Alexander provides a coherent framework for understanding the sense of *4QZodiacal Physiognomy*, but upon closer scrutiny his interpretation of the text is problematic on several accounts.

First, Alexander does not really explain the terms "house of light" and "house of darkness," or "pit of darkness" for that matter. Like other scholars, he simply asserts its sectarian meaning and assumes that it signals the dualistic nature of *4QZodiacal Physiognomy*.[63] This does not throw any light on the use of the word "house" (בית), the combination of which with "light" (אור) and "darkness" (חושך) occurs in no other text from Qumran and cannot just be taken as another example of Qumran dualism.[64]

Without the context of the *Two Spirits Treatise* there seems to be no reason to interpret the light and darkness terminology of *4QZodiacal Physiognomy* as dualistic. There are no clues in the extant text that the "house of light" and the "house of darkness" represent two fundamentally opposed, causal principles of reality and its constitutive elements. The light and darkness imagery cannot, therefore, simply be taken as an element of dualism in the text. The light and darkness phraseology of *4QZodiacal Physiognomy*, however, can be explained satisfactorily without adducing the *Two Spirits Treatise*. Following Albani, I argued in Chapter Three that the "house of light" and the "house of darkness" should be understood as references to the areas above and below the horizon between which the ascendant zodiacal sign is divided. Such a framework, however, does not in itself imply dualism.

Second, like most scholars who understand *4QZodiacal Physiognomy* against the background of the *Two Spirits Treatise*, Alexander provides no clarification for the realization of the numbers in the "house of light" and the "house of darkness." The only reflection on the number nine is that an uneven number excludes an even balance between light and darkness, thus causing someone to fall into either the domain of light or the domain of darkness. Stegemann, in addition, comments that the number nine is high enough to allow for the classification of various types of people alluded to in 1QS 3:13–14,[65] but this is problematic (see below). There is, however,

[62] Alexander, "Physiognonomy," 390-93.
[63] See e.g. Dimant, "Dualism at Qumran," 62-64.
[64] Cf. Bergmeier, *Glaube als Gabe*, 80; Albani, "Horoscopes in the Qumran Scrolls," 314.
[65] Cf. n. 46 above.

no further reflection on the question how the combination of these numbers is determined.

Third, Alexander does not elucidate the zodiacal information that 4Q186 1 ii 8–9 gives by saying that the individual is born "in the foot of *Taurus*" (ברגל השור). He assumes this means that the person was born when the sun was in *Taurus*, but it is not clear whether, and if so how, he relates this to the numbers allotted to the "house of light" and the "house of darkness."

The more general scholarly understanding of *4QZodiacal Physiognomy*, therefore, fails to account satisfactorily for the astrological framework of the text, the realization of the numbers in the "house of light" and the "house of darkness," or the zodiacal position "in the foot of *Taurus*" in 4Q186 1 ii 9, especially in relation to the other two elements.

However, the final point is decisive against the traditional interpretation that *4QZodiacal Physiognomy* is concerned with the division of the human spirit between light and darkness. And if this text does not deal with the human spirit there is no need to relate it to the *Two Spirits Treatise* as has been done.

Physiognomy, Spirit, Astrology, and Arithmetic

According to the more general understanding, *4QZodiacal Physiognomy* demonstrates that the human body and spirit are believed to be astrologically determined. Alexander argues that the physiognomic, spiritual, and astrological elements in the text are closely linked. However, on closer examination the understanding of the relationship between the latter two elements, i.e. human spirit and zodiacal birth sign, is not sustainable.

Many scholars maintain the idea that according to *4QZodiacal Physiognomy* the human spirit is somehow thought of as consisting of nine parts that can be divided between two entities, viz. the "house of light" and the "house of darkness." This notion, however, conflicts fundamentally with the idea that people's zodiacal birth sign determines each division of the human spirit astrologically. For example, according to Alexander:

> the clear link between physiognomy and astrology in the text makes it very likely that the complete text of 4Q186 differentiated only twelve human types – one for each sign of the zodiac.[66]

This implies that every possible division of the human spirit between light and darkness is linked to a different sign of the zodiac. However, if one follows this line of reasoning it is completely impossible for each zodiacal sign in the original text of *4QZodiacal Physiognomy* to be linked to a different type of human being in terms of his spirit (and body).

[66] Alexander, "Physiognonomy," 389.

Considering how many different combinations between parts of light and darkness can be made for a human spirit that is thought of as having nine parts, there are only eight such arrangements possible.[67]

"house of light" (ביתהאור)	"house of darkness" (ביתהחושך)
8	1
7	2
6	3
5	4
4	5
3	6
2	7
1	8

This is obviously not enough in relation to the number of twelve zodiacal signs. It is also questionable whether the number of eight possible divisions suffices in light of the many different types of people implied by 1QS 3:13–14, thus Stegemann, but as no numbers are mentioned in this text this remains unknown.

If one assumes that the zodiacal sign under which people are born determines in a distinguishing manner the physiognomy of people and the nature of their spirits, then there is a discrepancy between the numbers used in *4QZodiacal Physiognomy*. This discrepancy has not been noted before, but it poses a serious problem for the traditional interpretation that *4QZodiacal Physiognomy* envisages the division of the human spirit. If, on a scale of nine, the numbers mentioned in the text refer to the human spirit, only eight different types are possible, which is not in accordance with twelve zodiacal signs. A variable set of numbers makes no sense within this interpretation, because then the different human spirits could not be compared to each other. A recurrence of the same division of light and darkness does not make much sense either because the astrological influence of the zodiacal signs must be distinguishing in order to tell one type of person from another. In theory a different zodiacal sign implies a different physiognomy as well as a different configuration of the spirit in terms of light and darkness.

Not the Human Spirit in 4QZodiacal Physiognomy *(4Q186)*

If it is presumed that the human physiognomy, the human spirit, and the zodiacal birth sign are directly linked with each other, this means that something does not add up here. The physiognomic element is straightfor-

[67] It is irrelevant here to include the possibility that the scale included zero because the number of possible combinations would still be just ten. This is the case with Schmidt's proposal, which in the end has just ten different divisions of diurnal and nocturnal decans that are variable, and, hence, ten types of people. See Chapter Three n. 148.

ward and beyond suspicion. The remaining fragments of *4QZodiacal Physiognomy* make clear that different types of human bodies were described in the complete text. The astrological element seems less straightforward. Only in one entry is a clear reference to a zodiacal sign and a part of it preserved (4Q186 1 ii). However, comparison with other ancient texts that combine human physiognomies with zodiacal signs, most notably the Greek *zodiologia*, suggests that different types of bodies imply different signs of the zodiac.[68] The astrological element is, therefore, also a clear feature. This would set the minimum number of entries at twelve in the complete text of *4QZodiacal Physiognomy*, unless one suggests that not all signs were listed, which only begs the question as to the connection between the signs, the numbers, and the human spirit.[69] These considerations, therefore, suggest that the third element, viz. that of the human spirit, is the one that does not add up.

Like most scholars, Alexander assumes that the words רוח לו refer to the human spirit, translating it as "his spirit." However, not only can he not satisfactorily explain the terminology used and the realization of the numbers, but also, more importantly, he cannot account for the connection between the number of zodiacal signs and the number of alleged divisions of the human spirit that are possible on a nine-point scale. These considerations argue against the general understanding that *4QZodiacal Physiognomy* is concerned with the division of the human spirit between light and darkness. This means that on the basis of the text of *4QZodiacal Physiognomy* itself it seems unlikely that "spirit" (רוח) implies the human spirit, which entails that this text need not be related to the *Two Spirits Treatise* as has been generally assumed.

That the human spirit is not the object of concern in the text of *4QZodiacal Physiognomy* is possibly also suggested by the use of the construction רוח לו, which I translate as "there is a spirit for him." The occurrence of the construction רוח לו seems strange if what is meant is "his spirit," i.e. the human spirit of the described type of person. In the rest of the text such a possessive relationship is expressed, as is to be expected, by the use of attached suffixes.[70] The reader would, therefore, anticipate רוחו if

68 See e.g. Hippolytus, *Refutation of All Heresies* 4.15.4–27.2. See the section on physiognomics and astrology in Chapter Two.

69 Within the astrological framework of the ascendant interpretation, the total number of entries in the original text is, of course, much higher. If, however, one were to assume that the human spirit is meant within the astrological framework of the ascendant interpretation, the problem seems to be that many human spirits would have the same division of light and darkness, which diminishes the heuristic value of the text for telling one person from another.

70 See 4Q186 1 ii 5: ושוקיו and ואצבעות רגליו; 1 iii 5: ורואשו; 1 iii 6: ושניו; 1 iii 6–7: ואצבעות ידיו; 1 iii 7: ושוקיו; 1 iii 8: ואצבעות רגליו; 2 i 1: וע]יניו and וזקניו; 2 i 2: בת קולו and ושניו; 2 i 4: ידיו ואצבעות; 2 i 5: ושוקיו and וכפות רגליו.

"his spirit" were intended, which is also the regular way of referring to people's innate spirits.[71] Although the words רוח לו have in general been translated as "his spirit," it is important to point out that they do not say that exactly.[72] Strictly speaking, רוחלו occurs in *4QZodiacal Physiognomy* as a nominal construct meaning "there is a spirit for him." This suggests the possibility that another, external spirit is meant.

Moreover, one should allow for the possibility that the distinct construction רוח לו ("there is a spirit for him") was chosen precisely for the purpose of drawing the reader's attention to the fact that the human spirit is not meant as the object of reference in this case. In the physiognomic descriptions the suffix is immediately attached to the nouns, and that is also the case with the reference to the person's zodiacal sign in 4Q186 1 ii 9 (בהמתו, "his animal") following the occurrence of רוחלו ("there is a spirit for him") in 4Q186 1 ii 7. As already said, the reader would expect רוחו for "his spirit," but that is not the case. The construction רוחלו ("there is a spirit for him") seems thus to stand out in the text.[73]

ZODIACAL SPIRITS IN *4QZODIACAL PHYSIOGNOMY* (4Q186)

Whatever the exact sense, it is clear that *4QZodiacal Physiognomy* relates the "spirit" (רוח) in connection with certain numbers to the "house of light" and the "house of darkness." An explanation of this word should, therefore, also be able to account for the sense of this terminology and the establishment of the numbers.

In Chapter Three I have argued that Albani's ascendant interpretation accounts adequately for the realization of these numbers, and also that the combination of *melothesia* and *dodecatemoria* forms the astrological background of the division of the zodiacal sign in *4QZodiacal Physiognomy*, signaled by the words "in the foot of *Taurus*" (ברגל השור) in 4Q186 1 ii 9. This latter element demonstrates that *4QZodiacal Physiognomy* was part of a tradition in which the sign of *Taurus* was divided differently than in the tradition exemplified by the *Rhetorius-Teucer text*. An important consequence of this is that the number nine need not have been a set number for

[71] Cf. CD 3:3; 20:24; 1QS 2:14; 4:26; 6:17; 7:18.23; 9:15.18; 4Q279 5 5; 4Q416 7 3 (=4Q418 77 4); 4Q417 1 i 18 (=4Q418 43–45 i 14); 2 i 1.3; 4Q426 11 3; 11Q29 1. Cf. Sekki, *Meaning of* Ruaḥ *at Qumran*, 118-21, 123, who does not comment on the construction רוחלו.

[72] Allegro, DJD 5.89-91, was aware of the difficulty here; he translated "he has (of) spirit." But he did understand it as a reference to the human spirit. As did also Wise, "Horoscope Written in Code," 277-78, who translates "he possesses a spirit."

[73] The occurrence of ר]וח[לה ("he has a [sp]irit") in 4Q561 3 2 is too fragmentary to determine its exact meaning or to relate it to רוחלו in *4QZodiacal Physiognomy*.

all zodiacal signs in the entire, original text, which is also made clear by Rhetorius-Teucer's list of other zodiacal signs.[74]

If this interpretation is correct, then the numbers cannot stand by themselves but must be seen in relation to a specific zodiacal sign. Due to the ascendancy of a specific part of one of the twelve signs, the division of that sign's parts between the "house of light" and the "house of darkness" is realized. This connection to the sign identifies the division between light and darkness. By that I mean that the concrete zodiacal signs make it possible to distinguish between similar divisions of light and darkness. For according to the understanding of the text in this study, there would be many combinations of light and darkness that were repeated for the division of every zodiacal sign. This means that those combinations that were the same could only be distinguished if identified with one of the twelve signs.

Since *4QZodiacal Physiognomy* relates the numbers to a "spirit" (רוח), this reference should also be considered within the astrological framework of the text. If, therefore, the numbers listed in *4QZodiacal Physiognomy* are a result of the ascendant zodiacal sign, then the "spirit" should probably also be related to the zodiacal sign. For that reason, I suggest that the word רוח ("spirit") is used in *4QZodiacal Physiognomy* to refer to spirits that are related to the zodiacal signs. In other words, one should allow for the possibility that the spirits mentioned in the text are zodiacal spirits; one for each of the twelve zodiacal signs.

Angels and Stars as Animated Beings in Second Temple Period Judaism

During the Second Temple period, the sense of the word רוח ("spirit") developed and expanded to include different concepts of reference. *Inter alia*, it was used in the Dead Sea Scrolls for spirits, angels, and demons.[75] A similar semantic field can also be observed for the Greek word πνεῦμα ("spirit"), which takes on the sense of a supernatural spirit or intermediary.[76]

Various Jewish texts show that angels and spirits have a cosmological function bearing responsibility for the course of the sun, the moon, the

[74] See Chapter Three n. 85.

[75] Cf. Sekki, *Meaning of* Ruaḥ *at Qumran*, 145-71; Davidson, *Angels at Qumran*, 155-56; M. Mach, "Angels," in *Encyclopedia of the Dead Sea Scrolls*, 24-27, at 25. Sekki, *Meaning of* Ruaḥ *at Qumran*, 99, 145, states that רוח as human spirit has a consistently feminine gender, whereas רוח as demon or angel has a consistently masculine gender, and that, therefore, רוח in *4QZodiacal Physiognomy* must mean the human spirit. But Sekki, *Meaning of* Ruaḥ *at Qumran*, 145-47, 155-63, lists numerous examples where רוח in the feminine gender is used to refer to angels and demons. There is no reason why this should be restricted to the feminine plural form and cannot include the feminine singular. See also the Aramaic examples in 1Q20 20:16–20.26.28; 4Q197 4 i 13; 4Q560 1 ii 5–6.

[76] Cf. H. Kleinknecht, "*πνεῦμα* im Griechischen," *TWNT* 6:333-57, at 337; E. Sjöberg, "רוּחַ im palästinischen Judentum," *TWNT* 6:373-87, at 373-74.

planets, and the stars.[77] In *1 Enoch*, for example, Enoch is given a tour of the cosmos in which the angel Uriel, leader of the luminaries,[78] explains to him its workings. Towards the end of the journey Uriel says to Enoch:

> I have now shown you everything, and I have revealed everything to you so that you may see this sun and this moon and those who lead the stars of the sky and all those who turn them – their work, their times, and their emergences.[79]

The angels and spirits have been assigned their duties by God, as is made clear by the *Hodayot* text from Qumran Cave 1:

> You have stretched out the heavens for your glory. Everything [which it contains] you have [es]tablished according to your will, and powerful spirits, before they became h[oly] angels [...] eternal spirits in their realms: luminaries according to their mysteries, stars according to [their] circuits, [all the stormy winds] according to their roles, lightning and thunder according to their duties and well-designed storehouses according to [their] purposes [...] according to their secrets.[80]

Similar lists in which the angels control celestial, meteorological, and other processes of nature have been preserved in other texts.[81]

When Enoch passes his knowledge on to his son Methuselah regarding the law of the stars, he gives:

> the names of those who lead them, who keep watch so they enter at their times, who lead them in their places, in their orders, in their times, in their months, in their jurisdictions, and in their positions.[82]

Regarding the close link between angels and stars, Enoch can, therefore, say that he saw that "their motion is according to the number of angels."[83] The distinction is not always sharply made between angels and spirits controlling the celestial elements and being equal to them. These texts not only

[77] See e.g. Bietenhard, *Die himmlische Welt*, 25, 101-3; C.A. Newsom, "Angels: Old Testament," *ABD* 1:248-53, at 252; M. Mach, *Entwicklungsstadien des jüdischen Engelglaubens in vorrabbinischer Zeit* (TSAJ 34; Tübingen: J.C.B. Mohr [Paul Siebeck], 1992), 262-65; Davidson, *Angels at Qumran*, 314-15.

[78] *1 En.* 21:1–5; 72:1; 74:2; 75:3; 79:6.

[79] *1 En.* 80:1. Translations are from G.W.E. Nickelsburg and J.C. VanderKam, *1 Enoch: A New Translation* (Minneapolis: Fortress, 2004). Cf. *2 En.* 4.

[80] 1QH[a] 9:9–13. Translation from García Martínez and Tigchelaar, *DSSSE*, 159.

[81] Cf. *Jub.* 2:2; *1 En.* 60:14–22; *2 En.* 4–6; 11; 19. The items enumerated in these lists correspond with other lists in apocalyptic literature that catalogue revealed things. See M.E. Stone, "Lists of Revealed Things in the Apocalyptic Literature," in *Magnalia Dei: The Mighty Acts of God: Essays on the Bible and Archaeology in Memory of G. Ernest Wright* (eds. F.M. Cross, Jr., W.E. Lemke and P.D. Miller; Garden City, New York: Doubleday, 1976), 414-52.

[82] *1 En.* 82:10.

[83] *1 En.* 43:2.

demonstrate the connection between angels and stars, they also give expression to the concept of an animated, spirited universe.[84]

This latter aspect of animated stars is also suggested by some Second Temple period texts that seem to have been familiar with the notion that human beings after their death were to join the angels and were to be like stars.[85] Thus, in the Hebrew Bible the *Book of Daniel* expresses the idea of astral immortality:

> And the wise will shine like the brightness of the firmament, and those who lead the many to righteousness like the stars forever and ever.[86]

In various Jewish texts the afterlife of the righteous in community with the angels is expressed in terms of astral imagery.[87] Of course, this imagery may be merely metaphorical as in Daniel, but a passage from the *Similitudes* of *1 Enoch* seems to suggest another possibility:

> And I saw other lightnings and stars of heaven; and I saw that he called them by their names, and they listened to him. And I saw a righteous balance, how they are weighed according to their light, according to the breadth of their spaces and the day of their appearing. (And I saw how) their motion produces lightning, and their motion is according to the number of the angels, and they keep faith with one another. And I asked the angel who went with me and showed me what was hidden, 'What are these?' And he said to me, 'The Lord of the Spirits has shown you a parable concerning them; these are the names of the holy ones who dwell on the earth and believe in the name of the Lord of the Spirits forever and ever.'[88]

Enoch sees the stars of heaven and how they are weighed on a righteous balance according to the degree of their radiance, their magnitude, and their movement.

At first, this may be reminiscent of the data in ancient star catalogues where information regarding a star's brightness, magnitude, longitude, and latitude are listed. The continuation of the passage, however, reveals the interest in *1 Enoch* in these astronomical matters. When Enoch asks what these are, the angel responds that they are the names of the holy ones who dwell on earth. From *1 En.* 41:1–2 it seems clear that the holy ones are human beings who have passed judgment after the deeds of humanity are

[84] Cf. A. Scott, *Origen and the Life of the Stars: A History of an Idea* (Oxford: Clarendon, 1991), 56-62, 91-93; Mach, *Entwicklungsstadien des jüdischen Engelglaubens*, 173-84.

[85] For Greco-Roman traditions, see F. Cumont, *Lux Perpetua* (Paris: Paul Geuthner, 1949), 142-88.

[86] Dan 12:3. Cf. *1 En.* 39:4–5; 104:2. See Collins, *Daniel*, 393-94.

[87] Cf. Mach, *Entwicklungsstadien des jüdischen Engelglaubens*, 159-73; K.P. Sullivan, *Wrestling with Angels: A Study of the Relationship between Angels and Humans in Ancient Jewish Literature and the New Testament* (AGJU 55; Leiden: Brill, 2004), 30-31, 131-39.

[88] *1 En.* 43:1–4. Cf. Nickelsburg, *1 Enoch 1*, 529.

weighed in the balance and now reside in heaven.[89] In both passages the name of the Lord of the Spirits seems of central concern. In *1 En.* 43:4 the holy ones are those believing in the name, whereas in 41:2 the sinners are those denying the name. Assuming a close relationship between stars, angels, and righteous ones, one should allow for the possibility that *1 En.* 43:1–4 suggests that the astronomical measurements of the stars are related to the state of the righteous ones.[90] It is tempting to understand this in light of a remark by Pliny:

> We have stated that the stars are attached to the firmament, not assigned to each of us in a way in which the vulgar believe, and dealt out to mortals with a degree of radiance proportionate to the lot of each, the brightest stars to the rich, the smaller ones to the poor, the dim to those who are worn out; they do not each rise with their own human being, nor indicate by their fall that someone's life is being extinguished. There is no such close alliance between us and the sky that the radiance of the stars there also shares our fate of mortality.[91]

According to Pliny, there were many people in his time who regarded the stars as soul mates and believed that their fate was intertwined with the degree of brightness of their star. The stars seem to have been regarded as some sort of signpost signifying the life allotted to individual people.[92] It is possible that *1 En.* 43:1–4 expresses a notion similar to the element of the star's brightness in connection with people's fate in this passage from Pliny.

Against the background of these notions of angels performing cosmological functions and stars as animated beings, I suggest that *4QZodiacal Physiognomy* is familiar with the idea that angels or spirits also accompany the zodiacal signs so that these were believed to be animated beings, having a spirited nature. This interpretation is further strengthened by the *Testa-*

[89] See also *1 En.* 61:8.

[90] Cf. Stone, "Lists of Revealed Things," 428-31; J.J. Collins, "Journeys to the World Beyond in Ancient Judaism," in *Apocalyptic and Eschatological Heritage*, 20-36, at 28-29.

[91] Pliny, *Natural History* 2.6.28–29. Translation from H. Rackham, *Pliny: Natural History* (vol. 1; LCL 330; Cambridge, Massachusetts: Harvard University Press, 1949), 187-89.

[92] See also the later ridicule of the idea by Eusebius of Alexandria, PG 86,1.453–454, that in the time of Adam and Eve there were only two stars in heaven, and after the Flood eight stars for Noah and his family. Cf. Bouché-Leclercq, *L'astrologie grecque*, 386 n. 1; W. Gundel, *Sterne und Sternbilder im Glauben des Altertums und der Neuzeit* (Bonn: Kurt Schroeder, 1922; repr., Hildesheim: Georg Olms, 1981), 237-49; Cumont, *Lux Perpetua*, 172; Boll, Bezold, and Gundel, *Sternglaube und Sterndeutung*, 152-53. The idea of a close link between individual human beings and stars may have some affinity with Plato's *Timaeus*, where the demiurge creates souls equal in number to the stars. The demiurge created them from the remains of the elements that he used for creating the soul of the universe. Furthermore, those souls who lived well were able to return and dwell in their native stars. Cf. Scott, *Origen and the Life of the Stars*, 12-13.

ment of Solomon, which provides important evidence for the idea that the zodiacal signs and decans have spirits and demons that represented them.

The Testament of Solomon *and Zodiacal Spirits*

The *Testament of Solomon* demonstrates most clearly the concept of zodiacal spirits.[93] Various passages from this text show that spirits (πνεύματα) and demons (δαίμονες) were identified with signs of the zodiac.

The *Testament of Solomon* provides a wealth of information on magic and demonology, and it is not without reason that scholars have characterized some of its parts as a magical handbook. In the text Solomon is confronted with many demons and spirits who are subjugated by him. The narrative framework of these confrontations is set against the background of the building of the Temple in Jerusalem. After their subjugation by Solomon the demons are set to work on the building of the Temple. The purpose of the text is to provide the reader with knowledge of what spirits and demons of the air, the earth and the regions beneath the earth are set against men and how their attacks can be thwarted. The confrontations between Solomon and the demons adhere to a basic pattern, but this has not been preserved in all instances. A demon appears before Solomon and gives his

[93] The composition of the *Testament of Solomon* dates between the fourth–sixth century CE, but it contains older traditions that possibly go back to the beginning of the Common Era, especially *T. Sol.* 18 on the zodiacal and decanal demons. See e.g. C.C. McCown (ed.), *The Testament of Solomon* (UNT 9; Leipzig: J.C. Hinrichs, 1922); Gundel, *Dekane und Dekansternbilder*, 49-62; K. Preisendanz, "Ein Wiener Papyrusfragment zum Testamentum Salomonis," *Eos* 48/3 (1956): 161-67; R. Daniel, "The Testament of Solomon XVIII 27-28, 33-40," in *Festschrift zum 100-jährigen Bestehen der Papyrussammlung der österreichischen Nationalbibliothek: Papryus Erzherzog Rainer* (Vienna: Brüder Hollinek, 1983), 294-304; D.C. Duling, "Testament of Solomon," *OTP* 1:935-87; Alexander, "Incantations and Books of Magic," 372-75; H.M. Jackson, "Notes on the Testament of Solomon," *JSJ* 19 (1988) 19-60; D.C. Duling, "The Testament of Solomon: Retrospect and Prospect," *JSP* 2 (1988) 87-112; J. Harding and L. Alexander, "Dating the Testament of Solomon," posted May 28, 1999, at http://www.st-andrews.ac.uk/%7Ewww_sd/date_tsol.html; P.S. Alexander, "Jewish Elements in Gnosticism and Magic *c.* CE 70–*c.* CE 270," in *The Cambridge History of Judaism, Volume Three: The Early Roman Period* (eds. W. Horbury, W.D. Davies and J. Sturdy; Cambridge: Cambridge University Press, 1999), 1052-78, at 1068; Von Stuckrad, *Ringen um die Astrologie*, 394-420; P.A. Torijano, *Solomon the Esoteric King: From King to Magus, Development of a Tradition* (JSJSup 73; Leiden: Brill, 2002), 144-50; S.I. Johnston, "The *Testament of Solomon* from Late Antiquity to the Renaissance," in *Metamorphosis of Magic*, 35-49; T.E. Klutz, "The Archer and the Cross: Chorographic Astrology and Literary Design in the *Testament of Solomon*," in *Magic in the Biblical World: From the Rod of Aaron to the Ring of Solomon* (ed. T.E. Klutz; JSNTSup 245; London: T&T Clark, 2003), 219-44; Alexander, "Contextualizing the Demonology," 613-35; T.E. Klutz, *Rewriting the* Testament of Solomon*: Tradition, Conflict and Identity in a Late Antique Pseudepigraphon* (LSTS 53; London T&T Clark, 2005). I was unable to consult P. Busch, *Das Testament Salomos: Die älteste christliche Dämonologie, kommentiert und in deutscher Erstübersetzung* (TUGAL 153; Berlin: Walter de Gruyter, 2006). I thank Tobias Nicklas for drawing my attention to this title.

or her name, the star in which it resides and the name of the angel that can be used to avert its attack.

What is of interest here is that the spirits and demons are presented as inhabiting or being identical with zodiacal signs and constellations. Solomon's confrontation with the spirits and demons begins with his concerns about the health of the master workman's little boy at the time when the Temple of Jerusalem was being built. The cause of the boy's ill health is a demon named Ornias who sucks blood from the boy's right thumb. In *T. Sol.* 2:2 the demon Ornias, who has been subjugated by the boy using Solomon's magical ring, is forced to tell Solomon that he resides in the zodiacal sign *Aquarius*. After his subjugation Ornias must bring other demons before Solomon for interrogation. In some of these encounters with the demons explicit reference is made to the constellations that they reside in, such as in *T. Sol.* 4:6 with the female demon Onoskelis, in *T. Sol.* 5:4 with the demon Asmodeus, who was born of the union between a human mother and an angel, in *T. Sol.* 6:7 with the demon Beelzeboul, who is prince of the demons, and in *T. Sol.* 7:6 with the demon Lix Tetrax.[94] Finally, another clear connection between astrology and demonology is that the seven planets (the five planets Saturn, Jupiter, Mars, Venus, and Mercury, as well as the sun and the moon) also have their spirits. In *T. Sol.* 8:2–4 these seven spirits present themselves to Solomon as the seven heavenly elements (στοιχεῖα) that rule this world of darkness, saying that their stars in heaven may look small, but that they are named like gods.

The concept of zodiacal spirits and demons is demonstrated elaborately in *T. Sol.* 18. Here Solomon is confronted with thirty-six heavenly elements (στοιχεῖα), which are spirits calling themselves the world rulers of the darkness of this age. Like the other spirits before them, they appear before Solomon because he has authority over all the spirits of the air, the earth and the regions beneath the earth. *T. Sol.* 18 is a demonological catalogue that lists the names of the spirits of the zodiacal circle, the harm they cause to human beings, and the means for driving them away and curing people.

The thirty-six spirits represent the thirty-six decans, one of the subdivisions of the zodiac in ancient astrology.[95] Like the other spirits and demons before them, the thirty-six decanal spirits must tell Solomon who they are, what they do, and how their harm can be thwarted. Thus, the first decanal spirit (πνεῦμα) comes before Solomon:

[94] Cf. also *T. Sol.* 10:3; 15:4–6. For the possibility that *T. Sol.* 14:3 contains a reference to the zodiacal sign *Sagittarius*, see Klutz, "The Archer and the Cross, 232-38; Klutz, *Rewriting the* Testament of Solomon, 44-47.

[95] See the section on subdivisions of the zodiacal signs in Chapter Three.

> Then I, Solomon, summoned the first spirit, saying to him: 'Who are you?' And he replied: 'I am the first decan of the zodiacal circle (and) I am called Ruax. I make heads of men suffer pain and temples to throb. When I hear only: "Michael imprison Ruax," I withdraw immediately.'[96]

At the end of the interrogation Solomon refers to all the spirits as the thirty-six demons (δαίμονας) that plague humanity.

The astrological background of the demonology in *T. Sol.* 18 is that of decanal *melothesia*. According to the concept of *melothesia*, the different parts of the human body are under the influence of certain astrological entities. In the case of planetary *melothesia* the various parts of the body are distributed among the planets. In the case of zodiacal *melothesia* the signs of the zodiac have authority over specific parts of the human body. The concept of decans enabled astrologers to refine their system of zodiacal *melothesia* and thereby also their predictions or treatments. The human body was divided according to the number of thirty-six decans.[97] Thus, in *T. Sol.* 18 most of the decanal spirits of the zodiac are responsible for harm caused to a certain part of the human body.[98]

It is noteworthy that one of the manuscripts of the *Testament of Solomon* clarifies that the other spirits who were questioned before by Solomon were:

> spirits from *Aries* and *Taurus*, *Gemini* and *Cancer*, *Leo* and *Virgo*, *Libra* and *Scorpio*, *Sagittarius*, *Capricorn*, *Aquarius* and *Pisces*.[99]

Not only does this demonstrate that all the spirits and demons fall under the dominion of the twelve zodiacal signs,[100] but also that the signs themselves have spirits and are identical with them. It is, therefore, possible to speak of zodiacal spirits.

The *Testament of Solomon* thus demonstrates that in antiquity there were people who considered the signs of the zodiac and its subdivisions, the decans, to have or to be spirits. In addition to the notion in Jewish texts of the Second Temple period that angels controlled the movements of sun, moon, planets, and stars as well as the physical processes of nature, such as thunder, lightning, rain, and winds, there is thus evidence that the zodiacal signs and decans were imagined to be supernatural, spiritual beings or demons. The combination of angels, demonology, and zodiacal signs and decans that one encounters in the *Testament of Solomon* corresponds to

[96] *T. Sol.* 18:4–5.
[97] On these different forms of *melothesia* see Chapter Two n. 218.
[98] Cf. Klutz, "The Archer and the Cross," 243.
[99] *T. Sol.* 18:3 in Ms P. See McCown, *Testament of Solomon*, 51.12*.
[100] Cf. Torijano, *Solomon the Esoteric King*, 149.

similar ideas in some Greek magical papyri.[101] As animated beings, the zodiacal signs were considered to be astral spirits and they could be venerated as gods, something that already occurred in Babylonian astral magic where the stars functioned as mediators between people and the gods.[102]

In the *Testament of Solomon* the zodiacal and decanal spirits are clearly harmful beings, and in the Manichean *Kephalaia*, another late antique text, they are also considered as evil spirits.[103] Despite this negative view of the zodiacal spirits in these texts, a zodiacal spirit could apparently also be seen as a positive force to be summoned for aid in magical practices. Thus, there is an interesting magical amulet in Hebrew from the Cairo Genizah in which the zodiacal sign *Leo*, who has the angel Sarbeil appointed over him, is adjured to protect a woman in childbirth from harmful spirits and to drive away all kinds of male and female demons.[104] In light of the foregoing, one should allow for the possibility that *Leo* is considered here as an animated, personified being. In other words, it is possible that according to this medieval Jewish text *Leo* is a zodiacal spirit that can be summoned, just as the zodiacal and decanal spirits in the *Testament of Solomon*.

Zodiacal Spirits and Planetary Demons at Qumran

As a consequence of the astrological framework, the sense of the word "spirit" (רוח) in *4QZodiacal Physiognomy* should be understood in relation to the zodiacal signs. From the astrological context is appears likely that רוח concerns the spirit of the zodiacal sign. It is possible that the construction רוח לו ("there is a spirit for him") was intended to draw the reader's attention to the fact that not the human but the zodiacal spirit is meant, but apart from this the text is not more explicit.

Just as in the *Testament of Solomon* where the general word πνεῦμα ("spirit") is used for reference to the spirits of zodiacal signs and decans, so *4QZodiacal Physiognomy* uses the general word רוח ("spirit"). Such a sense is compatible with the development of the word's meaning in Second Temple period texts and the different concepts it conveys, similar to its Greek equivalent. I suggest, therefore, that in the three remaining instances where

[101] Cf. H.G. Gundel, *Weltbild und Astrologie in den griechischen Zauberpapyri* (MBPF 53; Munich: C.H. Beck, 1968), 16-20, 52-54.

[102] Cf. Boll, Bezold, and Gundel, *Sternglaube und Sterndeutung*, 97-99, 104-5, 125-26; Gundel and Böker, "Zodiakos," 543-47; Hübner, *Eigenschaften der Tierkreiszeichen*, 130; Reiner, *Astral Magic*.

[103] Cf. *Kephalaia of the Teacher* IV.27.14–29; XV.48.31–53; LXIX. See I. Gardner, The Kephalaia of the Teacher*: The Edited Coptic Manichaean Texts in Translation with Commentary* (NHMS 37; Leiden: E.J. Brill, 1995).

[104] T.-S. K 1.18. See L.H. Schiffman and M.D. Swartz, *Hebrew and Aramaic Incantation Texts from the Cairo Genizah: Selected Texts from Taylor-Schechter Box K1* (STS 1; Sheffield: Sheffield Academic Press, 1992), 69-79.

the text says that there is a spirit for the described type of person it refers to the zodiacal spirit.[105]

In Chapter Three I have presented *4QZodiacal Physiognomy* as an example of individual astrology at Qumran as opposed to the interest in general, mundane astrology in *4QZodiology and Brontology ar*. Although the manuscript of *4QZodiacal Physiognomy* is not a collection of actual horoscopes, the text does share the same genethlialogical concern of those texts, viz. individual people.

It is interesting to note that these two areas of astrological interest are possibly also reflected in the occurrence of individual and general spirits in an astrological sense. The zodiacal spirits in *4QZodiacal Physiognomy* that are of concern for individual people seem to have their counterparts in planetary spirits or demons who are set over entire regions and people.

Alexander Toepel has suggested that the trees mentioned in the Aramaic para-Danielic manuscripts *4QFour Kingdoms*$^{a-b}$ *ar* (4Q552–553) refer to planetary demons.[106] In this text four trees occur which are interrogated by someone, perhaps Daniel. The trees apparently represent personified figures. The first tree is asked to give his name. The tree answers that his name is Babel, upon which its interrogator says to him that he is the one who rules over Persia. In the remaining text two more trees are questioned, but their names have not been preserved, nor the countries over which they rule. For the second tree only some geographical information remains. It seems that a coastal area is implied as the area of dominance of this tree, because sea and harbor are mentioned.

Toepel plausibly argues that not only are the four trees connected with the four empires as their symbolic representations, but also that they are endowed with personality: they rise and move away, they speak, they have names and rule over their kingdoms. The astrological background here is that of four planets (Jupiter, Mars, Saturn, and Venus) who are connected with the four cardinal directions.[107] It is possible that these planets appear as spirits in the form of trees in the text of *4QFour Kingdoms*$^{a-b}$ *ar*.[108] If this interpretation of planetary demons is correct, this text forms an excellent counterpart with that of the zodiacal spirits in the text of *4QZodiacal Physiognomy* who rule over individual people.

105 Cf. 4Q186 1 ii 7; 1 iii 8; 2 i 6.

106 A. Toepel, "Planetary Demons in Early Jewish Literature," *JSP* 14 (2005) 231-38.

107 On chorographic astrology, see Bouché-Leclercq, *L'astrologie grecque*, 328-47; F. Cumont, "La plus ancienne géographie astrologique," *Klio* 9 (1909) 263-73; Honigmann, *Die sieben Klimata*; Boll, Bezold, and Gundel, *Sternglaube und Sterndeutung*, 9-10, 64-65, 157-58; Barton, *Ancient Astrology*, 179-85.

108 The notion of planetary demons also appears in *T. Sol.* 8 (see above).

People and their Zodiacal Signs and Spirits

In Hellenistic astrology the ascendant, being the part of the zodiacal sign ascending above the eastern horizon, determined people's nativities and was, therefore, the single most important element in ancient horoscopes. In antiquity this astrological idea was matched by the prime importance for people's fates and lives with which the zodiacal birth sign was accredited.[109]

It was believed that people were closely linked to their birth signs. This is demonstrated by the Greek *zodiologia* and by the notion of "zodiacal children."[110] The significance of the zodiacal birth sign is also expressed in the following passage from the Manichean *Kephalaia*:

> Before they die, they who will die are marked out by the stars and the signs of the zo[di]ac in the sphere. They are appointed for them; in the[m] are their births. And their root (i.e. people's fate) is bound up with their zodiacal signs; and they are compelled by them and brought to an equal judgement in accordance with their deeds and sins.[111]

This is a late antique text, but it expresses well an idea that has an older tradition.

According to the astrological framework of *4QZodiacal Physiognomy*, the text shares the concern of Greek astrology for the ascendant as the determining factor in people's nativities. The text expresses this in 4Q186 1 ii 8–9 by saying that the horoscope in which a certain type of person was born was "in the foot of *Taurus*" (ברגל השור). *4QZodiacal Physiognomy* demonstrates the notion that the zodiacal signs influence the shape and appearance of the human body, albeit that the signifying relationship is reversed in this text as opposed to the *zodiologia*. It thus shows that, at least with regard to the human body, people are closely linked with their zodiacal birth sign, or, more precisely, according to the way their sign was divided between the "house of light" and the "house of darkness."

The text of *4QZodiacal Physiognomy* augmented the importance of the zodiacal sign for people's fates by the notion of the zodiacal spirit. Like the zodiacal signs, these zodiacal spirits have a close relationship with human beings. The text refers to these spirits with the words רוח לו ("there is a spirit for him"). These zodiacal spirits are related to the types of people described as a consequence of their moment of birth under one of the twelve signs of the zodiac.

[109] Cf. Gundel and Böker, "Zodiakos," 582-85.

[110] See Chapter Two n. 225.

[111] *Kephalaia of the Teacher* XLVI.117.32–118.5. Translation from Gardner, Kephalaia of the Teacher, 124. Cf. XLVIII.122.11–15; CI.254.31–255.2.

If this interpretation is correct, this means that types of people not only shared the same zodiacal sign (*synastria*),[112] but also that they had a special connection with its zodiacal spirit. This relationship between the zodiacal spirits as supernatural beings and individual people becomes perhaps more understandable against the background of texts in which angels not only function on a macrocosmic level, but in which they are also imagined as having a close relationship with certain special individuals. In addition to angels set over nations,[113] there are examples of angels guarding individual human beings.[114] In the *Book of Tobit* the angel Raphael guides Tobit on his journey and safeguards him from danger.[115] In the *Testaments of the Twelve Patriarchs* various sons of Jacob have a guardian angel.[116] In the New Testament *Book of Acts* 12:15 the other disciples say that it is not Peter but his angel who stands at the gate.[117] These examples demonstrate that supernatural beings such as angels were believed to stand in a close relationship with certain individual people. One should, therefore, allow for the possibility that people were also believed to stand in a special relationship to the spirit of their zodiacal birth sign.

In the context of the decanal spirits in *T. Sol.* 18, Alexander speaks of these demons as being *synastroi* with people on the basis of their ascendant sign in their nativity.[118] Following this notion it seems possible to understand the zodiacal spirits in *4QZodiacal Physiognomy* as *synastroi* with people from their moment of birth onwards, i.e. both spirit and person shared the same zodiacal sign.

There are only twelve zodiacal spirits, one for each zodiacal sign. Due to the position of the ascendant at the moment of birth, the division of the zodiacal spirit differed between different types of people, although they shared the same zodiacal sign. *4QZodiacal Physiognomy* not only mentions the division between light and darkness, but also states what zodiacal sign is the birth sign of the types of people (4Q186 1 ii 9: "And this is his animal: *Taurus*"). The text identifies people's zodiacal birth sign and provides information with regard to the nature of the zodiacal spirit at the time of birth, thus differentiating between the two. The division of light and darkness served as an indication for the nature of the zodiacal spirits at that

[112] In ancient astrology the notion of *synastria* refers to the zodiacal signs that people shared as their birth sign. Cf. Gundel and Gundel, *Astrologumena*, 144, 283, 290.

[113] Cf. Mach, *Entwicklungsstadien des jüdischen Engelglaubens*, 257-62.

[114] K.E. Grözinger, "Engel III. Judentum," *TRE* 9:586-96, at 589-90; O. Böcher, "Engel IV. Neues Testament," *TRE* 9:596-99, at 597; Newsom, "Angels," 252.

[115] E.g. Tob 5:21; 6:2–9. Cf. H. Schüngel-Straumann, *Tobit* (HThKAT; Freiburg: Herder, 2000), 159-60; J.A. Fitzmyer, *Tobit* (CEJL; Berlin: Walter de Gruyter, 2003), 49, 160-61.

[116] H.W. Hollander and M. de Jonge, *The Testaments of the Twelve Patriarchs: A Commentary* (SVTP 8; Leiden: E.J. Brill, 1985), 191.

[117] Cf. Sullivan, *Wrestling with Angels*, 71-74.

[118] Alexander, "Contextualizing the Demonology," 633.

time. And the reference to the zodiacal sign identified for the reader which zodiacal spirit was meant in the case of a particular physiognomic type.

The Nature of Zodiacal Spirits Divided between Light and Darkness

Accordingly, the descriptions of the human body lead the reader of *4QZodiacal Physiognomy* to the various subdivisions of people's zodiacal signs and spirits between the "house of light" and the "house of darkness." The physiognomic purpose of the text was, therefore, to provide data concerning the division of people's zodiacal spirits between light and darkness, the exact zodiacal position at their birth, i.e. their horoscope, and, consequently, what their zodiacal sign was.

The division between the "house of light" and the "house of darkness" can be explained astrologically as the result of the ascendant's position vis-à-vis the eastern horizon, but it seems that in *4QZodiacal Physiognomy* this was taken in terms of the zodiacal spirit being divided between light and darkness. Understanding the phrase רוח לו ("there is a spirit for him") in this way aligns it with the ascendant interpretation, which explains the realization of the numbers in the "house of light" and the "house of darkness" as the result of the division of the zodiacal sign between the areas above and below the horizon. Dependent on the moment of birth, the ascendant divided the sign's parts above and below the horizon, and the zodiacal spirit was divided likewise.

The physiognomic interest of the text was, therefore, not only in providing its intended reader with the ascendant position of people's horoscopes, but also to reveal the division of people's zodiacal spirits between light and darkness. In Chapter Two it was mentioned that the text of *4QZodiacal Physiognomy* seems to suggest a semiotic relationship between, on the one hand, the physiognomies of the human bodies and, on the other hand, the division of numbers between light and darkness. More parts in the "house of light" than in the "house of darkness" seems to imply a more attractive appearance of the body, while, conversely, more parts in the "house of darkness" seems to entail a less attractive appearance.[119] This suggests that the division of the zodiacal signs and spirits exerts its influence upon the shape and appearance of the human body.

In Chapter Three a passage was adduced in which Ptolemy stated that those parts of the zodiacal sign beneath the earth were to be ignored in the determination of the horoscope.[120] It seems that those parts below the earth, in the "house of darkness," were not deemed influential for the power that

[119] See the section in Chapter Two on the semiotic relationship between the appearance of the body and the division of numbers between the "house of light" and the "house of darkness."

[120] See Chapter Three n. 172.

one's ascendant sign (horoscope) has. If in *4QZodiacal Physiognomy* a connection is made between the division of the zodiacal signs and spirits between the "house of light" and the "house of darkness" and the appearance of the human body, the assumption seems to be that the more parts of a zodiacal sign that have ascended, the more powerful the radiating influence of the sign and spirit on the people who were born at that moment. Such a connection made it possible to identify people's zodiacal signs and spirits by reading their bodies.

Thus, the "house of light" has a positive connotation and the "house of darkness" a negative one. This terminology in *4QZodiacal Physiognomy* need not necessarily be understood within the framework of (Qumran) dualistic thought. The positive connotation of light and the negative connotation of darkness suggested by the text were also known in other ancient astrological writings. This is shown by a Hellenistic astrological text ascribed to Hermes Trismegistos (*Liber Hermetis Trismegisti*). This text describes the various influences that sections of the zodiacal signs, when they are the ascendant, exert on human beings, and the text arranges the sections in light and dark parts. A variant of this tradition in Firmicus Maternus suggests the positive and negative influence upon human affairs, people's characters or spirits, and their bodies.[121]

The interest of *4QZodiacal Physiognomy* in the division of the zodiacal signs and their spirits seems to be governed by the need to have knowledge of the nature of people's zodiacal spirits. The division between light and darkness was a means for measuring this nature. The number of parts in the "house of light" and the "house of darkness" indicated the power and influence of the zodiacal spirits. With regard to individual people this may have been different each time on the basis of the astrological moment of birth. The zodiacal spirit's nature and its relationship to individual types of people were modified according to the circumstances at birth, *in casu* they depended on the division of the zodiacal sign. This means that the zodiacal spirit's nature could differ in just as many ways as the zodiacal sign could be divided. Each configuration thus corresponded to a type of person.

The text of *4QZodiacal Physiognomy* can be seen as an attempt to draw connections between different types of people and the natures of their zodiacal spirits. It structures and classifies these relationships in a list that connects the shape and appearance of the human body with the subdivisions of the signs and their spirits in the "house of light" and the "house of dark-

121 Firmicus Maternus, *Mathesis* 4.22. Cf. Bouché-Leclercq, *L'astrologie grecque*, 229-37; W. Gundel, *Neue astrologische Texte des Hermes Trismegistos: Funde und Forschungen auf dem Gebiet der antiken Astronomie und Astrologie* (ABAWPHK 12; Munich: Bayerischen Akademie der Wissenschaften, 1936), 50-73, 289-94; Hübner, *Grade und Gradbezirke*, 1:147-71.

ness." In this way, the text demonstrates an interest in knowledge of cosmic matters and relationships between heavenly elements and human beings. To have knowledge of these matters was to understand the power and influence of the signs and spirits upon people, which may have extended further than just the appearance of the body. The further relevance of this knowledge is explored in Chapter Five.

CHAPTER V

FUNCTIONS AND CONTEXTS OF PHYSIOGNOMIC AND ASTROLOGICAL LEARNING IN THE DEAD SEA SCROLLS AND SECOND TEMPLE PERIOD JUDAISM

INTRODUCTION

In the previous chapters, I have been concerned with the reconstruction and interpretation of the remains of the physiognomic catalogues preserved in *4QZodiacal Physiognomy* and *4QPhysiognomy ar*. In this final chapter the scope will be broadened to include questions about the possible function, context, and status of these physiognomic catalogues and the knowledge that they convey to the intended reader inside and outside the Qumran community.

Any extrapolation from physiognomic texts, such as the two lists from Qumran, to social functions and contexts must remain tentative. There are few references in the ancient literature that reveal in what setting, by whom, and how such texts and the learning contained in them were used, and these must be weighed carefully.[1] In Chapter Two some insight was gained concerning the social background of those involved in the cultivation of physiognomic learning in the Babylonian and Greco-Roman traditions. Although our knowledge remains limited, this may clarify some aspects of the social context of physiognomic and, for that matter, astrological learning in Second Temple period Palestine.

Before speculating in more detail about the social and cultural locus of the people transmitting, reading, and using texts such as *4QZodiacal*

[1] Compare this, for instance, with what is known about the authors of ancient Jewish dreams. See F. Flannery-Dailey, *Dreamers, Scribes, and Priests: Jewish Dreams in the Hellenistic and Roman Eras* (JSJSup 90; Leiden: Brill, 2004). Regarding the issue of people or social groups behind texts, cf. also M.E. Stone, "The Book of Enoch and Judaism in the Third Century B.C.E.," *CBQ* 40 (1978): 479-92, at 483: "Caution in such matters is wise, for the movement from tendencies of thought discerned in the analysis of texts to the positing of the existence of otherwise unattested social groups is fraught with peril. Yet, that danger is one that the scholar must brave if his analysis is conducted in terms that imply a sociological matrix for the development of ideas." See also the cautious remarks with regard to the Dead Sea Scrolls by P.R. Davies, "Sects from Texts: On the Problems of Doing a Sociology of the Qumran Literature," in *New Directions in Qumran Studies: Proceedings of the Bristol Colloquium on the Dead Sea Scrolls, 8–10 September 2003* (eds. J.G. Campbell, W.J. Lyons and L.K. Pietersen; LSTS 52; London: T&T Clark, 2005), 69-82.

Physiognomy and *4QPhysiognomy ar* some attention needs to be devoted to the type of knowledge offered by these texts. Should they be, as has been suggested, regarded as scientific writings? Can more be said about the nature of the learning transmitted by these texts? If *4QZodiacal Physiognomy* draws connections between different elements, such as the human body, zodiacal signs and spirits, and stones, is it possible to recognize a specific attitude behind the collection of these various aspects in one text?

Who was interested in these expressions of physiognomic and astrological learning? If, as has been suggested, for example, for the astronomical and geographical parts of Enochic literature, these texts from Qumran represent learned, scientific texts, then their authors or readers too were presumably part of a well-educated body of people in ancient Jewish society. How did people become familiar with physiognomic and astrological learning? Is it possible to be more precise and locate the education in and transmission of that knowledge with, say, priestly or secular scribes and scholars? Taking into account Babylonian and Hellenistic physiognomic and astrological traditions and the cultural exchange of ideas that went on in the Persian and Hellenistic period,[2] outside influences on Hellenistic-Early Roman period Palestine seem likely in the case of *4QZodiacal Physiognomy* and *4QPhysiognomy ar*, and scholars have suggested this. Is it possible to trace the cultural origin of the concepts operative in these Qumran texts to Babylonian or Hellenistic influence? Finally, inquiries into ancient Jewish science by scholars have focused primarily on material provided by *1 Enoch* and the lists of revealed things in apocalyptic literature. What can texts like the physiognomic catalogues from Qumran contribute to our understanding of ancient Jewish science?

Some scholars assume that *4QZodiacal Physiognomy* was of marginal interest to members of the Qumran community, while others think it was very important for the community, especially for its leadership. What was the status of physiognomic and astrological learning? With regard to the latter the controversial character of it surfaces again and again in different sources. Astrology was condemned and criticized by some, but also accepted and admired by others. How does this controversial nature bear on the presence of the physiognomic-astrological text *4QZodiacal Physio-*

[2] For cultural exchanges between Mesopotamia and Greece, see e.g. A. Kuhrt and S. Sherwin-White (eds.), *Hellenism in the East: The Interaction of Greek and Non-Greek Civilizations from Syria to Central Asia after Alexander* (Berkeley: University of California Press, 1987); W. Burkert, *The Orientalizing Revolution: Near Eastern Influence on Greek Culture in the Early Archaic Age* (RevAnt 5; Cambridge, Massachusetts: Harvard University Press, 1992); Dalley et al., *Legacy of Mesopotamia*; W. Burkert, *Babylon, Memphis, Persepolis: Eastern Contexts of Greek Culture* (Cambridge, Massachusetts: Harvard University Press, 2004).

gnomy in the collection of manuscripts from Qumran? Is the inverted and mixed writing of the manuscript in some way related to its status?

Finally, were people interested in the relationships between a person's body and his fate, and a person's body and his zodiacal sign and its spirit, just for the sake of it, or was there also a more practical relevance and application of that learning? If *4QZodiacal Physiognomy* is concerned with establishing the division between light and darkness of a person's zodiacal sign and its spirit in relation to the appearance of the human body, then the question may be raised whether there was any further practical relevance and function to that knowledge and, if so, in what way. Scholars have suggested that physiognomic learning may have functioned in the context of an admission procedure for the Qumran community, but if *4QZodiacal Physiognomy* assesses the zodiacal instead of the human spirit, is such a context still conceivable? Also, taking into account that the Qumran physiognomic catalogues are non-sectarian compositions, can a function and practical relevance for them be thought of outside of the Qumran community, i.e. in a non-sectarian context?

Ancient Science and Understanding Cosmic Connections

The manuscripts *4QZodiacal Physiognomy* and *4QPhysiognomy ar*, or *4QZodiology and Brontology ar* for that matter,[3] are invaluable evidence for a Jewish interest during the Second Temple period in the workings of nature and the cosmos. These texts complement in an important way the evidence from Enochic and apocalyptic literature.[4]

4QZodiacal Physiognomy and *4QPhysiognomy ar* demonstrate the existence and dissemination of certain forms of physiognomic and astrological knowledge in Palestine during the Hellenistic-Early Roman period. Modern definitions of science or distinctions such as "science" and "pseudo-science" may distort our understanding of the function and context of these texts in ancient Jewish society in Palestine.

Comparison with similar texts from Babylonian and Greco-Roman cultures, taking their socio-cultural background into account too, suggests that these two physiognomic lists from Qumran represent examples of ancient

[3] The *Treatise of Shem* is possibly another astrological example, but the dating to the first century BCE is far from certain. Cf. J.H. Charlesworth, "Treatise of Shem," *OTP* 1:473-86; A. Mengozzi, *Trattato di Sem e altri testi astrologici* (TVOa 7, LSc 1; Brescia: Paideia, 1997); J.H. Charlesworth, *Die Schrift des Sem* (JSHRZ-NF 2.9; Gütersloh: Gütersloher Verlaghaus, 2005).

[4] See Stone, "Lists of Revealed Things"; P.S. Alexander, "Enoch and the Beginnings of Jewish Interest in Natural Science," in *The Wisdom Texts from Qumran*, 223-43.

Jewish science.[5] Furthermore, the nature of the learning exhibited in these texts from Qumran can be characterized as an expression of the notion of cosmic sympathy. This is less clear with *4QPhysiognomy ar* because the extant fragments do not attest to the combination of other branches of learning than physiognomics, but seems an appropriate characterization for *4QZodiacal Physiognomy*.

Physiognomics and Astrology as Ancient Science

Within the historiography of science it is a debated issue whether the classification of physiognomic and astrological texts as scientific is correct. The methods and reasoning of ancient physiognomic and astrological learning do not compare with modern standards of scientific research, nor can it be assumed that, comparable to the modern period, a notion of natural science as a separate domain of intellectual inquiry was perceived.[6] This, however, does not invalidate characterizing these arts and their interest in the fabrics of reality as scientific. Science is not detached from social reality, it is a historically defined activity conducted by people in different contexts. What counts as scientific knowledge may differ over time and place depending on context.[7]

Like medicine, Greek culture characterized physiognomic and astrological learning as τέχναι (arts), indicating it to have been crafts requiring much practice and experience. These arts were regarded as conjectural bodies of knowledge in which the practitioner could end up with the wrong results, but without disqualifying the presuppositions of the arts themselves.[8] Risking an anachronism with the characterization "scientific," the term is helpful too in drawing attention to the learned character and possibly high status of the knowledge contained in the Qumran physiognomic texts. Like the τέχναι (arts) of astrology, physiognomics, and medicine in Greco-Roman society, the appropriation of these bodies of knowledge within Palestinian Judaism of the Hellenistic-Early Roman period would have presupposed practice and experience. This pursuit of knowledge required not only the

[5] Alexander, "Physiognonomy," 389; Alexander, "Beginnings of Jewish Interest in Natural Science," 230-31.

[6] It is possible that in Babylonia a mathematical astronomical interest may have existed purely for its own sake. Cf. Swerdlow, *Babylonian Theory of the Planets*, 174.

[7] On science in antiquity, see e.g. G.L. Irby-Massie and P.T. Keyser, *Greek Science of the Hellenistic Era: A Sourcebook* (London: Routledge, 2002); G.E.R. Lloyd, *Ancient Worlds, Modern Reflections: Philosophical Perspectives on Greek and Chinese Science and Culture* (Oxford: Clarendon, 2004). On the classification of ancient astrology and physiognomics as science, see e.g. Barton, *Power and Knowledge*; Rochberg, *The Heavenly Writing*.

[8] Cf. Pseudo-Aristotle, *Physiognomonica* 806a 15; Plutarch, *Sulla* 5.5–6; Ptolemy, *Tetrabiblos* 1.2.13 (1.3.18). Cf. Barton, *Power and Knowledge*, 2, 16 n. 57; Vogt, *Physiognomonica*, 307; Sassi, *Science of Man*, 179-83.

means to devote one's time to such an activity of learning, but also higher education.

The arts of physiognomics and astrology were a means to get a grip on reality in a systematic way. Physiognomics claimed to understand certain aspects of individual people by classifying types of people in a certain fashion. In Babylonian tradition the human body was one of the means through which the gods communicated their messages about people's fates. This notion was worked out in the huge catalogue of *Alamdimmû* that listed all sorts of possible signs in the appearance and shape of the human body and their meaning for people's fates. In Greco-Roman tradition physiognomics was a sophisticated means of codifying certain social values about types of people, their characters and the visibility of a character through a person's body. It was not only accompanied by a logico-semiotic reasoning, but also by a physiological foundation through the notion of humors and their mixture that was considered responsible for the configuration of people's bodies and characters.

Astrology and astronomy developed in such a way as to enable people to predict the movements of sun, moon, and planets, and also to predict, from their perceived character in relation to the zodiacal signs, their influences on earthly affairs and the lives of people.[9] Not only was there a sense of analogy between heavenly and earthly matters ("as above, so below"), but the notion also developed that the whole structure of reality was such that everything in it was intricately interwoven with everything else.[10] The four elements, for example, were believed to constitute the entire universe. This accounted for the connections between heavenly elements such as planets and zodiacal signs and types of people. The mixture of a planet had a distinct effect on the mixture of the human body.[11]

Cosmic Sympathy and the Understanding of Cosmic Correspondences in 4QZodiacal Physiognomy *(4Q186)?*

The combination of different elements of learning in *4QZodiacal Physiognomy* reveals a tendency to bring various branches of knowledge together

[9] Cf. Ptolemy, *Tetrabiblos* 1.1.1.

[10] Von Stuckrad, *Ringen um die Astrologie*; K. von Stuckrad, "Jewish and Christian Astrology in Late Antiquity – A New Approach," *Numen* 47 (2000): 1-40, does not sufficiently take this element into account, but he, rather one-sidedly, defines astrology as a matter of understanding the quality of time in terms of analogy and correspondence. Such a definition does not fully capture the sense of ancient astrology, which certainly also contained the notion of a more direct causation between heavenly and earthly elements. Cf. J. Dillon, "Plotinus on Whether the Stars are Causes," in *La science des cieux: Sages, mages, astrologues* (ed. R. Gyselen; ResOr 12; Bures-sur-Yvette: Groupe pour l'Étude de la Civilisation du Moyen-Orient, 1999), 87-92.

[11] Cf. Ptolemy, *Tetrabiblos* 3.12. See further the section in Chapter Two on the connection between physiognomics and astrology.

in one type of text. In addition, this sort of text may also indicate a particular attitude towards various aspects of the cosmos. The connection between physiognomics and astrology not only demonstrates the all-pervasive influence of the astrological worldview, but also shows that different things in the universe were seen as interrelated.[12]

In Greco-Roman tradition, Stoic philosophy formulated the notion of a cosmic harmony (συμπάθεια) between all elements in the universe in which everything had connections with everything else. This notion "is related to the idea of correspondences, which designated animals, plants and stones as sympathetic or antipathetic to particular conditions, which might or might not be seen to be caused by the heavenly bodies."[13] The Stoic idea of universal sympathy was the philosophical basis for Greco-Roman astrology in which the heavenly elements had causative force upon all earthly matters. The notion of cosmic sympathy between all elements in the universe makes the connections between different sorts of areas of learning understandable. In order to understand the nature and effect of specific stones, plants, or herbs in certain circumstances, one needed to understand their connections with other elements from the universe and how this in turn determined their mutual influence.[14] Astrology was not only related to physiognomics, but also to medicine and magic, illustrating the interrelatedness of everything.[15]

Nils Heeßel has argued recently that Babylonian scholarly tradition experienced a paradigm shift during the Persian and Hellenistic periods. He suggests that the Babylonian scholars moved away from the traditional framework of analogical thought to one of cosmic sympathy in which the universe is thought to be alive. The notion of cosmic sympathy goes further than a semantic correspondence between various elements. It claims that all things are interrelated and that there is not just one sense to be discovered behind every sign. According to Heeßel, Babylonian scholars needed to know the exact workings of things in different configurations to understand the relations between them. This explains how procedures of medicine, which involve the use of such things as plants and stones, are related to astrological conditions.[16]

The Late Babylonian text *LBAT* 1593 connects astrological circumstances with medical treatment, when it says to the reader:

[12] Cf. the section in Chapter One on magico-medicinal stones and Chapter Two n. 210.

[13] Barton, *Ancient Astrology*, 103.

[14] Cf. Barton, *Ancient Astrology*, 103-4; Barton, *Power and Knowledge*, 38; J.-P. Brach and W.J. Hanegraaff, "Correspondences," in *Dictionary of Gnosis and Western Esotericism* (ed. W.J. Hanegraaff et al.; vol. 1; Leiden: Brill, 2005), 275-79.

[15] Cf. Barton, *Ancient Astrology*, 185-97. See also Chapter Two n. 216.

[16] Heeßel, "Stein, Pflanze und Holz."

> you salve, feed, and fumigate the patient with the stone, herb, and wood (respectively).[17]

Heeßel suggests that a worldview of cosmic sympathy is behind this Late Babylonian text, a text that in another section also combines astrology and physiognomics.[18] Together with Greek astrological texts, it provides valuable comparative material for interpreting the interest in the connections between zodiacal signs, their spirits, human beings, and magico-medicinal stones in *4QZodiacal Physiognomy*.

4QZodiacal Physiognomy relates zodiacal signs and their spirits to human beings. This demonstrates an interest in cosmic matters and relationships between heavenly elements and human beings. The impression that a notion of cosmic sympathy may have determined the interest of the text is further strengthened by the mention of a sort of granite stone in 4Q186 1 ii 2. The reference to specific stones was possibly an important element in the text. Within the network of intricate relationships between heavenly elements and earthly matters the properties of stones were important, as is clear from Babylonian and Greek astrological texts.[19] One should allow for the possibility that in *4QZodiacal Physiognomy* stones were related to different zodiacal signs and the nature of their spirits, perhaps for magical or medicinal reasons. This suggests that a worldview of cosmic sympathy may be behind the combination of these different elements in this physiognomic-astrological text from Qumran.

The Social and Cultural Locus of Physiognomics and Astrology in Hellenistic-Early Roman Period Palestine

For Jewish society in Palestine during the Hellenistic-Early Roman period, we lack the evidence that is available for the social and cultural locus of physiognomic and astrological learning in Babylonian and Greco-Roman societies. Nevertheless, it seems likely that texts such as the Qumran physiognomic catalogues circulated only within a limited segment of society, restricted to an educated elite.

Learning the Physiognomic and Astrological Arts

In both Babylonian and Greco-Roman traditions, the arts or sciences of physiognomics and astrology were the domain of intellectual elites: the *āšipu* ("magician-exorcist"), philosopher, doctor, or astrologer. These were

[17] Reiner, "Early Zodiologia," 424. For this text see also the section in Chapter Two on the connection between physiognomics and astrology.
[18] Heeßel, "Stein, Pflanze und Holz," 15-16.
[19] See the section in Chapter One on magico-medicinal stones.

the few people who had or were afforded the means to devote their time to something not of immediate practical value; education would have been a necessary prerequisite for this pursuit of knowledge.[20]

During the Hellenistic period physiognomic and astrological learning would have been transmitted in the Babylonian temples, probably under the control of *āšipu*-families. In the Greco-Roman world some basic aspects of astronomy were on the Greek school curriculum, but the level of education in this subject must not be exaggerated. The more advanced, mathematical aspects of astronomy were for specialists; individual teachers may have gathered some students around them.[21]

There is not much evidence for education in the sort of learning one finds in physiognomic and astrological handbooks. We do not know exactly how people became familiar with such learned knowledge and on what level. The subject matter may have been of interest to more people who perhaps picked up some general physiognomic or astrological notions. Interest, however, in the texts themselves, such as, for example, the text of the pseudo-Aristotelian *Physiognomonica*, would have been very limited. It is unlikely that such specialized, technical texts circulated widely across many levels of society. Of course, the main condition for access is advanced literacy, which already narrowed down considerably the potential readership in antiquity.[22]

[20] Cf. Gordon, "Hellenistic Magic and Astrology"; D. Konstan, "Conventional Values of the Hellenistic Greeks: The Evidence from Astrology," in *Conventional Values*, 159-76, at 160, 170; P. Magdalino, "The Byzantine Reception of Classical Astrology," in *Literacy, Education and Manuscript Transmission in Byzantium and Beyond* (eds. C. Holmes and J. Waring; MMed 42; Leiden: Brill, 2002), 33-57, at 34-37.

[21] See H.-I. Marrou, *Histoire de l'éducation dans l'Antiquité* (2 vols; Paris: Seuil, 1948; repr., 1981), 1:263-77; 2:45, 84-85; M.L. Clarke, *Higher Education in the Ancient World* (London: Routledge & Kegan Paul, 1971), 49-52; S.F. Bonner, *Education in Ancient Rome: From the Elder Cato to the Younger Pliny* (London: Methuen, 1977), 77-78; R. Cribiore, *Gymnastics of the Mind: Greek Education in Hellenistic and Roman Egypt* (Princeton: Princeton University Press, 2001), 3. On the more social and political dimensions of education in Greek and Roman antiquity, see also Y. Lee Too (ed.), *Education in Greek and Roman Antiquity* (Leiden: Brill, 2001); R.F. Hock, "The Educational Curriculum in Chariton's *Callirhoe*," in *Ancient Fiction: The Matrix of Early Christian and Jewish Narrative* (eds. J.A. Brant, C.W. Hedrick and C. Shea; SBLSymS 32; Leiden: Brill, 2005), 15-36.

[22] On ancient literacy see e.g. W.V. Harris, *Ancient Literacy* (Cambridge, Massachusetts: Harvard University Press, 1989); A.I. Baumgarten, *The Flourishing of Jewish Sects in the Maccabean Era: An Interpretation* (JSJSup 55; Leiden: Brill, 1997), 114-36; J.L. Crenshaw, *Education in Ancient Israel: Across the Deadening Silence* (New York: Doubleday, 1998), 29-49; C. Wilcke, *Wer las und schrieb in Babylonien und Assyrien: Überlegungen zur Literalität im Alten Zweistromenland* (BAWPHK.S 6; Munich: C.H. Beck, 2000); A. Millard, *Reading and Writing in the Time of Jesus* (BibSem 69; Sheffield; Sheffield Academic Press, 2000); C. Hezser, *Jewish Literacy in Roman Palestine* (TSAJ 81; Tübingen: Mohr Siebeck, 2001); Cribiore, *Gymnastics of the Mind*, 147-59, 162-64, 180-83; P.S. Alexander, "Literacy among Jews in Second Temple Palestine: Reflections on the Evidence from Qumran," in *Hamlet on a Hill: Semitic and Greek Studies Presented to Professor T. Muraoka on the Occa-*

One possibility is that knowledge was gained through reading the handbooks. Polemo said that one could not put all physiognomic learning in one book, but for some people this may have been enough to know about the subject, even if in an inadequate way. Extracts may perhaps have been available in books that people could have picked up in passing, just like Aulus Gellius with the bundle of *mirabilia* books he came across in Brundisium.[23]

Another possibility is that people who wanted to know the physiognomic art more seriously learned it under the guidance of someone who was an expert in it, either in a one-to-one relationship or within a group of students. A large part would have been transmitted orally, as the texts alone did not suffice.[24] The situation for astrology would not have been much different.[25] Individual readers of the handbooks would not get very far. Vettius Valens referred to pupils he instructed further about his book. Most likely, people would have learned the different aspects of the astrological art through direct communication in an apprenticeship with a master astrologer.[26]

It is difficult to determine who exactly may have become familiar with learned knowledge of physiognomic and astrological arts in Second Temple period Judaism and how this would have happened.[27] Presumably, scholarly scribes and teachers would have been the sort of people to have access

---

sion of his Sixty-Fifth Birthday (eds. M.F.J. Baasten and W.Th. van Peursen; OLA 118; Leuven: Peeters, 2003), 3-24.

[23] See Chapter Two n. 160.

[24] On the importance of direct, oral instruction, the "living voice," see L. Alexander, "The Living Voice: Scepticism towards the Written Word in Early Christian and in Graeco-Roman Texts," in *The Bible in Three Dimensions: Essays in Celebration of Forty Years of Biblical Studies in the University of Sheffield* (eds. D.J.A. Clines, S.E. Fowl and S.E. Porter; JSOTSup 87; Sheffield: Sheffield Academic Press, 1990), 221-47; J. Mansfeld, "Galen, Papias, and Others on Teaching and Being Taught," in *Things Revealed: Studies in Early Jewish and Christian Literature in Honor of Michael E. Stone* (eds E.G. Chazon, D. Satran and R.E. Clements; JSJSup 89; Leiden: Brill, 2004), 317-29.

[25] The reference to the Babylonian priest Berossos who founded a school on the Greek island of Cos does not shed much light on the exact nature of this school. See Vitruvius, *On Architecture* 9.6.2. Regarding the schools of "Chaldean astrologers," cf. Gordon, "Hellenistic Magic and Astrology," 132.

[26] See Barton, *Ancient Astrology*, 134-39.

[27] Similar questions may also be asked regarding the emergence of an interest in medicine in the third-second centuries BCE, as indicated by the physicians that Ben Sira refers to (Sir 10:10; 38:7.12.15, although we know nothing of the state of their learning) and the interest in medicinal knowledge in the *Book of Tobit* or *Jub.* 10:10–13 (cf. B. Kollmann, "Göttliche Offenbarung magisch-pharmakologischer Heilkunst im Buch Tobit," *ZAW* 106 [1994]: 289-99). Medicinal knowledge and treatment skills were passed on in a direct, practical context, just like other crafts in antiquity. In *Jub.* 10:10–13 medicine is instructed orally to Noah, who then also writes it down in a book that he gives to his son Shem for further transmission. For the learning of skills and practical insight in ancient medicine besides the knowledge in books, see e.g. O. Temkin, "Greek Medicine as Science and Craft," *Isis* 44 (1953): 213-25. For the mainly oral transmission of craft knowledge, see Long, *Openness*.

to texts such as *4QZodiacal Physiognomy* and *4QPhysiognomy ar*, being interested in the knowledge they contained and perhaps also responsible for copying them. Although the localization of this interest with learned people is plausible, it remains rather vague.

There is little concrete evidence for education in Palestine during the Hellenistic and Early-Roman period.[28] Wisdom to manage oneself in various areas of life vis-à-vis the aristocratic upper class and the rural lower class could be passed on by individuals such as Ben Sira, who, around the turn of the third–second centuries BCE, probably provided private instruction at his home. Teachers like Ben Sira would have transmitted most of their knowledge orally. Actual texts played a minor role in the transmission of tradition through education.[29] This was probably also the case later at Qumran during the first century BCE until the first century CE.[30]

There are no clear data for institutionalized schools in this period. Writing and scribal training, for example, may have been in the hands of both priestly scribes and secular scribal guilds, the latter perhaps being responsible for administrative and business documents of ordinary citizens and merchants, but this is no more than conjecture.[31] Moreover, ancient Jewish scribes could perform different functions in different contexts, which would entail different levels of training. One needs to be careful positing the existence of a well-defined, homogenic class or group of scribes in Palestine during the Hellenistic-Early Roman period. Sometimes scribes would have been no more than copyists, while in other instances there could have been overlap between scholarly and scribal activity.[32] Nonetheless, contrary to Greco-Roman societies where the aristocratic elites did not have a high regard for teachers and scribes, writing and learning would have had more

[28] Cf. Hezser, *Jewish Literacy*, 39-109. On Jewish upper-class education in Alexandria, see Crenshaw, *Education in Ancient Israel*, 11-15.

[29] Cf. O. Wischmeyer, *Die Kultur des Buches Jesus Sirach* (BZNW 77; Berlin: Walter de Gruyter, 1995), 174-200; J.L. Crenshaw, "The Primacy of Listening in Ben Sira's Pedagogy," in *Wisdom, You Are My Sister: Studies in Honor of Roland E. Murphy, O.Carm., on the Occasion of His Eightieth Birthday* (ed. M.L. Barré; CBQMS 29; Washington, DC: Catholic Biblical Association of America, 1997), 172-87; Crenshaw, *Education in Ancient Israel*, 228-30, 269-72; R. Doran, "The High Cost of a Good Education," in *Hellenism in the Land of Israel* (eds. J.J. Collins and G.E. Sterling; CJAS 13; Notre Dame, Indiana: University of Notre Dame, 2001), 94-115; D.M. Carr, *Writing on the Tablet of the Heart: Origins of Scripture and Literature* (New York: Oxford University Press, 2005), 201-14.

[30] Alexander, "Literacy among Jews in Second Temple Palestine," 20-21.

[31] Crenshaw, *Education in Ancient Israel*, 112-13.

[32] Cf. C. Schams, *Jewish Scribes in the Second Temple Period* (JSOTSup 291; Sheffield: Sheffield Academic Press, 1998). For comparative data on scribes of early Christian literature, cf. K. Haines-Eitzen, *Guardians of Letters: Literacy, Power, and the Transmitters of Early Christian Literature* (New York: Oxford University Press, 2000).

social standing in Second Temple period Palestine, but their status should not be exaggerated.[33]

We do not know whether sciences such as astronomy/astrology or physiognomics were on an educational or scribal curriculum, but it is very doubtful. Scholars assume that ancient Jewish education in reading and writing was primarily centered on Torah study, but that instruction in certain general skills such as to count, weigh, measure, and calculate may have been taught by parents or relatives.[34] It is also plausible that such skills were taught in practice whenever certain crafts required such knowledge. Be that as it may, one should allow for the possibility that expertise in physiognomic and astrological learning circulated with individual, scholarly scribes or teachers from Jewish society who presumably attracted some students. Whether these scholars came from a priestly or secular milieu is impossible to determine; neither can be ruled out. Such people would have been responsible for the transmission of learned physiognomic and astrological texts in Second Temple period Palestine. Familiarity with these arts was probably due to direct or indirect contacts with Babylonian and/or Hellenistic culture.[35]

Alien Wisdom

In the case of the physiognomic and astrological learning exhibited by the Qumran catalogues, it is likely that we are dealing with the appropriation of "alien wisdom" in Hellenistic-Early Roman period Judaism.[36]

With regard to scientific learning in different strands of Enochic literature – *1 Enoch* as a whole as well as its different parts being composite texts – scholars have rightly pointed to a Mesopotamian background for the astronomical aspects of the *Astronomical Book* (*1 En.* 72–82).[37] For certain

[33] Cf. Crenshaw, *Education in Ancient Israel*, 178-79; Alexander, "Literacy among Jews in Second Temple Palestine," 17.

[34] Cf. Hezser, *Jewish Literacy*, 73-74. Consider, however, also the pedagogical function of the metrological number and measure notations in the *Aramaic Levi Document* 31-47. See H. Drawnel, *An Aramaic Wisdom Text from Qumran: A New Interpretation of the Levi Document* (JSJSup 86; Leiden: Brill, 2004), 280-93.

[35] On Hellenism in Palestine, see e.g. Hengel, *Judentum und Hellenismus*; Collins and Sterling, *Hellenism in the Land of Israel.*

[36] Cf. A. Momigliano, *Alien Wisdom: The Limits of Hellenization* (Cambridge: Cambridge University Press, 1975); Gordon, "Hellenistic Magic and Astrology," 132; Alexander, "Beginnings of Jewish Interest in Natural Science," 232.

[37] See e.g. O. Neugebauer, "The 'Astronomical' Chapters of the Ethiopic Book of Enoch (72 to 82)," in The Book of Enoch *or* 1 Enoch*: A New English Edition* (M. Black; SVTP 7; Leiden: E.J. Brill, 1985), 386-414, at 387, 394-95; Albani, *Astronomie und Schöpfungsglaube*, 155-72; Ben-Dov, *Astronomy and Calendars*. For the Mesopotamian background of other features too, see J.C. VanderKam, *Enoch and the Growth of an Apocalyptic Tradition* (CBQMS 16; Washington, DC: Catholic Biblical Association of America, 1984); H.S. Kvan-

elements of cosmography and geography in the *Book of the Watchers* (*1 En.* 1–36), however, a Greek background besides a Mesopotamian one is also possible.[38] A Babylonian origin for the transmission of scientific ideas into Second Temple period Judaism is further strengthened by Qumran calendar texts that use elements of a Babylonian lunar system.[39] It seems probable that this westward transmission of scientific ideas from Babylonia occurred through Aramaic sources.[40] If these available sources indicate predominantly a Babylonian origin for much of the learned knowledge in Second Temple period texts, then the question is how the physiognomic texts from Qumran fit in this picture.

The Aramaic physiognomic catalogue *4QPhysiognomy ar* may have a Babylonian background, but, apart from its language, the text provides no conclusive evidence for this. It is possible that Mesopotamian examples stand behind the text, but, as has been argued in Chapter Two in the section on the literary dependency of the physiognomic catalogues, a Hellenistic origin cannot be excluded.

If the modified ascendant interpretation for the Hebrew physiognomic catalogue *4QZodiacal Physiognomy* is correct, then this text presents significant evidence for a Hellenistic origin and transmission of certain scientific ideas into Palestine during the Hellenistic-Early Roman period. The text's concern with the ascendant part of the zodiacal sign at the moment of birth points decisively in the direction of Greek astrology. Even more, if *4QZodiacal Physiognomy* stands in the tradition that merges *melothesia* and *dodecatemoria*, it is possible that Jews in Palestine appropriated Hellenistic astrology through Egyptian channels, perhaps through Alexandria. The oldest example of this tradition is Teucer, who was active in Egypt, being most likely of Greek descent, probably during the first century BCE.[41] Amidst much evidence for the Babylonian origin of scientific ideas in Second Temple period Judaism, then, one should allow for the possibil-

---

vig, *Roots of Apocalyptic: The Mesopotamian Background of the Enoch Figure and the Son of Man* (WMANT 61; Neukirchen-Vluyn: Neukirchener Verlag, 1988).

[38] Cf. Nickelsburg, *1 Enoch 1*, 61-62, 279-89; K. Coblentz Bautch, *A Study of the Geography of 1 Enoch 17-19: "No One Has Seen What I Have Seen"* (JSJSup 81; Leiden: Brill, 2003), 231-57.

[39] Cf. Ben-Dov and Horowitz, "Babylonian Lunar Three."

[40] Cf. Greenfield and Sokoloff, "Astrological and Related Omen Texts," 202; Alexander, "Beginnings of Jewish Interest in Natural Science," 238-39. J. Ben-Dov, "Mesopotamian Science in West Semitic Sources: The Case of Qumran," in *The Interpretations of Ancient Astral Science* (eds. H. Falk and D. Brown), *forthcoming*, has recently argued that concrete evidence for an Aramaic medium is lacking for the spread of scientific astronomy from Babylonia to Greece and Egypt, which seems to have been rendered immediately into Greek. In the case of celestial divination, Aramaic sources may have transmitted Babylonian lore. Cf. also Chapter Two nn. 23 and 205. I am grateful to Jonathan Ben-Dov for sending me his Berlin paper.

[41] See Chapter Three nn. 77, 219.

ity that *4QZodiacal Physiognomy* testifies to a Greco-Egyptian background. What to us may seem enigmatic terminology – the "second column," or the "house of light" and the "house of darkness" – possibly represents an effort on the part of an ancient Jewish author to translate and appropriate foreign learning and unfamiliar words and concepts.

Although a Jewish interest in matters such as physiognomics and astrology may already date to the Persian period, like Enochic astronomy,[42] the Hellenistic period is more likely, at least as far as the astrological notions in *4QZodiacal Physiognomy* are concerned. The Qumran manuscripts are the oldest Jewish texts available on this matter, dating to the second half of the first century BCE, and there is no need to push the date back much further. This temporal limitation, however, does not imply a limitation in geographic region. It is possible that physiognomic learning was also received from Mesopotamian culture. Cultural contacts during the Hellenistic-Early Roman period were not limited to one geographic area. Thus, it is possible for *4QPhysiognomy ar* to have a Mesopotamian background, while *4QZodiacal Physiognomy* may have a Greco-Egyptian one.

Martin Hengel argues that these texts were an expression of a rationalistic interest in Hellenistic sciences by the Essenes. He limits this interest to the early period of the Qumran community around the middle of the second century BCE.[43] The date, however, of the manuscripts seems at odds with this limitation. Those responsible for writing or copying the texts during the second half of the first century BCE apparently regarded them worthwhile enough to do so, which demonstrates an interest in the subject matter they convey. Although it is doubtful whether the interest in matters of astrology and physiognomics should be characterized in terms of rationalism, it seems likely that it presupposes openness to the scientific notions that were current in surrounding cultures. Given the non-sectarian character of these texts, however, they not only demonstrate an interest in these matters on the part of the Qumran sect, but also of Judaism at large.

Ancient Jewish Science and Revealed Things

Scholars have regarded Jewish interests in natural science and the cosmos and its workings as an integral part of Enochic literature as well as of other apocalyptic texts in which lists of revealed things play a central role. That literature would evince an attitude according to which esoteric knowledge concerning man, nature, and the cosmos did not remain hidden for all of mankind but was revealed to some special individuals through heavenly

[42] Alexander, "Beginnings of Jewish Interest in Natural Science," 231-32, 240.

[43] Hengel, *Judentum und Hellenismus*, 432-38. See also M. Hengel, "Qumran and Hellenism," in *Religion in the Dead Sea Scrolls* (eds. J.J. Collins and R.A. Kugler; SDSSRL; Grand Rapids, Michigan: Eerdmans, 2000), 46-56, at 54.

mediation. Interest in these matters in apocalyptic literature has led scholars to see the people behind it as part of a well-educated body of people from Jewish society, possibly active in and/or around the Jerusalem Temple, which entertained speculations about the cosmos and humankind's position in it. Around the turn of the third–second centuries BCE, people transmitting texts like the Enochic *Astronomical Book* and the *Book of the Watchers* may have been in opposition with a wisdom teacher like Ben Sira who was critical of investigations into realms of knowledge that had not been revealed for all to see, as opposed to the revealed exoteric wisdom transmitted by himself.[44]

In their efforts to understand ancient Jewish scientific interests scholars have mainly focused on the material provided by *1 Enoch* and the lists of revealed things in apocalyptic texts. This is not the place to raise the issues of origins, developments, and attitudes towards science in Second Temple period Judaism in depth, but two brief comments should be made when the physiognomic and astrological lists from Qumran are taken into account.

First, although it is possibly an anachronism to argue for a separate domain of scientific inquiry in Second Temple period Judaism, one should allow for the possibility that the physiognomic and astrological texts from Qumran were not framed by a religious interest in divine, eschatological judgment, as is the case with the Enochic and apocalyptic material.[45] Although the original, complete texts may have given these lists in a revelatory, apocalyptic framework, this seems unlikely.[46] The Qumran physio-

[44] Although these different attitudes have been related to social groups, their exact identities, social positions, and mutual connections remain debated issues. See e.g. Stone, "Lists of Revealed Things"; Stone, "Book of Enoch"; M.E. Stone, *Scriptures, Sects and Visions: A Profile of Judaism from Ezra to the Jewish Revolts* (Cleveland: Collins, 1980), 27-47; M.E. Stone, "Enoch, Aramaic Levi and Sectarian Origins," *JSJ* 19 (1988): 159-70; M. Himmelfarb, *Ascent to Heaven in Jewish and Christian Apocalypses* (New York: Oxford University Press, 1993), 72-94; R.A. Argall, 1 Enoch *and Sirach: A Comparative Literary and Conceptual Analysis of the Themes of Revelation, Creation and Judgment* (SBLEJL 8; Atlanta: Scholars Press, 1995); Nickelsburg, *1 Enoch 1*, 65-67; Alexander, "Beginnings of Jewish Interest in Natural Science"; B.G. Wright III, "Putting the Puzzle Together: Some Suggestions concerning the Social Location of the Wisdom of Ben Sira," in *Conflicted Boundaries in Wisdom and Apocalypticism* (eds. B.G. Wright III and L.M. Wills; SBLSymS 35; Atlanta: Society of Biblical Literature, 2005), 89-112; R.A. Horsley, "The Politics of Cultural Production in Second Temple Judea: Historical Context and Political-Religious Relations of the Scribes Who Produced *1 Enoch*, Sirach, and Daniel," in *Conflicted Boundaries*, 123-45; A.Y. Reed, *Fallen Angels and the History of Judaism and Christianity: The Reception of Enochic Literature* (Cambridge: Cambridge University Press, 2005), 24-83.

[45] Stone, "Lists of Revealed Things"; Stone, "Book of Enoch," 487; Nickelsburg, *1 Enoch 1*, 37-42; Alexander, "Beginnings of Jewish Interest in Natural Science," 234-36, 239; Reed, *Fallen Angels*, 67.

[46] In addition, it is unnecessary to invoke a religious orientation of the subject matter in order to understand it. Cf. the discussion of the interpretation of *4QZodiacal Physiognomy* within the dualistic framework of the *Two Spirits Treatise* discussed in Chapter Four. Moreover, what would be the religious orientation of the selenodromion and brontologion in

gnomic and astrological lists are more elaborate and explicit than the apocalyptic lists of revealed things and more complex than the simple arithmetical scheme in the *Astronomical Book* of *1 Enoch*, although the astronomical material in this latter text might very well have existed independently before being incorporated into the Enochic framework.[47]

Second, the physiognomic and astrological texts from Qumran demonstrate that during the Hellenistic-Early Roman period, Jews in Palestine were interested in contemporary scientific knowledge, and not just in "outdated" forms of Mesopotamian astronomy as in the *Astronomical Book.*

Michael Stone argued that the transmission of the *Astronomical Book* was "a deliberate act of archaism" by people who "faithfully preserved and cultivated views of natural science that were already antiquated at the time they were committed to writing." This attitude was fostered either by "the conscious rejection of Greek science" or by "the creation of a social context into which such science did not penetrate."[48] According to Stone the tradents of the *Astronomical Book* can be characterized as separatist, sectarian circles that kept themselves apart from contemporary culture, its learning and natural science. Moreover, the circles that studied the Enochic tradition were linked in some way to those people who founded the Qumran community.[49]

Such an interpretation of ancient Jewish scientific interests, however, ignores the evidence provided by the physiognomic and astrological texts from Qumran. These texts demonstrate that contemporary scientific notions, such as the zodiac and the connection between physiognomics and astrology, found their way into Jewish society during the Hellenistic-Early Roman period. These two elements represent developments that post-date the older astronomical science in MUL.APIN and the *Astronomical Book.*[50] At

---

4QZodiology and Brontology ar? *Pace* García Martínez and van der Woude, "Groningen Hypothesis," 522-23.

[47] With regard to the possibility of an independent existence of the lists of revealed things, see Stone, "Lists of Revealed Things," 418-19.

[48] Stone, "Enoch, Aramaic Levi," 164. Concerning theological motives for the reception of older Mesopotamian astronomy in the *Astronomical Book*, see Albani, *Astronomie und Schöpfungsglaube*, 273-350.

[49] Stone, "Enoch, Aramaic Levi," 164-66. Cf. Alexander, "Beginnings of Jewish Interest in Natural Science," 239-40. Alexander, "Beginnings of Jewish Interest in Natural Science," 234, characterizes the group or groups behind Enochic literature as innovators and modernists who were open to the scientific thought of their day and attempted to integrate novel ideas in a new form of Judaism. But he does not raise the issue that a type of "outdated" Mesopotamian science was picked up. Cf. Reed, *Fallen Angels*, 68-69.

[50] The zodiac was not developed before the fifth century BCE. It does not figure in the earlier Mesopotamian text MUL.APIN, which was the source for the *Astronomical Book*, see Chapter Three. The connection between astrology and physiognomics occurs first in Late Babylonian sources (*Esoteric Babylonian Commentary* and *LBAT* 1593) and Hellenistic astrology, see Chapter Two.

Qumran these "modern" ideas were apparently transmitted alongside "older" forms of science such as in the *Astronomical Book*.[51] Such a situation demonstrates that in this respect the Qumran community cannot be described as isolationist or ignorant of what was going on in contemporary Babylonian and Hellenistic sciences. Regardless of the motivations of the initial transmitters of the *Astronomical Book* sometime in the fourth–third centuries BCE, it seems that during the first century BCE some people in Jewish society had an open mind for some of the more contemporary scientific developments.

THE STATUS OF PHYSIOGNOMIC AND ASTROLOGICAL LEARNING IN THE QUMRAN COMMUNITY AND SECOND TEMPLE PERIOD JUDAISM

Scholars disagree on the status of the physiognomic lists within the collection of the Dead Sea Scrolls. The discussion centers on *4QZodiacal Physiognomy*. According to some scholars its astrological learning was controversial and of minor interest to the Qumran community. Therefore, it is unlikely that this interest reflects any acceptance of astrology or its practice within the community.[52] With the publication of all the Dead Sea Scrolls, and particularly wisdom texts such as *Instruction* and *Mysteries*, other scholars have drawn attention to various references to times of birth, the position of stars, and horoscopes in the wisdom texts. The references are couched in general terms and do not explicate any technical astrological notions, but it seems likely that they demonstrate the acceptance of certain astrological notions and concepts within these texts, probably in connection with ideas on predestination.[53]

There is no indication that physiognomic learning in Hellenistic-Early Roman period Judaism was opposed in the way that astrology was by some people. Acceptation of the basic premise that the body signifies character seems implied by the instances of physiognomic consciousness in Jewish sources discussed in Appendix II. It is, therefore, likely that physiognomic learning was not as controversial as astrological learning, or at least did not receive as much attention as astrology did in the sources at our disposal.

[51] Some of the Qumran calendars show that Late Babylonian methods of lunar schemes were partly known and recognized. Moreover, the lunar pattern in these texts has no bearing on the cult or the cycle of priestly courses in some of the other calendar texts, see Ben-Dov and Horowitz, "Babylonian Lunar Three."

[52] Cf. Lehmann, "New Light on Astrology"; Lange, "Essene Position on Magic"; Albani, "Horoscopes in the Qumran Scrolls," 315-24.

[53] Cf. E. Tigchelaar, "Your Wisdom and Your Folly: The Case of 1-4QMysteries," in *Wisdom and Apocalypticism*, 67-88; E. Tigchelaar, "Astronomie en kalenders; horoscopen en astrologie," in *Fragmenten uit de woestijn*, 167-83, at 177-80; Kister, "From Ben Sira to *Mysteries*"; Schmidt, "'Recherche son thème de géniture.'"

There are no Jewish texts extant that explicitly criticize physiognomic learning, although Ben Sira cautions his readers not to depend blindly on people's looks.[54] Astrology, however, like magic, attracted much more attention, and attitudes towards astrology varied between favorable and unfavorable.

Ambivalent Attitudes: The Controversial Status of Astrology

The art of astrology has always enjoyed an ambivalent and controversial status. In antiquity astrology attracted its critics,[55] but it was also held in high esteem and could be called "the queen of sciences."[56] While there are both affirmative and unfavorable comments, the controversial status of physiognomic learning is less pronounced,[57] possibly because it was less in the center of political attention than astrology was. Scholars have shown that in the Roman Empire astrological knowledge functioned politically within the highest networks of power, being also a potentially subversive science that needed to be controlled.[58] Regarding physiognomic learning it has been argued that it functioned politically on a more local level of power as a means of social control.[59] Clear evidence, however, equivalent to that for Roman society, is lacking for the political function of astrology or physiognomics in Hellenistic-Roman period Palestine.[60]

With regard to astrology an ambivalent attitude is attested in Jewish sources, demonstrating its controversial status. This is illustrated for exam-

[54] Sir 11:2. See Appendix II.

[55] Cf. Bouché-Leclercq, *L'astrologie grecque*, 570-93; Gundel and Gundel, *Astrologumena*, 99-102, 188-89, 296-99; A.A. Long, "Astrology: Arguments Pro and Contra," in *Science and Speculation: Studies in Hellenistic Theory and Practice* (eds. J. Barnes et al.; Cambridge: Cambridge University Press, 1982), 165-92; Barton, *Ancient Astrology*, 52-57; Barton, *Power and Knowledge*, 50-54, 65.

[56] Philo, *On the Preliminary Studies* §50. Philo is ambivalent in his appreciation. See e.g. von Stuckrad, *Ringen um die Astrologie*, 224-53, who argues that Philo opposed fatalistic (horoscopic) astrology, but did not deny astrology's signifying value regarding general events on earth. See Philo, *On the Creation of the Cosmos* §§58–59a (cf. Chapter Two n. 175). See also Gundel and Gundel, *Astrologumena*, 180-83.

[57] Cf. Chapter Two n. 7.

[58] See F.H. Cramer, *Astrology in Roman Law and Politics* (MAPS 37; Philadelphia: American Philosophical Society, 1954); Barton, *Power and Knowledge*, 27-94; W. Orth, "Astrologie und Öffentlichkeit in der frühen römischen Kaiserzeit," in *Kommunikation in politischen und kultischen Gemeinschaften: Stätten und Formen der Kommunikation im Altertum V* (eds. G. Binder and K. Ehlich; BAC 24; Trier: Wissenschaftlicher Verlag Trier, 1996), 99-132; A. Schmid, *Augustus und die Macht der Sterne: Antike Astrologie und die Etablierung der Monarchie in Rom* (Cologne: Böhlau, 2005).

[59] See Barton, *Power and Knowledge*, 95-131.

[60] Von Stuckrad, *Ringen um die Astrologie*, 102-33, 141-58, argues that astrological considerations were prominent in political decisions made by Alexander Jannaeus, Herod the Great and Bar Kokhba. However, apart from adducing stellar imagery and referring to astrologically significant planetary positions at certain moments, he fails to provide convincing evidence for astrological reasoning behind the political motives of these three leaders.

ple by the debate on the origin of astrology. Most ancient astrological sources state that astrology originated in Egypt, while some acknowledge a Mesopotamian origin.[61] On the one hand, there were Jewish writers, such as Artapanus (third–second century BCE), Pseudo-Eupolemus (prior to the first century BCE), and Josephus (first century CE), who contributed to this debate by claiming a Jewish origin for astrological learning with Enoch or Abraham. On the other hand, sources such as the Enochic *Book of the Watchers* (8:3) and *Jubilees* (8:3) traced astrological learning back to fallen angels (the Watchers) illicitly revealing secret knowledge to humankind who sinned because of this learning.[62] The first position demonstrates pride in having a great ancestor teaching astrology to other peoples, while the second position shows wariness of its potential sinfulness in locating it with fallen angels who sin by disclosing astrological teachings to people.[63]

It is not possible, or desirable, to define a monolithic stance on astrology within Hellenistic-Early Roman period Judaism.[64] In different contexts different attitudes will have existed, but each source needs to be evaluated on its own account first.[65] Considering, for example, the people behind Enochic literature, one should allow for the possibility that the appropriation of alien wisdom like astrology was initially considered favorable, but

[61] Cf. Gundel and Gundel, *Astrologumena*, 9-51.

[62] Cf. also the critical stance of *Jub.* 12:16–18.

[63] Cf. Bouché-Leclercq, *L'astrologie grecque*, 577-78; Gundel and Gundel, *Astrologumena*, 51-59; C. Böttrich, "Astrologie in der Henochtradition," *ZAW* 109 (1997): 222-45; G.H. van Kooten, "Enoch, the 'Watchers,' Seth's Descendants and Abraham as Astronomers: Jewish Applications of the Greek Motif of the First Inventor (300 BCE-CE 100)," in *Recycling Biblical Figures: Papers Read at a NOSTER Colloquium in Amsterdam, 12-13 May, 1997* (eds. A. Brenner and J.W. van Henten; STAR 1; Leiden: Deo, 1999), 292-316; Nickelsburg, *1 Enoch 1*, 188-201; A.Y. Reed, "Abraham as Chaldean Scientist and Father of the Jews: Josephus, *Ant.* 1.154-168, and the Greco-Roman Discourse about Astronomy/Astrology," *JSJ* 35 (2004): 119-58; A.Y. Reed, "Heavenly Ascent, Angelic Descent, and the Transmission of Knowledge in 1 Enoch 6–16," in *Heavenly Realms and Earthly Realities in Late Antique Religions* (eds. R.S. Boustan and A.Y. Reed; Cambridge: Cambridge University Press, 2004), 47-66.

[64] This also applies to rabbinic Judaism. L. Wächter, "Astrologie und Schicksalsglaube im rabbinischen Judentum," *Kairos* 11 (1969): 181-200, construes strong opposition between astrology and Jewish religion and argues that rabbinic Judaism in Palestine was not really affected by astrological teachings. Lehmann, "New Light on Astrology," argues that astrology was a sectarian phenomenon at variance with normative Judaism. Both approaches fail to take account of the diversity of positions, and thereby downplay evidence for astrological interests. They also give too much weight to certain theological positions, but these do not necessarily result in the absence or marginalization of astrology within society.

[65] Cf. J.H. Charlesworth, "Jewish Interest in Astrology during the Hellenistic and Roman Period," *ANRW* II 20.2 (1987), 926-50; Von Stuckrad, *Ringen um die Astrologie*; Von Stuckrad, "Jewish and Christian Astrology in Late Antiquity"; M. Bar Ilan, "Astrology in Ancient Judaism," in *The Encyclopaedia of Judaism: Volume V, Supplement Two* (eds. J. Neusner, A.J. Avery-Peck and W.S. Green; Leiden: Brill, 2004), 2031-37. See also W. Hübner, *Zodiacus Christianus: Jüdisch-christliche Adaptationen des Tierkreises von der Antike bis zur Gegenwart* (BKP 144; Meisenheim: Anton Hain, 1983).

came to be regarded more critically at a later stage during the Hellenistic period.[66] If this model is appropriate for explaining the textual development of certain parts of the Enochic corpus, this does not, however, imply a linear development because later tradents of this literature may have chosen to emphasize the former, more positive stance.

Given the quantity, distribution, and nature of the sources at our disposal, it is not possible to draw detailed and definitive conclusions regarding the development of attitudes towards astrology. Scholars, however, should not simply lend more authority to sources or stages of tradition that exhibit a critical attitude when assessing the status of astrological learning at Qumran and beyond. The people behind the text of *Jubilees* may have been disinclined to accept lunar calculations for calendrical purposes, but this did not stop people at Qumran, where many copies of *Jubilees* have been found (implying its authoritative status for the community),[67] from engaging in such matters. Texts like *Jubilees* may have been authoritative at Qumran for certain matters, such as chronology, but not for others. Similarly, for the astrological learning in *4QZodiacal Physiognomy* or *4QZodiology and Brontology ar* it seems likely that these texts, together with the astrological references in texts like *Instruction* and *Mysteries*, represent more than just a critical interest in their subject matter, demonstrating a varied interest in astrology at Qumran.

Inverted and Mixed Writing and Scribal Strategies of Secrecy[68]

When considering texts as cultural artifacts in themselves,[69] it is important to reflect on the inverted and mixed writing of *4QZodiacal Physiognomy* and its significance concerning the status of the text at Qumran. Why was *4QZodiacal Physiognomy* written or copied in the manner that it was?[70]

[66] Cf. Alexander, "Beginnings of Jewish Interest in Natural Science," 234-36; Reed, *Fallen Angels*, 42-44, 60.

[67] The alleged authoritativeness is based on two arguments: 1. the number of copies; 2. the purported references to or quotations from the *Book of Jubilees* in CD 16:3–4 and 4Q228. On the first quotation, cf. most recently D. Dimant, "Two 'Scientific' Fictions: The So-Called *Book of Noah* and the Alleged Quotation of *Jubilees* in CD 16:3-4," in *Studies in the Hebrew Bible*, 230–49.

[68] See also Popović, "Physiognomic Knowledge in Qumran and Babylonia," 166-76; M. Popović, "The Human Body, Astrology, and Writing in Code: Physiognomic Divination in the Dead Sea Scrolls," paper read at the CNRS Symposium *Divination et écriture, écriture de la divination*, Paris, December 9, 2005. The proceedings are to be published by J.-J. Glassner.

[69] Cf. Haines-Eitzen, *Guardians of Letters*; Alexander, "Literacy among Jews in Second Temple Palestine."

[70] The fact that the Aramaic physiognomic list *4QPhysiognomy ar* was not written in a cryptic manner does not render the following discussion obsolete. Similar to the use of the Cryptic A script, it is difficult to assess the reasons why some texts were written in this script and others of similar genre were not. See also the Introduction n. 30.

It is possible that the manner of writing was meant to convey a sort of scribal pride. It seems hardly the work of a beginning scribe. The writing is executed with a fine hand, which concurs with the effort it must have cost to write in the direction from left to right, different from the usual right to left.[71] Together with the use of characters from various scripts, the inverted writing of the entire manuscript can be considered a distinguishing mark that indicates scribal craftsmanship and perhaps scholarly pride. These considerations may suggest that this manuscript was the work not only of a skilled scribe but also perhaps of a scholarly scribe.

In addition to the possibility of an intended magical effect,[72] the inverted and mixed writing possibly represents a scribal strategy of secrecy, limiting the availability of and accessibility to the physiognomic-astrological learning in the text.[73] The manner of writing prevents people from easily reading the text and quickly obtaining knowledge of its subject matter.[74] Some scholars have suggested that the use of inverted and mixed writing is a literary style to restrict knowledge of the text's contents to a few initiates within the Qumran community, such as the *Maskil*.[75] The in-

[71] People's names were sometimes written backwards – Ναδυοι (=Ἰουδάν [Judah]); מולש (=שלום [Salome]) – which may be the result of confusion between Greek and Hebrew. See T. Ilan, *Lexicon of Jewish Names in Late Antiquity: Part I, Palestine 330 BCE – 200 CE* (TSAJ 91; Tübingen: Mohr Siebeck, 2002), 31, 115, 250. This phenomenon, however, does not explain the inverted writing in an entire text such as in *4QZodiacal Physiognomy*.

[72] See the section on magico-medicinal stones in Chapter One.

[73] Epiphanius, *Panarion* 19.4.3, transmitting information ascribed to Elxai, gives an intriguing example of unintelligible words written in Greek that most likely render an Aramaic sentence when read from right to left (the inversion may have been there in the hypothetical Aramaic text). According to Elxai nobody should search for the meaning of these words. The sentence has effect in the (translated) Greek book. Perhaps it would have been readily understandable to Aramaic readers, but the question why it was written in an inverted way in the original text remains, even if the prohibition of not searching for its meaning was inserted by the translator. See G.P. Luttikhuizen, *The Revelation of Elchasai: Investigations into the Evidence for a Mesopotamian Jewish Apocalypse of the Second Century and its Reception by Judeo-Christian Propagandists* (TSAJ 8; Tübingen: J.C.B. Mohr [Paul Siebeck], 1985), 100-3, 124-25. Of course, this is not exactly the same as the text of *4QZodiacal Physiognomy*, but it may present an illustration of the mystery surrounding inverted writing.

[74] It took some time before modern scholars understood how to read the text, as is clear from Allegro's letters home. See Brown, *John Marco Allegro*, 29-30. This difficulty was later downplayed by Allegro. See Allegro, *The Dead Sea Scrolls: A Reappraisal*, 57.

[75] See Carmignac, "Les Horoscopes," 199; Hengel, *Judentum und Hellenismus*, 436; D. Mendels, "Hellenistic Utopia and the Essenes," in *Identity, Religion and Historiography: Studies in Hellenistic History* (D. Mendels; JSPSup 24; Sheffield: Sheffield Academic Press, 1998), 420-39, at 435-36 (originally published in *HTR* 72 [1979]: 207-22); Lichtenberger, *Studien zum Menschenbild*, 144; Charlesworth, "Jewish Interest in Astrology," 939-40; Alexander, "Physiognonomy," 391. For the *Maskil*, see Chapter Four n. 27.

K. von Stuckrad, *Frömmigkeit und Wissenschaft: Astrologie in Tanach, Qumran und frührabbinischer Literatur* (EH 23/572; Frankfurt: Peter Lang, 1996), 123-24, argues that the encoding of *4QZodiacal Physiognomy* may be a reaction to the Roman imperial edicts prohibiting the practice of astrology and especially genethlialogy. His argument is not convincing and misunderstands *4QZodiacal Physiognomy* as a horoscope.

verted and mixed writing thus indicates that "the contents of the text were not intended for everybody, and that uttermost care was taken to keep them accessible only to a very few experts."[76]

Others argue that this was done because of the controversial nature of the contents and because astrology would have been valued negatively within the Qumran community.[77] According to Albani "the mixture of various sorts of scripts probably gave the impression of impurity to Jewish readers" and the cryptic script "could hardly be a sign of the special value of the contents of this document."[78] This assumption, however, is contradicted by epigraphic evidence from Masada. Among three hundred and one small ostraca, some are inscribed with characters of the paleo-Hebrew script in combination with Greek letters, such as *alpha* and *beta*.[79] In addition, quite a few ostraca of another eighty tags with specific names combine characters from square, paleo-Hebrew, and Greek scripts.[80] These epigraphic examples from Masada make it clear that the use of various scripts need not have signaled a notion of impurity to Jews in antiquity. This suggests that it need not have been the possibly controversial character of the learning contained in *4QZodiacal Physiognomy* that caused the concealment.

It may, I suggest, rather have been the high status accredited to the kind of knowledge transmitted by the text that was the reason for the manner in which it was written. Whatever the attitudes at times of some people towards the arts of physiognomics and astrology may have been, submitted to writing in technical lists they unmistakably also represented higher forms of learning, handled by educated people. The utilization of the writing techniques of inverted writing and mixed scripts was a scribal means to limit the accessibility to and availability of this expert knowledge to those who were suitable to understand and use it. Literacy at Qumran may have been widespread, but must not be exaggerated. The inverted and mixed writing of *4QZodiacal Physiognomy* can thus have been effective enough to discourage insufficiently skilled and knowledgeable members of the community from taking account of its contents. In the context of the Qumran community this probably means that leadership figures like the *Maskil* were

[76] García Martínez, "Magic in the Dead Sea Scrolls," 30.

[77] Cf. e.g. Albani, "Horoscopes in the Qumran Scrolls," 318-21.

[78] Albani, "Horoscopes in the Qumran Scrolls," 320-23. See also n. 52 above.

[79] Y. Yadin and J. Naveh, *The Aramaic and Hebrew Ostraca and Jar Inscriptions* (vol. 1 of *Masada: The Yigael Yadin Excavations 1963-1965, Final Reports*; Jerusalem: Israel Exploration Society/The Hebrew University of Jerusalem, 1989), 13, 16. See also Naveh, "Aramaic Magic Book from Qumran," 254 n. 7.

[80] Yadin and Naveh, *Aramaic and Hebrew Ostraca*, 13, 17-19, 41-42. According to Naveh these shards might have been used as "tokens [...] in the food-rationing system at besieged Masada." (18)

the ones with access to and in possession of the learned knowledge of *4QZodiacal Physiognomy*.[81]

The technical compendium *4QZodiacal Physiognomy* demanded a certain level of knowledge of those who were suitable to handle its learning, analogous to the demands expressed in the Babylonian *Esagil-kīn-apli Catalogue* or the prohibitions in Greco-Roman astrological treatises not to transmit knowledge to the uninitiated,[82] which in turn affected the way the manuscript was written.

The manner of writing may, therefore, demonstrate some form of information control on the dissemination of certain forms of learning within the community of Qumran.[83] Accordingly, the possession of this prized piece of learning bestowed and confirmed power and prestige. The anthropologist Hugh Urban has studied the ways and strategies that people in the Śrīvidyā Indian Tantra and the Rectified Scottish Rite of French Freemasonry traditions use to conceal or reveal certain valued information, which in turn bestows status on the one controlling that knowledge. Urban argues that the concept of secrecy is best understood "in terms of its *forms* or *strategies* – the tactics by which social agents conceal or reveal, hoard or exchange, certain valued information. In this sense, secrecy is a discursive strategy that transforms a given piece of knowledge into a scarce and precious resource, a valuable commodity, the possession of which in turn bestows status, prestige, or symbolic capital on its owner."[84] One can perhaps say that the written form of *4QZodiacal Physiognomy* signified the cultural capital that the text's learning represented and that possession of the manuscript bestowed symbolic capital on the people in possession of it.

VALUE AND FUNCTION OF LEARNED KNOWLEDGE

The knowledge in the physiognomic and astrological lists may have been valued in itself as a piece of speculative, scientific learning about man and certain cosmic elements without any further more practical relevance.[85] It is possible that this knowledge was regarded as a valuable commodity in itself, bestowing status on those in possession of it.[86]

[81] Cf. Alexander, "Literacy among Jews in Second Temple Palestine," 18-19.

[82] See Chapter Two nn. 58, 158. Besides being literary *topoi*, one should allow for the possibility that in some texts such statements may also indicate a real concern with guarding knowledge from other people in one way or another.

[83] Cf. Chapter Two n. 62.

[84] Urban, "The Torment of Secrecy," 210.

[85] Cf. Alexander, "Physiognonomy," 389.

[86] Apart from the status that the lists of revealed things may give to the one claiming to be in possession of the speculative knowledge enumerated in them, their function and role in

Pierre Bourdieu's ideas concerning "cultural capital," and social distinctions may be helpful in understanding the social value of knowledge here.[87] Education can be seen as an investment in cultural capital, but also as a mechanism reinforcing the social position of those already in possession of the necessary dispositions to enter into it. The cultivation of different forms of knowledge can function as a criterion for social distinctions between people. Education and knowledge bestow status. Tacitly they signal, confirm, and regulate existing relations of power by the inclusion or exclusion of people from certain circles or institutes of society.

Within this context of "cultural capital" the learning of *4QZodiacal Physiognomy* and *4QPhysiognomy ar* may be appreciated as a prized piece of knowledge signaling and confirming the status of those having access to and possessing it. These texts perhaps objectified the speculative, scientific interests of some elite members of Hellenistic-Early Roman period Jewish society or of the Qumran community. The pursuit and possession of that knowledge may have confirmed that elite status.

In addition to such an appreciation of the social value of the learning contained in *4QZodiacal Physiognomy* and *4QPhysiognomy ar*, it is also possible to venture some hypotheses on a more practical application of the knowledge conveyed by these texts in different contexts.

With regard to the possible function of *4QPhysiognomy ar* scholars have not made any suggestions, but regarding *4QZodiacal Physiognomy* it has been argued that it possibly functioned as part of a physiognomic test applied by leadership figures like the *Maskil* during the procedure of admission into the Qumran community.[88] As has been made clear in Chapter Two, distinction should be made between the texts and the knowledge contained in them when inquiring about the practical application. The actual texts need not have been used in an actual physiognomic practice, whatever that may be. One should, nevertheless, allow for the possibility that such catalogue texts could function as works of reference, whether at the moment of a physiognomic diagnosis or at another moment, for example, during education or study. Be that as it may, the knowledge conveyed by the texts may have served some practical purpose. In addition, another distinction to

---

apocalyptic literature are not entirely clear. Cf. Stone, "Lists of Revealed Things," 427, 443; Himmelfarb, *Ascent to Heaven*, 79.

[87] See e.g. P. Bourdieu, *Distinction: A Social Critique of the Judgement of Taste* (trans. R. Nice; London: Routledge & Kegan Paul, 1984); P. Bourdieu, "The Forms of Capital," in *Handbook of Theory and Research for the Sociology of Education* (New York: Greenwood, 1986), 241-58. Cf. N. Veldhuis, *Elementary Education at Nippur: The Lists of Trees and Wooden Objects* (Diss. Groningen, 1997), 137-46, for the use of Bourdieu's concept of "cultural capital" to explain the social uses of the science of writing in Old Babylonian education.

[88] See Chapter Four n. 57 and below in this chapter.

be made applies to the possible contexts in which the physiognomic and astrological learning may have functioned. Given that the two catalogues are non-sectarian compositions, one should acknowledge a non-sectarian context alongside a sectarian one.

THE PREDICTIVE FUNCTION OF PHYSIOGNOMIC LEARNING IN *4QPHYSIOGNOMY AR* (4Q561)

It seems that the purpose of physiognomic divination in *4QPhysiognomy ar* is to discern the future and fate of the described individuals. In the discussion in Chapter Two about the functions of Babylonian physiognomic learning, it was remarked that it could have been used as a tool of social control exercising power over the social make-up of different circles such as the royal court or priests at temples. It remains, however, unclear whether and, if so, how the physiognomic omens functioned in such procedures.

Besides a mere intellectual interest in the subject matter and the social value it may have had for people, one can only speculate about the practical purpose of the learning listed in *4QPhysiognomy ar*. Its applied value is difficult to assess.

The predictive value of physiognomic learning in *4QPhysiognomy ar* may have been of interest in a divinatory practice in which people consulted a diviner or physiognomist in order to know what their future had in store for them. An example that comes to mind in another context is that of the family of Britannicus who sought the advice of a *metoposcopus* to inquire about the boy's fate as future emperor.[89] It is also possible that in the context of Qumran knowledge of people's future may have been relevant in understanding their situation. Perhaps the text of *4QPhysiognomy ar* had an eschatological focus or was given one by its readers. The descriptions of the human body may have been understood to somehow reveal certain eschatological information or indications, somewhat similar perhaps to the revelatory function of the description of Noah in *1 En.* 106 (see Appendix II).

THE DIAGNOSTIC FUNCTION OF PHYSIOGNOMIC-ASTROLOGICAL LEARNING IN *4QZODIACAL PHYSIOGNOMY* (4Q186)

In Chapter Four, I concluded that the concern of *4QZodiacal Physiognomy* is the discernment of the nature of people's zodiacal spirit. The division between the "house of light" and the "house of darkness" indicates this nature. This division is astrologically the result of the ascendant zodiacal

[89] Suetonius, *Lives of the Caesars: Titus* 2. Cf. Evans, *Physiognomics*, 52.

sign's position in relation to the eastern horizon (Chapter Three), but in this text it is consequently the zodiacal spirit that is divided between light and darkness.

Similar to the observation of the body in ancient medicine, physiognomic learning has a diagnostic function here.[90] According to *4QZodiacal Physiognomy* the human body signifies the division between light and darkness of people's zodiacal spirits, which have a close relationship with them since the moment of their birth. The diagnosis of people's bodies reveals the character of their zodiacal spirits. What was the importance of this knowledge? What could be the practical relevance and function of knowing someone's zodiacal sign, its position at birth, and the division of the zodiacal sign and its spirit between light and darkness? Apart from a scientific interest in cosmic sympathy between zodiacal signs, their spirits and human beings, and the classification of this knowledge in a catalogue, one should allow for the possibility of a more practical, diagnostic application of the knowledge contained in *4QZodiacal Physiognomy*.

The Importance of Knowing the Nature of People's Zodiacal Spirits

It was important to know what the nature was of the zodiacal spirit that attended someone because it could be a potentially dangerous and harmful being. The appreciation of zodiacal spirits in *4QZodiacal Physiognomy* is, I suggest, more in line with the negative evaluation of them in the *Testament of Solomon* and the Manichean *Kephalaia* than with the positive image in the Cairo Genizah amulet in which *Leo* is adjured.[91] The *Testament of Solomon* speaks of the thirty-six decanal spirits as "the thirty-six demons that plague humanity."[92] Zodiacal spirits could attack people, cause illnesses or other calamities and inconveniences during life.

Before too negative a picture of the zodiacal spirits emerges, one should bear in mind that the influences exerted by the zodiacal signs were of various kinds. Some of these were beneficial, some were less beneficial, and others not at all beneficial.[93] The zodiacal spirits were, just like the zodiacal signs, responsible for what happened to people during life and, therefore, also for people's illnesses or bad luck.

As indicated by *4QZodiacal Physiognomy*, the nature of someone's zodiacal spirit is modified according to the position of the zodiacal sign at the moment of birth. The number of parts within the "house of light" and the "house of darkness" serve, I suggest, to reveal the more or less beneficent or maleficent nature of people's zodiacal spirits. More parts of light would

[90] Cf. Chapter Two n. 109.
[91] See the section on the *Testament of Solomon* and zodiacal spirits in Chapter Four.
[92] *T. Sol.* 18:42.
[93] Cf. Hübner, *Eigenschaften der Tierkreiszeichen*, 201-38.

have been indicative of a beneficial character, while more parts of darkness of a maleficent one. One should allow for the possibility that more than just two opposing characters were indicated, that the situation was more nuanced. It was not a matter of simply beneficent or maleficent. For example, zodiacal spirits with more parts of light could still cause inconveniences, but these would have been of a less harmful nature than those caused by spirits with less parts of light. In other words, to the readers of the text of *4QZodiacal Physiognomy* the division of light and darkness indicated, I suggest, the degree to which zodiacal spirits were potentially harmful; whether a spirit could be really harmful or not very. The more light, the less harmful, the more darkness, the more harmful an attack by a zodiacal spirit would have been.

Practical Applications of Physiognomic-Astrological Learning in *4QZodiacal Physiognomy* (4Q186)

Regarding the practical application, people could have used the knowledge listed in *4QZodiacal Physiognomy* as a diagnostic tool during a physiognomic inquiry. The physiognomic diagnosis was believed to determine people's horoscopes and the nature of their zodiacal signs and spirits.

Knowledge of the nature of people's zodiacal spirit could be relevant in various contexts. Here I suggest two possible contexts, not mutually exclusive, for the diagnostic value of the physiognomic-astrological knowledge in *4QZodiacal Physiognomy*. First, knowing the nature of zodiacal spirits could have been important in a more general magico-medicinal context. Second, in line with earlier proposals, this knowledge could have been relevant in a sectarian context to control and regulate admission to the Qumran community.

A Magico-Medicinal Context: Diagnosing the Zodiacal Culprit

In a magico-medicinal context knowledge of a zodiacal sign's character could have been helpful to determine treatments. People could consult a medical astrologer, *magus*, or doctor with complaints of illness or bad luck for which their zodiacal spirit might be responsible, in which case astrological circumstances needed to be taken into account.[94] In antiquity, however, many people would not have known their time of birth or zodiacal

[94] For the connection between astrology and medicine (iatromathematics), see e.g. Bouché-Leclercq, *L'astrologie grecque*, 517-42; F. Cumont, *L'Égypte des astrologues* (Brussels: Fondation Égyptologique Reine Élisabeth, 1937), 93-94, 128-29, 171-74; A.-J. Festugière, *La révélation d'Hermès Trismégiste, I: L'astrologie et les sciences occultes* (2d ed.; CEA.EtB; 1950; repr., Paris: Les Belles Lettres, 1983), 123-86; Reiner, *Astral Magic*, 43-60; Sassi, *Science of Man*, 177-83.

sign. It was, therefore, important to discover these in order to identify the zodiacal culprit.

In this situation the medical astrologer could resort to a physiognomic examination diagnosing which zodiacal spirit was causing the affliction and what its nature was. In other words, knowledge of the zodiacal sign's character would have determined the diagnosis of the disease's nature; we would say, whether a disease was of benign or malicious character.

In terms of pneumatology, this understanding of diseases being caused by external spirits is matched by other texts that identify the cause of illnesses to be external and demonic, such as *T. Sol.* 18.[95] Other texts from Qumran also identify external spirits as the cause of diseases and afflictions, demonstrating that the spirits attacking people were believed to be of varying sorts. For example, the fragmentary Aramaic text 4Q560 (*4QExorcism ar*), which was originally some sort of demonic catalogue, listed adjurations against demons that cause pregnancy or childbirth problems, inflict various illnesses, and disturb people's sleep by dreams.[96] Although this text, as far as can be determined from its remains, does not speak of zodiacal spirits, one should allow for the possibility that zodiacal spirits were one kind of spirit imagined by members of the Qumran community and other Jews in Hellenistic-Early Roman period Palestine.

Magico-Medicinal Countermeasures and Apotropaic Stones

On the basis of his knowledge about the identity and nature of the zodiacal spirit, the medical astrologer could determine which appropriate magico-medicinal countermeasures to take to cure the patient and end the afflictions that are bothering him or her.

Knowledge of which zodiacal sign was responsible for what illness enabled astrologers, magicians, or doctors to take appropriate measures for curing people. For example, in one of his writings the second-century CE physician Galen ridicules a certain Pamphilus for claiming to use the thirty-six sacred herbs of demons and decans from a Hermetic text and also for his use of incantations and spells when gathering these herbs.[97] Galen's criticism demonstrates that at least some people believed astrology, demono-

[95] This sort of external cause of disease is in line with Babylonian medicine but not with Greek medicine, where illness is understood to have an internal cause resulting from the imbalance of the mixture (κρᾶσις) of humors. Cf. Chapter Two n. 188.

[96] See Penny and Wise, "By the Power of Beelzebub"; Alexander, "Wrestling Against Wickedness," 329-30; Naveh, "Aramaic Magic Book from Qumran"; García Martínez, "Magic in the Dead Sea Scrolls," 26-28. Cf. also the Cave 4 fragments of the Damascus Document that deal with the treatment of skin diseases and speak of a spirit that enters the body, see Chapter Two n. 187.

[97] Galen, *De simplicium medicamentorum temperamentis ac facultatibus* 6 pr = Kühn XI 796-98. See Barton, *Power and Knowledge*, 53-54.

logy, and medicine to be connected with each other in an intricate way. In addition, Galen's comments about Pamphilus are also interesting because he ridicules him for relying on books and having no real acquaintance with the herbs he is talking about. This suggests that books could play an important role as a source of learning and shows that "inadequate" knowledge, though contested by some, could have value for others.

The connection between astrology, demonology, and medicine, as well as the harmful and dangerous character of zodiacal spirits, is also demonstrated by the *Testament of Solomon*. *T. Sol.* 18 lists the names of decanal demons, the harm they cause to humans, and the means for driving them away and curing people. Many of the decanal spirits control a certain part of the human body (*melothesia*), inflicting various illnesses and injuries.[98]

During their interrogation by Solomon, however, these spirits not only tell him who they are and what they do, but also how they can effectively be defeated. In some cases, they can be subdued by calling upon the name of an angel. For example, if the first spirit hears "Michael, imprison Ruax," he tells Solomon that he retreats immediately. But in other cases, the demons advise Solomon to write down the angelic name or some other name on different surfaces such as papyrus, wood or ivy leaves, which should be hung around the neck, attached to doorways, or heaped up in a pile. Also, more complicated instructions are given to prepare herbs and ointments. For example, the sixteenth demon tells Solomon:

> I am called Katrax. I inflict incurable fevers on men. If anyone wants to regain health, let him pulverize coriander and rub it on his lips, saying, 'I adjure you by Zeus, retreat from the image of God,' and I retreat immediately.[99]

Alexander rightly argues for the magical and medicinal purpose of the catalogue in *T. Sol.* 18 that combines demonology with astrology:

> Demonology becomes a more rational, more predictable phenomenon. If a client comes to a magus complaining of illness or ill luck the magus can take cognizance of which star or which decan is in the ascendant at this point in time, and thus identify from the numerous demons the probable demonic culprit and apply the appropriate angelic restraint. Alternatively he could discover through the client's nativity which star or decan was in the ascendant at the time of his birth, which demon is *synastros* with him and therefore likely to be causing him problems. [...] Indeed, the magus can practice apotropaic medicine. Since there is a predictable system, he

[98] See the section on the *Testament of Solomon* and zodiacal spirits in Chapter Four.

[99] *T. Sol.* 18:20. Translation from Duling, "Testament of Solomon," 979. On the name Katrax, cf. Daniel, "Testament of Solomon XVIII 27-28, 33-40," 304.

> can identify from which quarter the attack is likely to come and offer amulets and incantations to defend against it.[100]

Although harmful, people believed that the influence exerted by zodiacal and decanal spirits could be stopped. It is possible that books and texts on these matters could have served some practical purpose. The text of *4QZodiacal Physiognomy* could have been used in a magico-medicinal practice that sought to establish the exact nature of the zodiacal spirit that caused a certain affliction.

A magico-medicinal context for *4QZodiacal Physiognomy* is supported by the mention in 4Q186 1 ii 2 of a specific sort of granite stone (אבן צונם). Although the text is fragmentary, it is very suggestive. *4QZodiacal Physiognomy* may have associated certain stones with zodiacal signs and spirits. These stones may have been listed and used for purposes of magico-medicinal treatment or as a preventive, apotropaic element. There are Babylonian and Greco-Roman astrological texts that list stones in connection with the signs of the zodiac, and also give instructions for medical treatment. A magical purpose is possibly also indicated by the way that the words אבן צונם ("a granite stone") are written. These were not written in an inverted manner like the rest of the text, but in the regular order from right to left. This manner of writing may suggest its magical power.[101]

A Sectarian Context: Physiognomic Control over People's Entrance into the Qumran Community and the Fight against Evil Spirits

The possible use of physiognomics as a means for exercising social control has been discussed in Chapter Two. Analogous to some Greco-Roman examples, scholars have suggested that *4QZodiacal Physiognomy* was used as a means to manage and regulate the entrance of new members into the Qumran community.[102] Whether physiognomic learning was really applied and, if so, in what manner, is impossible to determine, but it is evident from the ancient sources that it was understood or at least imagined as a real possibility. One should, therefore, bear in mind the option that a physiognomic test was applied during the procedure of admission into the Qumran community. However, instead of the human spirit being evaluated, it was, I

[100] Alexander, "Contextualizing the Demonology," 632.

[101] See the section on magico-medicinal stones in Chapter One and the section on cosmic sympathy above in this chapter.

[102] In addition to the Babylonian and Greco-Roman contexts, mention should also be made of the medieval Jewish tradition that credited the circles of Merkavah mystics with the practice of physiognomics to guard their secrets by only sharing them with those deemed eligible. Cf. P.S. Alexander, "The Historical Setting of the Hebrew Book of Enoch," *JJS* 28 (1977): 156-80, at 168; Schäfer, "Fragment zur Metoposkopie und Chiromantik," 85-86; Alexander, "Physiognonomy," 392; J.R. Davila, *Descenders to the Chariot: The People behind the Hekhalot Literature* (JSJSup 70; Leiden: Brill, 2001), 60-67, 72-73.

suggest, the nature of a person's zodiacal spirit that may have determined whether someone qualified for group membership.

The physiognomic-astrological learning of *4QZodiacal Physiognomy* concerning the connections between people, their horoscopes and the nature of their zodiacal spirits may have been instrumental in exercising control over the composition of the Qumran community. Considering the potentially dangerous and harmful nature of the zodiacal spirits, they could pose a threat not only to individuals, but also, through individual people, to a whole group or community. Knowledge of whether people's zodiacal spirits were liable to cause serious harm could have determined whether an individual would be allowed to become a member of a social group such as the community of Qumran. Those persons whose zodiacal spirits had more parts in the house of light were presumably judged to be better off. They had to fear less from serious afflictions inflicted upon them by their zodiacal spirits. An individual person could, therefore, have been deemed eligible to join the community. In this case the community had also less to fear from attacks by his zodiacal spirit, which could not only harm him, but also others within the community through him.

As has been discussed in Chapter Four, Alexander argued for the use of physiognomic learning in controlling entrance into the Qumran community, as a divinatory tool in the eschatological fight between the community and the sons of darkness. The major difference between my understanding of the text of *4QZodiacal Physiognomy* and that of Alexander is that he regards the reference to רוח ("spirit") to mean the human spirit, whereas I take it to refer to the zodiacal spirit. This latter understanding, however, would fit even better in the world-view of the Qumran community, according to which the sons of light were battling the evil spirits of Belial and had to defend themselves against their attacks upon them. It is possible that when zodiacal spirits had more parts of darkness than light they were considered to be potential demonic minions of Belial.

Again, in terms of Qumran pneumatology, the zodiacal spirits could be seen as another sort of spirit or demon inhabiting the world and possibly causing trouble for the members of the Qumran community.[103] Therefore, adapting Alexander's interpretation, I suggest that the knowledge in *4QZodiacal Physiognomy* was used in a sectarian context like that of the Qumran community that wished to guard itself against demonic attacks. The physiognomic-astrological knowledge was used as the justification for a pre-emptive strike, so to speak, by denying entry into the community to people whose zodiacal spirits were found upon physiognomic inquiry to be

[103] Cf. n. 96 above in this chapter. See also P.S. Alexander, "The Demonology of the Dead Sea Scrolls," in *Dead Sea Scrolls After Fifty Years*, 331-53.

potentially too dangerous or maleficent. Such a pre-emptive, defensive function of *4QZodiacal Physiognomy* finds support in two manuscript copies, 4Q510 (4QShir[a]) and 4Q511 (4QShir[b]), which contain songs that were meant to be recited by the *Maskil* as defensive measures against attacks by different sorts of demons upon the community.[104]

It is not necessary to assume that physiognomic inquiry was only limited in scope as a preventive measure taken before people were admitted to the community. Sectarian writings of the Qumran community such as the *Rule of the Community* demonstrate that the group was preoccupied with disciplining members, keeping them within the group, and out of the hands of the sons of darkness, because it was a real possibility and threat that members of the community could wander off the right track. Physiognomic divination could, therefore, have been used both as a preventive measure, which regulated membership of the group and prevented wrong people and their zodiacal spirits from entering and threatening the community, as well as a diagnostic tool, similar to the magico-medicinal context, to determine the kind of treatment and cure for community members attacked by zodiacal spirits of a less harmful nature.

[104] Cf. Alexander, "Wrestling Against Wickedness," 319-24; García Martínez, "Magic in the Dead Sea Scrolls," 21-23.

APPENDIX I

NOTES AND COMMENTS ON READINGS IN *4QZODIACAL PHYSIOGNOMY* (4Q186) AND *4QPHYSIOGNOMY AR* (4Q561)

4QZODIACAL PHYSIOGNOMY (4Q186)

Notes and Comments on Readings in 4Q186 1 i

4Q186 1 i 4: The small fragment carrying 1.4 first appears in PAM 41.314 and is joined in PAM 41.804 with 4Q186 1. The first letter is probably *he*: two legs and part of the head are visible. Two down strokes, the right one of which curves slightly to the left, follow *he*. These strokes might have been part of *ʾalep*, but this is not clear (see also PAM 42.616).

4Q186 1 i 7: With Allegro, I reconstruct the last letter of this line as *qop*.[1] The long down stroke curves to the left ("s"-shaped). This feature makes it probable that *qop* should be read here. Wise translates "Anyone, the ha[ir of whose head] shall be," most likely reconstructing ש]ער ראושו.[2] Although the reconstruction is interesting, the reading is paleographically improbable. The left down stroke extends further down below the right arm than is usual with *šin*/*śin*. Also, one would expect to see remnants of the right arm of *śin*.

4Q186 1 i 8: The second word, סגלגלים, "round," is entirely recorded in paleo-Hebrew characters. Allegro reads ווֹגֿלגלים, "*rounded*,"[3] but paleo-

[1] Allegro, DJD 5.88. See also García Martínez and Tigchelaar, *DSSSE*, 380.

Apart from three minor instances, the recent presentation (transcription and translation) of *4QZodiacal Physiognomy* by N. Gordon agrees completely with Allegro's *DJD* edition and does not add anything new; see D.W. Parry and E. Tov (eds.), *Additional Genres and Unclassified Texts* (DSSR 6; Leiden: Brill, 2005), 220-23. Therefore, I will not refer individually to this publication in the following discussion. The three minor differences are in 4Q186 1 ii 7: בבור/בבוֹר; 1 iii 3: "and his teeth are …"/"and his teeth are elevated"; 2 ii 7: the additional suggestion that גולש could mean "flowing."

[2] Wise, "Horoscope Written in Code," 277.

[3] Allegro, DJD 5.88-89 (the italics are Allegro's and indicate the uncertainty of the translation). Note that in his preliminary publication ("Astrological Cryptic Document," 292-93), Allegro gives a table of the scripts used in *4QZodiacal Physiognomy* that differs from the one in the final publication (Allegro, DJD 5.90). In the preliminary publication Allegro understands the disputed letter to be paleo-Hebrew *samek*, but suggests that it should be recognized as a *waw* (=ו?). In the final publication the table has been adapted and the identification of a *samek* has been replaced by a clear *waw*. But note that both tables are ordered al-

Hebrew *samek* is clearly visible and there is no need to revert to another reading as this one makes sense.[4] In addition, Allegro is uncertain of reading *gimel*. It is not clear whether he understands it to be an example of a paleo-Hebrew or a cryptic character. It is unnecessary to assume that *gimel* is written in a cryptic script.[5] The character is perfectly understandable as paleo-Hebrew *gimel*. It is comparable with *gimel* in some of the biblical manuscripts that are written entirely in paleo-Hebrew characters, especially 11QpaleoLev[a].[6] Strugnell proposes reading סנלגלים<ו>, assuming a leap by the copyist from *waw* to *samek* in paleo-Hebrew, a suggestion adopted by most translators.[7]

4Q186 1 i 9: The final letters are written continuously, seemingly as one word,[8] but it is also possible that the scribe forgot to separate two distinct words.[9] The reading of the first three letters is clear, but various readings have been proposed for the traces of characters after שאר.

Allegro reads two separate words, שאר הׄיׄשׁן, but he gives no translation for the last word ("the *flesh* of [...").[10]

According to Carmignac the traces of the letters that follow should be considered as forming one word with שאר, which he, contrary to Allegro (שְׁאֵר), understands to mean "remnant, rest" (שְׁאָר). He considers the fourth and fifth letters to be either *yod* or *waw*, although the fifth letter might also just be an ink spot. The remains of the sixth letter might have belonged to *he*, *dalet*, or *reš* (שאר־י(ו)י(ו)ה/ד/ר). If one reads two separate words, then Carmignac tentatively proposes reconstructing שאר י(ו)ר]ח ("the rest of the moo[n (mon[th)"), but if one word should be read, perhaps one could recon-

phabetically, and that the position of the sign for *waw/samek* remains the same in both publications, namely the alphabetic position of *samek*.

4 The reading סנלגלים is generally accepted. See Carmignac, "Les Horoscopes," 200; Delcor, "Recherches sur un horoscope," 299; Licht, "Legs as Signs," 19; Strugnell, "Notes en marge du volume V," 274; García Martínez and Tigchelaar, *DSSSE*, 380. Von Stuckrad, *Frömmigkeit und Wissenschaft*, 119 n. 419, states that Carmignac proposed reading *zayin* instead of *waw*, but this is clearly not the case.

5 It does not, for example, correspond in any way with the Cryptic A sign for *gimel* (cf. Pfann, "Introduction," 527), nor can it be related to the still undeciphered Cryptic B.

6 In this latter document the paleographic development of *gimel* has resulted in a subtle variation regarding the tilt of the letter as it is suspended from the upper line, because the scribe apparently had a tendency to let the bottom of the letter swing forward. Cf. R.S. Hanson, "Paleography," in *The Paleo-Hebrew Leviticus Scroll (11QpaleoLev)* (D.N. Freedman and K.A. Mathews; Winona Lake, Indiana: American Schools of Oriental Research, 1985), 15. This same trend is observable in *4QZodiacal Physiognomy*. Here the scribe has swung the bottom of *gimel* even more to the left, and the leg of the first *gimel* curves more inwards to the left than does that of the second *gimel*.

7 Strugnell, "Notes en marge du volume V," 274. See e.g. Maier, *Texte vom Toten Meer*, 2:135; Wise, "Horoscope Written in Code," 277; García Martínez and Tigchelaar, *DSSSE*, 381.

8 שאררׄוׄשׁ, see PAM 41.804; 42.616; 43.344; 43.438.

9 See also 4Q186 1 ii 8: אש˺הואה.

10 Allegro, DJD 5.88-89.

struct שארי(י)ה]ם ("[the]ir remains"), assuming that שאר is used in the plural. However, in his translation Carmignac distinguished two words and leaves the second one untranslated ("the rest of").[11]

Strugnell thinks Allegro's reconstructed *he* is nearly impossible, and that *šin* is not very certain. He suggests reading שאר רושן, but admits that the three final letters remain doubtful.[12]

Wise, finally, seems to combine Carmignac's interpretation of שאר and Strugnell's reading of the final three characters. He translates "but the rest of [his] head is not [...]."[13]

A tear beneath l.9 that runs upward right through the strokes of ink after שאר hampers any reading of the final letters of this line. The manuscript also seems to have suffered a crinkle at this point. To the left of the tear the tip of a head is clearly visible, either from paleo-Hebrew *reš* or square script *waw* or *yod*. Beneath this head a trace of ink that must be the end of a stroke is still observable. In PAM 41.804 the two elements are clearly not connected, which would seem to rule out the possibility of *waw* or *yod*, but PAM 42.616 is less clear and it even seems as if the head has part of the leg attached to it. Two elements are clearly discernable to the right side of the tear. First, one sees a small trace of ink, and, second, below the trace a down stroke with a stroke to the left on top is patently visible. Again, neither element is clearly connected in PAM 41.804, but this is not so clear in PAM 42.616.

Adopting Strugnell's reconstruction, I assume that the first letter is paleo-Hebrew *reš*. It consists of three elements. The first part is the head at the left side of the tear that must have been connected with the second element, which is the trace of ink to the right side of the tear. This was the connection between the upper stroke of the head and the down stroke of the leg.[14] The third element is the small trace of ink to the left side of the tear and below the head. The manuscript has crinkled causing the displacement of the leg of paleo-Hebrew *reš* diagonally underneath the head. This also resulted in the fourth element, the second letter *waw*, moving slightly lower. The last stroke of ink could be the left edge of paleo-Hebrew *šin*.

11 Carmignac, "Les Horoscopes," 201, 205. The *waw* in י(ו)רנח does not make much sense, which is perhaps why Carmignac did not incorporate it in his translation. See also Maier, *Texte vom Toten Meer*, 2:135 "und nicht sonstiges," although he adds "Fleisch" in brackets.

12 Strugnell, "Notes en marge du volume V," 274. See also García Martínez and Tigchelaar, *DSSSE*, 380-81: שאר רושנו, "and not the flesh of [his] head [...]"; Ejrnæs, Holst and Müller, *Dødehavsskrifterne*, 496.

13 Wise, "Horoscope Written in Code," 277.

14 Cf. 4Q186 1 ii 7: האור, where the head is elongated and pointed (completely different from the paleo-Hebrew *reš* in 1 i 8: רחבים).

Reading רֹ֯וׄשׁנו differs from the full spelling וראושו in 4Q186 1 iii 5, but this does not speak against this reconstruction.[15]

Notes and Comments on Readings in 4Q186 1 ii

The small fragment containing ll.1–4 first appears in PAM 41.314 and is joined in PAM 41.804 with 4Q186 1 ii.[16]

4Q186 1 ii 1: Most scholars read טמא, "unclean, impure,"[17] except Carmignac and Delcor who read טבא, "good."[18] The first reading is to be preferred. In PAM 42.616 it is clear that the scribe drew part of the down stroke that is attached to the horn of *mem* and descends to the left. The word טמא stands too isolated to determine its function. Perhaps it characterizes some state of the described individual, but this is not clear.

4Q186 1 ii 3: Allegro reads עז,[19] but the second letter might also be *waw*. There seems to be a small crinkle in the leather below the second character, or it may be that a small part of the surface has fallen off. This makes it difficult to determine whether the down stroke continued further down.

Allegro suggests reading עזעש. He identifies it with the constellation mentioned in Job 9:9 and 38:32,[20] but he does not explain the meaning of

[15] Admittedly, there are only a few cases where one and the same manuscript has alternative spellings. See 4Q403 1 ii 24 (רוש, לראשי); 1 ii 34 (ראוש); 4Q418 9+ 5 (ברושכה); 43–45 i 1 (ראש); 126 ii 7 (ראוש). In biblical manuscripts only in the Isaiah[a] scroll from Cave 1; see for the form without *ʾalep* (רוש) Isa 40:21; 41:26; 48:16.

[16] Note that at the top of 4Q186 1 ii a stroke of ink is still visible in PAM 40.615 and 41.314, but that it vanishes in PAM 41.804 when the other fragment is joined. What is left is a v-shaped crack, the outline of which is already unmistakably visible in PAM 40.615. Perhaps the fragment broke loose, or maybe it did not belong here in the first place. If it does belong here there would not have been enough room to attach the four-line fragment at the top of 4Q186 1 ii. In PAM 41.804 this fragment crosses right over the v-shaped crack, exactly where the small piece of leather was located in the previous PAM photographs. If we look only at PAM 40.615 and 41.314, the stroke of ink seems to be in a strange place. It is too close to the line below it to be part of another line. It seems like the upper stroke of *bet* or *reš*, but it is too high to belong to the same line of the *nun* below it (unless it were a scribal correction?).

[17] Allegro, DJD 5.88-89; Strugnell, "Notes en marge du volume V," 274; Maier, *Texte vom Toten Meer*, 2:135; Wise, "Horoscope Written in Code," 277; García Martínez and Tigchelaar, *DSSSE*, 380-81.

[18] Carmignac, "Les Horoscopes," 201-2; Delcor, "Recherches sur un horoscope," 299-300.

[19] Allegro, DJD 5.88-89.

[20] Allegro, DJD 5.90. For עָשׁ in Job, *BHS* (ed. G. Gerleman) proposes reading עַיִשׁ וּבָנֶיהָ instead as in 38:32, but without any manuscript evidence. The identification of the constellation mentioned in the two passages in Job, however, is somewhat doubtful (see A. de Wilde, *Das Buch Hiob: Eingeleitet, übersetzt und erläutert* [OTS 22; Leiden: E.J. Brill, 1981], 142-44, who sides with those identifying עיש/עש with *Aldebaran*). The most common identification seems to be with the *Great Bear* (see D.J.A. Clines, *Job 1-20* [WBC 17; Dallas, Texas: Word Books, 1989], 231), but some propose an identification with *Leo* (see H. Strauß, *Hiob: 2. Teilband, 19,1-42,17* [BKAT 16/2; Neukirchen-Vluyn: Neukirchener, 2000], 339).

his reconstructed reading איש עיש. Interestingly, the *Peshiṭta* translates עש in Job 9:9 and עיש in Job 38:32 with *ʿjūta*, which is used of the *Hyades*, a distinct group of stars in the head of the constellation *Taurus*, with the red star *Aldebaran* as its brightest member (cf. the Greek Λαμπαδίας).[21] As Allegro does not provide a translation for איש עיש ("a man of *Hyades*"?), it is difficult to surmise what he has in mind with his suggested reconstruction עי]ש. How is a reference to עש/עיש as a distinct section of the zodiacal *constellation Taurus* in 4Q186 1 ii 3 related to the indication of a specific part of the zodiacal *sign Taurus* in 4Q186 1 ii 9 (ברגל השור)?[22] There is no other textual evidence in which the *Hyades* are referred to as the "foot of *Taurus*" (רגל השור, 4Q186 1 ii 9).[23] Allegro's reconstruction does not seem to elucidate our understanding of *4QZodiacal Physiognomy*, and is to be rejected.

Wise translates "[And] anyone [whose] eyes are," probably reconstructing איש עי]ניו.[24] The latter reconstruction, however, seems improbable for syntactical reasons. One would expect איש אשר יהיו עי]ניו (or, at least, איש אשר עי]ניו), similar to the form of the phrase used in 4Q186 1 i 7 (ואיש אשר יהיה קׄ) and in *The Book of the Reading of the Hands by an Indian Sage*.[25] This could mean that it is not necessarily a bodily feature that is referred to in 4Q186 1 ii 3.

If, however, one reads ען, a reconstruction such as איש עו]ר is possible, which would mean that the individual described in 4Q186 1 ii is a blind person.[26] Another possibility could be that line 3 introduced a certain type of character. If one reads ען, one might reconstruct איש עו]ל ("an unjust man"),[27] or איש עׄו]רמה ("a shrewd man").[28]

[21] Cf. G. Schiaparelli, *Die Astronomie im Alten Testament* (Gießen: J. Ricker [Alfred Töpelmann], 1904), 53-56; M. Albani, "'Kannst du die *Sternbilder* hervortreten lassen zur rechten Zeit ...?' (Hi 38,32): Gott und Gestirne im Alten Testament und im Alten Orient," in *Das biblische Weltbild und seine altorientalische Kontexte* (eds. B. Janowski and B. Ego; FAT 32; Tübingen: J.C.B. Mohr [Paul Siebeck], 2001), 181-226, at 204-5.

For the position of *Aldebaran* in the head of the constellation *Taurus*, see e.g. Ptolemy, *Tetrabiblos* 1.9.3, and, more specifically, the star catalogue in Ptolemy, *Almagest* 7.5[.23].

[22] Allegro, of course, did not see this problem. Translating ברגל השור as "on the Festival of Taurus," he does not understand it as a reference to a part of the zodiacal sign *Taurus*.

[23] There is, however, evidence that עש (as the *Hyades*) was interpreted as the head of *Taurus*. In the Babylonian Talmud the term עש is explained as being יותא, and יותא is said by some to mean the tail of the Lamb (*Aries*), and by others the head of the Calf (*Taurus*). See *b. Ber*. 58b: מאי עש? אמר רב יהודה: יותא. מאי יותא? אמרי לה: זנב טלה, ואמרי לה: רישא דעגלא. The passage continues by arguing that those who identify עש with the tail of *Aries* hold the more reasonable position.

[24] Wise, "Horoscope Written in Code," 277.

[25] See the section in Chapter One on the beginning of the physiognomic entries.

[26] Lev 21:18; 11QT[a] 45:12.

[27] 4Q417 2 i 7 (*4QInstruction*[c]) says that an unjust man should not be considered as one who helps (ואיש עול אל תחשוב עוזר).

[28] Cf. 4Q525 23 5 (*4QBeatitudes*): באנשי ערמה.

If the text is arranged according to physiognomic criteria, it seems more likely that a bodily feature is being described rather than a character trait. If, in addition, אבן צונם in 1.2 represents the end of a previous account, it follows that 1.3 would be the beginning of another account in 4Q186 1 ii. Perhaps 1.3 mentioned first the head and then commented on the eyes by stating the person was blind. Two lines further on the thighs are described as long and slender.[29] This would mean, however, that a person whose zodiacal spirit has more parts in the "house of light" than in the "house of darkness" (6:3) is a blind person. The text, unfortunately, is too fragmentary to enhance our understanding of this matter.

4Q186 1 ii 4: This line is difficult to understand because of the poor condition of the manuscript here. Allegro, Delcor and Maier do not even provide a translation of this line.[30] It is hard to determine whether the body is the object of description in this line. One can only assume this on the basis of the next line, 1.5, which begins with the thighs. The presupposition is that 1.5 continues a bodily description from the previous line(s).

The beginning of 1.4 is not preserved. Allegro transcribes the first visible signs as]··ות.[31] The final letter is paleo-Hebrew *taw*, and the third letter is square script *waw*, with part of the head still visible in PAM 42.616. Of the second letter, a down stroke and a base are extant. The first letter is only present in a small trace of ink[32] and is in itself not indicative. Wise reads the second letter as *kap*, translates "and lo]ng," and presumably reconstructs ואר]וכות.[33] Wise takes this to be a description of the eyes mentioned at the end of 1.3 according to his reconstruction.[34] Wise's reconstruction at the end of 1.3, however, is syntactically problematic (see above). If one accepts Wise's reading for 1.4, it is not possible to determine which part of the body is described in this line as being long.[35] Perhaps his teeth or the fingers of his hands?

[29] In 4Q186 1 iii 5 the head is mentioned and two lines down the thighs are described as thick and hairy.

[30] Allegro, DJD 5.89; Delcor, "Recherches sur un horoscope," 309; Maier, *Texte vom Toten Meer*, 2:135.

[31] Allegro, DJD 5.88-89.

[32] See PAM 41.804; 42.616.

[33] Wise, "Horoscope Written in Code," 277.

[34] Wise, "Horoscope Written in Code," 277. At the end of 1.4 Wise reads the word "fix[e]d" as a qualification of the eyes ("fixed eyes"), which means that the description of the eyes begins in 1.3 and runs all the way to the end of 1.4 according to his reading. In *The Book of the Reading of the Hands by an Indian Sage* a person's eyes are described as long, ועיניו ארוכות, see Scholem, "Physiognomy," 491.5.

[35] Instead of reading *kap*, one could also read the second letter as *bet* and reconstruct וארכ]ובות, "the knees." But there are two objections, based on literary considerations, against this reading. First, the physical descriptions run from head to foot (the *a capite ad calcem* principle, however, need not be a strict criterion. In other physiognomic texts this order is not always rigorously followed). In 1.5 the thighs are the objects of description, but one would

Allegro reconstructs the next word as וה[נ]ה̊.[36] The first two letters are clearly visible. The first could be *waw* or *yod*, and the second letter is *he*. The manuscript is damaged and the surface layer of the leather is partly missing. The right part of the roof and upper part of the right leg of *he* are therefore lacking. This damaged condition makes it impossible to reconstruct the continuation of the word; only a few dots are visible in this damaged section. When the top layer of the leather is present again one can discern a small horizontal stroke that seems to have a curve upwards at the left, like the upper stroke (the sting) of *bet*, *dalet*, *kap*, *mem* or *reš*, but it is very vague.[37] There are two problems with Allegro's reconstruction. First, the gap seems too large for one letter (even if it were paleo-Hebrew *nun*), and, second, final *he* is difficult to read, since the curve seems to go upward.

In his preliminary publication Allegro gave another reading than in the final edition of the last word in 1.4. In the preliminary publication Allegro reads נק̊י̊[ו]ת, "clean, clear." He tentatively suggests that the second letter is used cryptically for *qop*, although it is clearly paleo-Hebrew *ṣade*.[38] But in his final publication Allegro does not maintain this suggestion. He then reads נצ̊י̊[ו]ת, a *nip'al* from נצה meaning "to be pressed, lean," but does not provide a translation.[39] Carmignac suggests that the third letter is not *yod*, but paleo-Hebrew *reš*. He reads נצר[ו]ת, a *qal* feminine passive participle plural from the root נצר, "to watch, keep, guard", referring to Isa 48:6 (נְצֻרוֹת, "hidden things") as an exact parallel for this form.[40]

The reading by Carmignac is to be preferred. The first letter, however, need not necessarily be square script but could be paleo-Hebrew *nun*.[41] The down stroke of paleo-Hebrew *mem* is tilted to the left in *4QZodiacal Physiognomy*,[42] and the same may apply to paleo-Hebrew *nun*. I suggest that this final word in 1.4 is entirely written in paleo-Hebrew characters. The

---

expect this to precede an account of the person's knees. Secondly, as far as one can tell, the individual's body parts are always followed by a personal suffix (See 4Q186 1 ii 5: ושוקיו and ואצבעות רגליו, 1 iii 5: וראושו, 1 iii 6: ושניו, 1 iii 6–7: ואצבעות ידיו, 1 iii 7: ושוקיו, 1 iii 8: ואצבעות רגליו, 2 i 1: וע[י]ניו and וזקניו, 2 i 2: ובת קולו and ושניו, 2 i 4: ואצבעות ידיו, 2 i 5: ושוקיו and וכפות רגליו).

[36] Allegro, DJD 5.88-89. But he does not take it up in his translation. Wise, "Horoscope Written in Code," 277, translates "but th[e]y" (וְהֵנָּה). See also F. García Martínez and A.S. van der Woude (eds.), *De rollen van de Dode Zee: Ingeleid en in het Nederlands vertaald: Deel 2* (Kampen: Kok, 1995), 498, "en [z]ij"), but García Martínez and Tigchelaar, *DSSSE*, 380-81, translate "and s[e]e" (וְהִנֵּה).

[37] See PAM 41.314; 41.804; 42.616.

[38] Allegro, "Astrological Cryptic Document," 292-94.

[39] Allegro, DJD 5.88-90. See also García Martínez and van der Woude, *De rollen van de Dode Zee*, 498, "schraal (?)" ("lean, meagre, poor").

[40] Carmignac, "Les Horoscopes," 202. See also García Martínez and Tigchelaar, *DSSSE*, 380. Delcor, "Recherches sur un horoscope," 299-300, wrongly ascribes the transcription נצורות to Carmignac.

[41] *Pace* Carmignac, "Les Horoscopes," 202, who refers to 4Q186 1 ii 6: מנ and 1 ii 9: עני as examples of a similar square script *nun*.

[42] See 4Q186 1 i 8: רחבים; 2 i 2: ממעֹנט.

second and third letters are paleo-Hebrew *ṣade* and *reš*. The fourth letter, filling the gap, is paleo-Hebrew *waw*. In PAM 41.804 one can observe a small stroke to the left of the gap and to the right of the paleo-Hebrew *taw*. This perhaps represents the horizontal stroke of paleo-Hebrew *waw*. The fifth and final letter is clearly paleo-Hebrew *taw*.

The sense of the final word in 4Q186 1 ii 4, נצרות, is hard to determine. Carmignac proposes the interpretation that, following the sequence of the description of the human body, it refers to the genital area of the body, which is kept secret ("gardées secrètes").[43] But if a part of the body were referred to one would expect a suffix attached to נצרות.[44] Wise presumably agrees with Carmignac's reading, but he translates נצרות as "fix[e]d," a reference to "fixed eyes."[45] García Martínez and Tigchelaar translate "secrets," leaving open to what it refers.[46] One could perhaps think of revealed secrets or a secretive character?[47]

In the context of a physiognomic description, Carmignac's suggestion is possible. In Babylonian as well as in medieval Jewish physiognomic tradition the penis is the object of physiognomic inquiry.[48] And in Greek *zodiologia* as well as the Mandean *Book of the Zodiac* the secret or private parts are also referred to.[49] Again, the text is too fragmentary to establish the sense of the final part of this line.

[43] Carmignac, "Les Horoscopes," 202-3. Carmignac quotes part of a *zodiologion* that he found in Starcky, "Un texte messianique araméen," 65, which speaks of moles on the secret parts of a person born in the zodiacal sign *Aries*: "he has moles on the eyes, on the breast, and on the secret parts" (ἐλαίας ἔχει ἐπὶ τῆς ὄψεως, ἐπὶ τοῦ στήθους καὶ ἐν τῷ κρυπτῷ, see CCAG 12.174.14–15). Cf. also Delcor, "Recherches sur un horoscope," 299-300.

Ptolemy, *Tetrabiblos* 3.13.15, mentions injuries and diseases of the secret parts (κρυπτῶν τόπων) caused by the planet Mars. Firmicus Maternus, *Mathesis* 5.3.38, says that if Saturn is positioned in *Scorpio* he causes tireless pains in the concealed and private parts (*absconsorum et latentium locorum assiduos dolores*).

[44] Cf. n. 35 above.

[45] Wise, "Horoscope Written in Code," 277. But it is improbable that the eyes are being referred to at the end of l.3 (see above), which weakens Wise's interpretation of נצרות.

[46] García Martínez and Tigchelaar, *DSSSE*, 381.

[47] Cf. e.g. Ptolemy, *Tetrabiblos* 2.2.10: τὰ πολλὰ κρύπτοντες.

[48] In the Babylonian physiognomic omen series *Šumma alamdimmû* the tenth tablet is devoted to descriptions of the penis and testicles (X:64–125), see Böck, *Die babylonisch-assyrische Morphoskopie*, 122-27. There is an astrological-physiognomic text in the Cairo Genizah (T.-S. NS 252:2) that gives descriptions of the male genitals, see Gruenwald, "Jewish Physiognomic," 317-19. For a comparison between some Babylonian omens and this Cairo Genizah text, see Böck, *Die babylonisch-assyrische Morphoskopie*, 67. The Babylonian series *Šumma alamdimmû* also pays attention to the vagina in the subseries devoted to the woman, see Böck, *Die babylonisch-assyrische Morphoskopie*, 165. It seems that Greco-Roman physiognomic tradition did not regard the sex organs as objects of physiognomic inquiry, but see the occurrence in the Anonymous Latin author, *De physiognomonia liber* §85: *qui virilia habent magna laneaque, stolidi sunt* ("those who have large and hairy testicles are stupid"), and also the Greek *zodiologia* in n. 49.

[49] CCAG 4.159.12; 160.1–2, 18–19, 31–32; 162.1; 163.3–4; 166.15; 167.1, 23–24; 168.23; 169.10; 10.102.8–9; 103.26–27; 105.11–12; 108.15–16; 109.28–29; 114.25; 117.3; 118.19;

4Q186 1 ii 6: Practically no scholar accepts Allegro's translation "and he is of the Second Vault."[50] Allegro points to עמודי שמים in Job 26:11,[51] and asserts that this terminology from Job is explained as "vaults" in *1 En.* 18:3. But, as in Job 26:11, the translation "pillars of heaven" seems better. In Enochic cosmology the universe is imagined as a building. The earth has its foundations, and the heavens are supported by the four winds functioning as pillars that are set at the ends of the earth.[52] The word עמוד is, therefore, best translated by "pillar" or "column."[53]

The architectural sense of the word עמוד is clearly attested in the Hebrew Bible.[54] It is used for pillars supporting the roofs of structures such as the tabernacle and the Temple.[55] The term עמוד is furthermore employed to describe the divine presence, either of God or of the angel of God, in a pillar of cloud (עמוד ענן) or a pillar of fire (עמוד אש) during the Exodus and the Wandering through the Wilderness.[56] A third distinction is the term's metaphorical sense. The metaphorical references to the pillars of earth and heaven in Job 9:6, 26:11, and Ps 75:4 reflect a cosmology in which the universe is structured as a building,[57] but it is doubtful whether this imagery is behind the usage of עמוד in *4QZodiacal Physiognomy*.

Occurrences of עמוד in the Dead Sea Scrolls fall primarily within the category of architectural usage,[58] while the rest are mainly analogous to the

---

119.29; 183.21–184.1; 12.176.6–7; 178.8–9; 181.28; 184.25–26. Drower, *Book of the Zodiac*, 9, 13, 30, 33, 35.

[50] Allegro, DJD 5.89-90. See also von der Osten-Sacken, *Gott und Belial*, 187: "Gewölbe"; García Martínez and van der Woude, *De rollen van de Dode Zee*, 498: "firmament."

[51] See Strauß, *Hiob*, 109; De Wilde, *Buch Hiob*, 249. Allegro also refers to Job 9:6. The "pillars of the earth" (see also Ps. 75:4), however, are different from the "pillars of heaven," see Clines, *Job 1-20*, 230.

[52] See Nickelsburg, *1 Enoch 1*, 276, 284-85. See also Coblentz Bautch, *Geography of 1 Enoch 17-19*, 35, 102-3.

[53] See also Dupont-Sommer, "Deux documents horoscopiques," 241.

[54] Cf. also D.N. Freedman, B.E. Willoughby and H.-J. Fabry, "עמוד *'ammûḏ*," *TWAT* 7:204-9.

[55] See the description of the tabernacle in Ex 26:32.37; 27:10–12.14–17; 35:11.17; 36:36.38; 38:10–12.14–15.17.19.28; 39:33.40; 40:18. See the description of the Temple in 1 Kgs 7:15–22.41–42. Pillars are also used for King Solomon's palace in Jerusalem, the House of the Forest of the Lebanon (see 1 Kgs 7:2–3.6). The architectural sense of the supportive nature of these pillars is vividly clear in the story of Samson's destruction of the Philistine temple of Dagon (see Judg 16:25–26.29).

[56] See Ex 13:21–22; 14:19.24; 33:9–10; Num 12:5; 14:14; Deut 31:15; Ps 99:7; Neh 9:12.19.

[57] A cosmic sense is also apparent in some occurrences of the verb עמד, see Josh 10:13; Hab 3:11; Ps 148:6.

[58] See 1QM 5:10; 3Q15 4:1; 6:1; 11:3; 4Q403 1 i 41 (perhaps the supporting pillars of the most exalted dwelling, עמודי משא זבול רום רומים, are a reference to the pillars of heaven? For זבול, see Isa 63:15, and Hab 3:11. Cf. also *b. Ḥag.* 12b where *Zebul* is one of the seven heavens. But see C. Newsom, "4QShirot 'Olat HaShabbath[d]," in *Qumran Cave 4.VI: Poetical and Liturgical Texts, Part 1* [eds. E. Eshel et al.; DJD 11; Oxford: Clarendon, 1998], 253-92, at

second sense describing the divine presence.[59] The reference in 4Q204 1 viii 29 (*1 En.* 18:11) to the pillars of fire (עמודׄני נורא) might be metaphorical in that it refers to the Watchers.[60]

A metaphorical use of עמוד is also attested in rabbinic literature.[61] Certain individuals are identified as being a pillar. Rabbi Johanan ben Zakkai is addressed as "light of Israel, right pillar (עמוד הימיני), and strong hammer." Abraham is called "pillar of the world" (עמודו של עולם), and the same is said about the righteous. These metaphorical examples of עמוד are related to the architectural sense. They express the supportive nature of certain individuals in a cosmic sense.

The meaning of עמוד in the phrase והואה מן העמוד השני ("and he is from the second column") in 4Q186 1 ii 6 is difficult to align with one of the various senses in which the word עמוד is used in the Hebrew Bible, the Dead Sea Scrolls, and rabbinic literature. It seems to be unique and particular to *4QZodiacal Physiognomy*.

4Q186 1 ii 7: Scholars transcribe the last word of this line in two ways. The first two letters are clearly Greek *beta*, but the two last letters are not so clear. The third letter can be either *waw* or *yod*, and the fourth letter is read either as *reš* or *taw*. Allegro reads בבוׄר, "in the Pit of," and asserts that the last letter is cursive *reš* used cryptically.[62] But Carmignac proposes reading בבית, "in the house of," and thinks that the last letter is a partly erased paleo-Hebrew *taw*.[63] Strugnell considers both readings to be paleographically doubtful. Carmignac's reading is not supported by a close study of the surface of the text, which shows no trace of loss, but Allegro's reading of cursive *reš* seems equally unlikely. Nonetheless, Strugnell is inclined to accept Carmignac's reading because it probably corresponds to the text of the author, whatever the exact reading of the text written by the scribe may have been.[64] It is not possible to give a clear identification of

---

276); 11Q13 3:10; 11Q19 10:11; 30:9; 31:9; 34:2–3.15; 35:10; 42:11. See also the Aramaic *New Jerusalem* texts 1Q32 1 1–2; 4Q554 1 iii 22; 5Q15 1 ii 4; 2 4–5; 11Q18 9 2; 11 6.

[59] See 4Q365 6a i 9; 4Q470 3 5; 4Q504 6 10; 4Q505 128 2 (?); 4Q506 126 2.

[60] Cf. Coblentz Bautch, *Geography of 1 Enoch 17-19*, 129-33; Nickelsburg, *1 Enoch 1*, 287.

[61] See U. Wilckens, "στῦλος," *TWNT* 7:734.

[62] Allegro, DJD 5.89-90. See also García Martínez and van der Woude, *De rollen van de Dode Zee*, 498; Maier, *Texte vom Toten Meer*, 2:136; Alexander, "Physiognonomy," 387; G. Vermes, *The Complete Dead Sea Scrolls in English* (London: Allen Lane The Penguin Press, 1997), 358; Ejrnæs, Holst and Müller, *Dødehavsskrifterne*, 496.

[63] Carmignac, "Les Horoscopes," 203. See also Dupont-Sommer, "Deux documents horoscopiques," 241; Delcor, "Recherches sur un horoscope," 300.303; A. Marx, "Y a-t-il une prédestination à Qumrân," *RevQ* 6/22 (1967): 163-81, at 179; Von der Osten-Sacken, *Gott und Belial*, 187 n. 2; Wise, "Horoscope Written in Code," 277; Schmidt, "Astrologie juive ancienne," 128 n. 9; Albani, "Horoscopes in the Qumran Scrolls," 285-86; García Martínez and Tigchelaar, *DSSSE*, 380.

[64] Strugnell, "Notes en marge du volume V," 274-75.

the last letter. One might suggest that the scribe used a cryptic letter from some unknown script, but there is no basis for such an argument and it does not help much. Given the fact that the issue cannot be decided on paleographic grounds, I accept Carmignac's reading on literary grounds. Given the other occurrences of בית,[65] I understand the final word in l.7 as בבית.

4Q186 1 ii 8: Part of the upper horizontal stroke and left down stroke of paleo-Hebrew *ḥet* is still visible.[66]

4Q186 1 ii 9: Allegro transcribes the third word as עני, "he will be poor,"[67] whereas Dupont-Sommer reads ענו, "he will be humble."[68] Both readings are possible.[69] I am in favor of the latter and understand it as an indication of character.

Notes and Comments on Readings in 4Q186 1 iii

Several fragments have been joined together to form 4Q186 1 iii. From PAM 40.615 it is clear that the central fragment (containing columns i and ii) preserves (part of) the last letters of ll.8–9 and also the bottom margin of column iii. In PAM 41.314 a large fragment is joined to the third column. This fragment contains four lines (ll.6–9), preserves the left part of the third column for ll.6–8, and also has the right part of column iv (ll.6–8). Finally, in PAM 43.438 a small fragment, which appears separately from PAM 41.804 onwards, is placed above the left part of column iii as ll.4–5 because it appears to have a left margin.[70]

4Q186 1 iii 4: The third character is problematic. The first letter is paleo-Hebrew *waw*, the second letter Greek *alpha*, while the fourth letter is paleo-Hebrew *he*. Allegro reads וא·ה and does not give a translation. The third letter is legible, but Allegro places a question mark over this character in

[65] See 4Q186 1 ii 7: בבית; 1 iii 8: בבית; 1 iii 9: מבית.

[66] See PAM 41.804; 42.616.

[67] Allegro, DJD 5.89. See also Wise, "Horoscope Written in Code," 277; García Martínez and Tigchelaar, *DSSSE*, 380-81.

[68] Dupont-Sommer, "Deux documents horoscopiques," 241. See also Vermes, *Complete Dead Sea Scrolls*, 358; Ejrnæs, Holst and Müller, *Dødehavsskrifterne*, 496.

[69] Cf. Strugnell, "Notes en marge du volume V," 275; Maier, *Texte vom Toten Meer*, 2:136.

[70] In addition, Allegro, DJD 5.91, suggests that the small fragment 4Q186 3 possibly belongs above the left part of 4Q186 1 iii. In *DJD* 5.Plate XXXI, the fragment is presented separately, but on PAM 43.438 it appears joined with 4Q186 1 iii as the left part of ll.2–4. This join, however, seems unlikely. First, the left margin of ll.2–3 is out of line with that of ll.4–8, because it stands ca. 1.0 cm to the right of the left margin of ll.4–8. Second, it is improbable that the designation "beautiful" (יפי) in 4Q186 3 3 is part of a description in which other qualifications such as "terrifying" and "protruding teeth" (מיראות and רומות ושניו, 4Q186 1 iii 6) appear. Of course, it is possible that 4Q186 1 iii 4 is the end of a previous account, thereby resolving the problem of opposing qualifications, but because it is impossible to determine this, the fragment is best treated separately.

his table of the scripts used in *4QZodiacal Physiognomy*.[71] Carmignac, however, reads the third character as a damaged Greek letter *beta*, and wonders why Allegro does not do the same instead of supposing it to be an unknown character. Consequently, Carmignac reads ואבה and translates "et il consentira (?)."[72] But Allegro's unwillingness to read the third character as a Greek *beta* is understandable. First, this character evidently differs from the other occurrences of Greek *B* in *4QZodiacal Physiognomy*.[73] It is more rounded and, more importantly, it lacks an upper "belly,"[74] resembling Greek minuscule *beta*. However, and secondly, the minuscule writing system does not predate, in its definitive form, the eighth century.[75] Although *4QZodiacal Physiognomy* makes use of different scripts, it seems unlikely that within the same script variant characters were also used to express the same letter. The third letter remains, therefore, unidentified.

4Q186 1 iii 5: Two small strokes of ink are visible above the word ואצבעות in 1.6.[76] Wise reads "[whose] ey[es]," which he connects with the beginning of the next line as fear-inspiring eyes.[77] It seems, however, paleographically impossible to assume one of the strokes represents the leg of *ʿayin* because this runs diagonally and not straight, as is the case with the two strokes of ink in 4Q186 1 iii 5. It is perfectly possible that the subject's eyes were mentioned in 1.5, but the manuscript provides no material support for this assumption.

4Q186 1 iii 6: Allegro reads מיראות, but thinks it is used "perhaps erroneously for מילאות."[78] Licht reads מוראות,[79] but *yod* is clearly distinguishable from *waw* in this case.[80] Carmignac derives מיראות from מרא, "to be fat," as a possible qualification of the cheeks.[81] Gordis, however, under-

[71] Allegro, DJD 5.89, 90. See also García Martínez and van der Woude, *De rollen van de Dode Zee*, 498; Maier, *Texte vom Toten Meer*, 2:136; Wise, "Horoscope Written in Code," 277; Vermes, *Complete Dead Sea Scrolls*, 358.

[72] Carmignac, "Les Horoscopes," 203-4, 205. See also Delcor, "Recherches sur un horoscope," 307; García Martínez and Tigchelaar, *DSSSE*, 382-83.

[73] Cf. 4Q186 1 i 8: רחבים; ii 7: בבית (twice); iii 8: בבית.

[74] It is clear from PAM 42.616 that the leather is sufficiently intact to determine that this character is not damaged in the sense that it might originally have had an upper "belly." This observation was confirmed by use of the microscope at the Dead Sea Scrolls laboratory of the IAA, September 22, 2005.

[75] Cf. E.M. Thompson, *An Introduction to Greek and Latin Palaeography* (Oxford: Oxford University Press, 1912; repr., New York: Burt Franklin, 1964), 103, 218; B.A. van Groningen, *Short Manual of Greek Palaeography* (Leiden: A.W. Sijthoff, 1940), 33.

[76] See PAM 41.314; 41.804; 42.616; 43.344; 43.438.

[77] Wise, "Horoscope Written in Code," 277.

[78] Allegro, DJD 5.89, 90.

[79] Licht, "Legs as Signs," 20. See also von der Osten-Sacken, *Gott und Belial*, 187 n. 3.

[80] Cf. ושניו in the same line.

[81] Carmignac, "Les Horoscopes," 204. See also Delcor, "Recherches sur un horoscope," 307; Vermes, *Complete Dead Sea Scrolls*, 358.

stands מיראות to be *piʿel* participle of ירא, "causing fear, frightening, awesome," possibly used to describe the subject's eyes.[82]

Allegro does not provide a translation for the sentence ושניו רומות לאבר, but comments that רומות לאבר is a phrase "presumably indicating a meaning opposite to על סרכמה of the teeth of the more favored individual of f.2, i 3; so perhaps here = 'lying askew' or the like."[83] Carmignac suggests, first, to understand רומות in an active sense, and, second, a confusion in hearing occurred between *ʾalep* and *ʿayin* (לעבר, "à côté, de travers"): "ses dents (sont) poussées de travers."[84] Other scholars do not emend לאבר and try to make sense of "the wing." Delcor takes ושניו רומות לאבר to mean that the teeth of the person are raised like the tip of a wing ("les dent sont élevées 'en aileron'").[85] Dupont-Sommer and Delcor read דומות instead of רומות and propose that the teeth resemble a wing.[86] Nebe, however, rightly remarks that *reš* and *dalet* are clearly distinguishable in *4QZodiacal Physiognomy*. Also, if אבר had been meant figuratively one would expect כאבר. Firstly, with Segal, the passive participle רומות of the verb רום should be understood as describing "a more or less permanent state as the result of a verbal action."[87] According to Nebe, the same is true for the active participle ויושבות in 4Q186 2 i 3. Secondly, אבר is אֱבַר, "Aussenseite." It is derived from the noun בר (*barr*) with *ʾalep-prostheticum*, according to Nebe. He finds support for this interpretation in Aramaic אבר(א)י and אבריתא. Nebe, therefore, translates ושניו רומות לאבר as "und seine Zähne stehen nach aussen."[88] I follow Nebe's interpretation.

4Q186 1 iii 7: Allegro does not show (by his transcription עֿבות) that in the first occurrence of עבות in this line the letter *ʿayin* is not written at all

[82] Gordis, "Document in Code," 38-39. See also Maier, *Texte vom Toten Meer*, 2:136; Wise, "Horoscope Written in Code," 277; García Martínez and Tigchelaar, *DSSSE*, 382-83; Ejrnæs, Holst and Müller, *Dødehavsskrifterne*, 496.

[83] Allegro, DJD 5.89, 90. See also Dupont-Sommer, "Deux documents horoscopiques," 243; M. Philonenko, "Deux horoscopes Qoumrâniens: Identification des personnages," *RHPR* 65/1 (1985): 61-66, at 64; García Martínez and van der Woude, *De rollen van de Dode Zee*, 498; Maier, *Texte vom Toten Meer*, 2:136; Vermes, *Complete Dead Sea Scrolls*, 358; García Martínez and Tigchelaar, *DSSSE*, 382-83; Ejrnæs, Holst and Müller, *Dødehavsskrifterne*, 496.

[84] Carmignac, "Les Horoscopes," 204-5.

[85] Delcor, "Recherches sur un horoscope," 308.

[86] Dupont-Sommer, "Deux documents horoscopiques," 243 n. 1; Gordis, "Document in Code," 39.

[87] M.H. Segal, *A Grammar of Mishnaic Hebrew* (Oxford: Clarendon, 1927; repr., 1978), 159-160, §§332-33.

[88] G.W. Nebe, "אבר in 4Q186," *RevQ* 8/30 (1973): 265-66. See also von der Osten-Sacken, *Gott und Belial*, 187; Wise, "Horoscope Written in Code," 277.

by the scribe in the manuscript.[89] One has to correct the scribe in this instance and read <ע>בות.[90]

Furthermore, Allegro reconstructs [ש]ער,[91] but part of the left down stroke of *šin* is distinguishable.[92] Therefore, I transcribe שׂער.

4Q186 1 iii 9: Notice that מבית instead of בבית is written here. Whether it is significant that מן instead of ב is used is not clear.[93] Contrary to the other reference to the "house of light" or the "house of darkness," the word is written in square script, except for a paleo-Hebrew *taw*.[94]

At the end of this line, Allegro transcribes ואיש, "And a man…,"[95] but Strugnell states that this reading is materially impossible, "mais le fragment devra être gratté au verso, puis lu par transparence, avant d'aboutir à un déchiffrement certain."[96] García Martínez and Tigchelaar only transcribe ו..ש ("And…").[97] Most scholars, however, do not provide a translation at all for the last part of l.9.[98] I think Allegro's reading is problematic. If the second letter is *ʾalep*, the left down stroke is strangely tilted to the right. Also, if the third letter is *yod*, the down stroke seems to stand at too sharp a diagonal. It is different from ואיש in 4Q186 1 i 7. From the photographs it is impossible to determine the reading of the letters between *waw* and *šin* with certainty.[99]

Nevertheless, the fact that the last word in this line begins with *waw*-conjunctive is significant because it indicates that a new element is introduced subsequent to the numbers allotted to the "house of darkness" and the

[89] Cf. Allegro, DJD 5.89.

[90] See also Carmignac, "Les Horoscopes," 205; Strugnell, "Notes en marge du volume V," 275; García Martínez and van der Woude, *De rollen van de Dode Zee*, 498; Maier, *Texte vom Toten Meer*, 2:136; Wise, "Horoscope Written in Code," 277; Vermes, *Complete Dead Sea Scrolls*, 358; García Martínez and Tigchelaar, *DSSSE*, 382.

[91] Allegro, DJD 5.89. See also Carmignac, "Les Horoscopes," 206; Licht, "Legs as Signs," 20; Delcor, "Recherches sur un horoscope," 307.308; García Martínez and Tigchelaar, *DSSSE*, 382; Ejrnæs, Holst and Müller, *Dødehavsskrifterne*, 496.

[92] See PAM 41.314; 41.804; 42.616; 43.344; 43.438.

[93] See also 4Q186 3 1.

[94] Cf. Allegro, *The Dead Sea Scrolls: A Reappraisal*, 57: "Having deciphered one column including a particularly puzzling phrase, it was encouraging to find another piece in a further purchase which contained the same phrase written, rather carelessly for the coder, in 'clear' Hebrew, confirming the decipherment."

[95] Allegro, DJD 5.89, 90. See also Carmignac, "Les Horoscopes," 206; Licht, "Legs as Signs," 20; Delcor, "Recherches sur un horoscope," 307-9; Ejrnæs, Holst and Müller, *Dødehavsskrifterne*, 496.

[96] Strugnell, "Notes en marge du volume V," 275.

[97] García Martínez and Tigchelaar, *DSSSE*, 382-83.

[98] Cf. Bergmeier, *Glaube als Gabe*, 78; Maier, *Texte vom Toten Meer*, 2:136; Wise, "Horoscope Written in Code," 278; Vermes, *Complete Dead Sea Scrolls*, 358; García Martínez, *Dead Sea Scrolls Translated*, 456.

[99] See PAM 41.314; 41.804; 42.616.

"house of light."[100] Whether this new element is the horoscope of the described person, as in 4Q186 1 ii 8, cannot, unfortunately, be determined.

Notes and Comments on Readings in 4Q186 1 iv

On the basis of the few words preserved in column iv, not much can be said regarding the content of this part of the *4QZodiacal Physiognomy*.

4Q186 1 iv 6: Most scholars do not translate שמה, but a few scholars interpret it as "there."[101]

4Q186 1 iv 7: Perhaps something inside (תוך) a specific body part was indicated.[102]

Notes and Comments on Readings in 4Q186 2 i

4Q186 2 i 1: Allegro transcribes סרכ֯·[ע]י֯נ֯י֯ו֯, and translates "order. His [ey]es."[103] Strugnell, however, reads סרכמ֯[ו]ה ע]יניו.[104] A small dot of ink is visible following *kap* and preceding the gap.[105] This might be part of the down stroke of *mem*.[106] I therefore follow Strugnell's reading.

The beginning of this line is a continuation of a previous column that must have ended with על.[107] Perhaps סרכמ֯[ו]ה refers to the subject's eyes as well ordered as opposed to crooked.

The reading and understanding of the word following the second lacuna are difficult, and most translations add a question mark or leave a blank space. Allegro reads ה֯ג֯מ֯יות, compares it with Aramaic גומרא, "glowing coals," but adds that "the 'gentilic' form is strange unless it presupposes an adjective *גומרי 'glowing,'" which is "possibly here an error for גומרות."[108] Strugnell, however, notes that *gimel* is not very probable, *he* is impossible, and מנמריות, "speckled" (from נמר, "to give a checkered/striped appearance"), should probably be read. But he has no better explanation for the ending

[100] Cf. 4Q186 1 ii 8: חה.

[101] Cf. Carmignac, "Les Horoscopes," 206; Delcor, "Recherches sur un horoscope," 310; García Martínez and Tigchelaar, *DSSSE*, 383.

[102] Cf. for the use of תוך in relation to body parts, for example, *b. Neg.* 6:8: הפה תוך העין תוך האוזן תוך החוטם תוך; T.-S. K 21.88 2/א 8: ובתוך פדחתו (Gruenwald, "Jewish Physiognomic," 310).

[103] Allegro, DJD 5.91. See also Wise, "Horoscope Written in Code," 278; Vermes, *Complete Dead Sea Scrolls*, 358; García Martínez, *Translated*, 456.

[104] Strugnell, "Notes en marge du volume V," 275. See also Maier, *Texte vom Toten Meer*, 2:136; García Martínez and Tigchelaar, *DSSSE*, 382-83; Ejrnæs, Holst and Müller, *Dødehavsskrifterne*, 496.

[105] Cf. PAM 41.804; 42.616.

[106] Cf. 4Q186 2 i 6: סרכמה.

[107] Cf. 4Q186 2 i 3.6: על סרכמה.

[108] Allegro, DJD 5.91. See also Vermes, *Complete Dead Sea Scrolls*, 358; Wise, "Horoscope Written in Code," 278: "light (?)."

-יות.[109] It is more probable the letter preceding the broken *mem* is *nun* instead of *gimel*. This is because the vertical stroke of the letter directly following the lacuna is too close for it to be *gimel*. There seems to be no space for the left down stroke of *gimel*.[110] Furthermore, there are other examples of *nun* in this manuscript where the upper part of the vertical stroke curves to the right.[111] Strugnell is also correct to note that *he* is impossible for the trace of ink to the left of the gap. For *he* one would expect a horizontal stroke to extend to the right, but such is evidently not the case. It could, however, be the upper right corner of *mem*, and Strugnell's reading *mem* is, therefore, possible. I assume the text attempts to specify the color of the eyes as being between black and a somewhat lighter shade of black.

In *The Secret of Physiognomy*, a person's yellow eyes are specified as being between light yellow and reddish: מי שעינו צהובה בין ירקות לאדמות (perhaps orange is described?).[112] The construction בין...ל is equivalent to בין...ובין used in *4QZodiacal Physiognomy*, and also occurs in *4QPhysiognomy ar*. The translation "between"[113] is to be preferred either to "both,"[114] or "neither/nor,"[115] which is expressed by the use of לוא...ולוא in 4Q186 2 i 3–4. Perhaps the word מנמריות, "speckled," is used in *4QZodiacal Physiognomy* as a reference to the stripes that are visible in the iris and that are better observed in a somewhat light-colored iris. If so, it is reasonable to assume that in 4Q186 2 i 1 it is used to differentiate between pitch black and a lighter color of black. In ancient physiognomics the eye received a lot of attention, and many specifications as to color were made.[116]

4Q186 2 i 2: At the left side of this line, Allegro transcribes]·מם, reading two paleo-Hebrew *mem*s, but he does not provide a translation.[117] Most scholars do not translate the beginning of the second line. Presumably a qualification of the beard, mentioned in l.1, is given here. Maier, therefore, reconstructs "und sein Bart str[ähnig(/gesprenkelt?)]," while Wise suggests "whose beard is sp[arse]."[118] They presumably read the *pu'al* participle

[109] Strugnell, "Notes en marge du volume V," 275. See also García Martínez and van der Woude, *De rollen van de Dode Zee*, 499. Maier, *Texte vom Toten Meer*, 2:136, gives both possibilities: "gefleckt/glühend." García Martínez and Tigchelaar, *DSSSE*, 382-83, translate Strugnell's reading ("striped"), but mistakenly transcribe מגמריות.

[110] Cf. 4Q186 1 ii 5: רגליו; 1 ii 9: ברגל; 1 iii 8: רגליו; 2 i 2: תרגל; 2 i 5: רגליו.

[111] See 4Q186 2 i 1: בין and וזקנו.

[112] See Scholem, "Physiognomy," 493.4.

[113] Strugnell, "Notes en marge du volume V," 275. See also Maier, *Texte vom Toten Meer*, 2:136; García Martínez and Tigchelaar, *DSSSE*, 383.

[114] Allegro, DJD 5.91. See also Vermes, *Complete Dead Sea Scrolls*, 358.

[115] Wise, "Horoscope Written in Code," 278.

[116] See Polemo 1.106.19-170.22F; Anonymous Latin, *De physiognomonia liber* §§ 20-43.

[117] Allegro, DJD 5.90-91.

[118] Maier, *Texte vom Toten Meer*, 2:136; Wise, "Horoscope Written in Code," 278.

ממועט. But the lacuna hardly provides enough space for three letters and a blank space preceding the next word והואה. Perhaps Maier and Wise reconstruct the trace of ink preceding the lacuna as the upper part of the left down stroke of *ʿayin* and assume a reading ממעט for the *puʿal* participle of מעט. A reconstruction ממׄע]ט seems possible.

The third word, תרגל, presents a difficulty because this form is unknown in Hebrew or Aramaic.[119] Allegro makes sense of it by recourse to "Arab *rajila* 'of a quality between lankness and crispness or curliness,'" and translates "and it is curly," taking והואה as a reference to the beard mentioned in l.1.[120] Some scholars do not offer a translation for תרגל,[121] while others follow Allegro.[122] According to the classical Arabic-Arabic dictionary *Lisān al-ʿarab*, the phrase *šaʿr rajal* (or *rajil* or *rajl*) is explained as "hair between lankness and curliness." The translation "wavy" seems apt.[123]

Following ובתקולו, Allegro reads עניה ("And the pitch of his speech [?] is subdued").[124] But Maier and Vermes seem to read ענוה instead.[125] Both readings seem possible.[126]

4Q186 2 i 4: Allegro reads ממ֯י֯ל֯·[, and translates "and he is."[127] Strugnell, however, reads ממולד֯, and, because of an additional trace of ink, suggests ממולד֯ו֯, in which case the following word, אצבעות, need not be supplemented by *waw*-conjunctive.[128] Wise translates "but is well built" (perhaps reading ממילא, "he is filled," in the sense of well built?).[129] With Strugnell, I read the third letter as *waw*. Following *lamed* in ממול the manuscript is too mutilated to discern with certainty whether the traces of ink belong to one or two letters. It is possible to discern the down stroke of *dalet*, but the trace of ink Strugnell understands as the tip of *waw* might also be part of *dalet*.[130] Nevertheless, Strugnell's reading is plausible, either with damaged or reconstructed *yod*. García Martínez and Tigchelaar translate

[119] It also occurs in 4Q561 4 2.

[120] Allegro, DJD 5.91.

[121] Maier, *Texte vom Toten Meer*, 2:136; Vermes, *Complete Dead Sea Scrolls*, 358.

[122] Wise, "Horoscope Written in Code," 278; García Martínez and Tigchelaar, *DSSSE*, 383; Ejrnæs, Holst and Müller, *Dødehavsskrifterne*, 496.

[123] *Lisān al-ʿArab* (Beirut: Dār Iḥyāʾ al-Turāṯ al-ʿArabī, 1988), s.v. *rjl.* I thank Fred Leemhuis for this information.

[124] Allegro, DJD 5.91. See also García Martínez and Tigchelaar, *DSSSE*, 382-83.

[125] Maier, *Texte vom Toten Meer*, 2:136; Vermes, *Complete Dead Sea Scrolls*, 358; Ejrnæs, Holst and Müller, *Dødehavsskrifterne*, 496.

[126] Wise, "Horoscope Written in Code," 278, translates "resonates," presumably understanding it as a *qal* singular feminine participle from the root IV ענה ("to sing, cry").

[127] Allegro, DJD 5.91. See also Maier, *Texte vom Toten Meer*, 2:136; Vermes, *Complete Dead Sea Scrolls*, 358; Ejrnæs, Holst and Müller, *Dødehavsskrifterne*, 496.

[128] Strugnell, "Notes en marge du volume V," 275. See also García Martínez and Tigchelaar, *DSSSE*, 382-83.

[129] Wise, "Horoscope Written in Code," 278.

[130] See PAM 41.804 and 42.616.

ממולדו as "and like that from his conception."[131] But I understand the preposition מן to have causative meaning ("because").[132] The person's physiognomic appearance, in this case being neither tall nor short, is said to be so because of his horoscope. This means that the physiognomic description is interrupted by a reference to the person's horoscope. But, more significantly, it signals awareness of the notion that the configuration of heavenly bodies at the moment of birth influences human appearance.

4Q186 2 i 6: A small stroke of *lamed* is visible, but Allegro does not suggest a reconstruction.[133] There is, however, enough space to reconstruct the beginning of this line as]ואצבעות רג[ל]יו[.[134]

4Q186 2 i 7: Allegro reads *nun* and *he*, joins two more fragments to the left of the remaining part of 4Q186 2 i 7–9, and reconstructs שמונה, "eight," in this line.[135] The reading *nun*, however, is impossible. The stroke of ink near the left edge of the leather makes a curve to the left.[136] It could be part of *samek*, *ʿayin* or *šin*. But *nun* is excluded because, first, a vertical stroke should have been visible on the extant leather of 4Q186 2 i 7, and, second, if Allegro's join of 4Q186 5 to the left of 4Q186 2 i 7–9 is accepted, a stroke of the base of *nun* should appear following *waw* in 4Q186 5 1.[137] Since this is evidently not the case, the impossibility of reading *nun* in 4Q186 2 i 7 militates against Allegro's arrangement of 4Q186 2 and 4Q186 5, and also his reading שמונה. The reading *he* seems strange, but it is not unusual for the legs of *he* to curve to the left in *4QZodiacal Physiognomy*.[138] Although the upper horizontal stroke seems to extend too far to the left, this is not conclusive evidence against reading *he*.

Together with 4Q186 5, Allegro placed 4Q186 4 to the left of 4Q186 2 i 7–9. Thus, he read 4Q186 2 i 7 as]עׄמוד השני שמונה וא[חת.[139] Strugnell then suggested the following reconstruction for 4Q186 2 i 7–8:[140]

6 ורוח ל[ו]

7]בבית האור מן ה[עמוד השני שמונה וא[חת]

8]בבית החושך וזה הואה [מולדו ילוד הו[אה עליו

131 García Martínez and Tigchelaar, *DSSSE*, 383.

132 Cf. B.K. Waltke and M. O'Connor, *An Introduction to Biblical Hebrew Syntax* (Winona Lake, Indiana: Eisenbrauns, 1990), 213: "The cause or means of a situation is marked by *mn*."

133 Allegro, DJD 5.91.

134 Cf. Bergmeier, *Glaube als Gabe*, 78; Maier, *Texte vom Toten Meer*, 2:136; Vermes, *Complete Dead Sea Scrolls*, 358; Wise, "Horoscope Written in Code," 278.

135 Allegro, DJD 5.91. See also Maier, *Texte vom Toten Meer*, 2:136; Vermes, *Complete Dead Sea Scrolls*, 358; Wise, "Horoscope Written in Code," 278; García Martínez and Tigchelaar, *DSSSE*, 382-83; Ejrnæs, Holst and Müller, *Dødehavsskrifterne*, 496.

136 See PAM 41.804; 42.616.

137 Cf. 4Q186 1 iii 9: ש]מׄונה.

138 See e.g. 4Q186 2 i 6: סרכמה.

139 Allegro, DJD 5.91. See also Maier, *Texte vom Toten Meer*, 2:136.

140 Strugnell, "Notes en marge du volume V," 275.

Most scholars accept this reconstruction.[141] Regarding the placement of 4Q186 4, however, this join is problematic for two reasons.

First, the leather is not so deformed as to allow enough room for Allegro's arrangement. The state of the manuscript certainly does not justify the large cut and separation of 4Q186 2 into two halves.[142] Had Allegro not made this cut, there would not even have been room for 4Q186 4, let alone 4Q186 4 and 4Q186 5.[143] The join is materially incorrect. This judgment was confirmed after studying the fragments at the Dead Sea Scrolls laboratory of the Israel Antiquities Authorities.[144]

Second, the resulting sentence is improbable.[145] Although the exact sense of העמוד השני is not certain, it is clear from the context in 4Q186 1 ii 6 that someone (והואה), most probably the person who is the subject of the preceding physiognomic description, is from the second column. According to Strugnell's reconstruction, however, it seems that the person's spirit (ורוח לו) is not only in the "house of light," but also from the second column. It is not clear what this means. Also, such a reading is evidently different from the use of העמוד השני in 4Q186 1 ii 6. Furthermore, in the two other extant examples of the section concerning the "house of light" and the "house of darkness," in 4Q186 1 ii 7–9 and 4Q186 1 iii 8–9, there is no interruption by another clause. As this reconstruction creates another difficulty instead of contributing to our understanding of the sense of the phrase העמוד השני, it is, therefore, to be rejected because it has no basis in the extant text of *4QZodiacal Physiognomy*. This fragment is best treated separately.[146]

[141] See also, although some with slight differences, Vermes, *Complete Dead Sea Scrolls*, 358; Wise, "Horoscope Written in Code," 278; García Martínez and Tigchelaar, *DSSSE*, 382-83; Ejrnæs, Holst and Müller, *Dødehavsskrifterne*, 496.

[142] Cf. Allegro, DJD 5.91: "The straight cut through the centre of the fragment has been made to facilitate arranging the pieces where the skin has warped." In PAM 41.314; 41.804; 42.616; 43.344 the second fragment of *4QZodiacal Physiognomy* is still intact.

[143] Cf. PAM 42.616; 43.344. See also Wise, "Horoscope Written in Code," 278. Nevertheless, Wise reads שמונה, "eight," accepting at least the join of 4Q186 5 to the left of 4Q186 2 i 7.

[144] September 22, 2005.

[145] Strugnell, unfortunately, does not give a translation for his reconstruction. Vermes, *Complete Dead Sea Scrolls*, 358, translates: "His spirit consists of eight (parts) [in the House of Light, of] the second Column, and one [in the House of Darkness," and García Martínez and Tigchelaar, *DSSSE*, 383, translate "[His] spirit has eight (parts) [in the house of light, in the] second column, and o[ne] [in the house of darkness]."

[146] Cf. Popović, "A Note," 635-38.

Notes and Comments on Readings in 4Q186 2 ii

4Q186 2 ii 5: In this case *4QZodiacal Physiognomy* has הוא instead of הואה. The latter orthography is according to the Qumran scribal practice and it occurs in the other six occurrences in the text.[147]

4Q186 2 ii 6: Allegro reads מ]עֹורב and understands it in the same way as מעורבים in 4Q186 1 i 9.[148] Strugnell states that *dalet* instead of *reš* should be read,[149] but this is incorrect.[150] A certain bodily feature is described as "mixed."[151]

4Q186 2 ii 7: This small fragment appears separately in PAM 42.616. The amount of space to the right of *gimel* suggests גולש/שלוג to be the final word of a column. In PAM 43.438 this fragment is joined as another line under 4Q186 2 ii 6. In addition to understanding the word as written in reverse order in accordance with the rest of the manuscript, Allegro suggests that it might not be "coded," similar to אב]צונם in 4Q186 1 ii 2. If read as גולש it might refer to a quality of the hair, either in the sense of "flowing hair" as poetically expressed in Song 4:1 and 6:5, or in the sense of "baldness." On the other hand, with Allegro one should allow for the possibility that the word is written in reverse manner in accordance with the rest of *4QZodiacal Physiognomy* as שלוג, and that it is related to the word שלג ("snow"). In this sense it may describe a physical characteristic, for example the skin, as being white or fair.[152] In *The Book of the Reading of the Hands by an Indian Sage* the sole of someone's foot is described as having the appearance of either red wine or snow, i.e. being a burgundy red or white (כתואר יין אדום או כתואר שלג).[153] The problem, of course, with this reading is that 4Q186 2 ii 7 has שלוג, not שלג, which is presumably the reason why Allegro thought the "non-coded" reading more probable here.[154] The reading שלוג remains possible, but its sense is not clear. From the context it is impossible to decide which bodily feature is described in this line, and also whether the word should be read "non-coded" or reversed.

147 See 4Q186 1 ii 6; 1 ii 8 (twice); 2 i 2; 2 i 3; 4 3. The table in Tov, *Scribal Practices*, 341, needs to be corrected accordingly.

148 Allegro, DJD 5.91.

149 Strugnell, "Notes en marge du volume V," 275.

150 In PAM 41.804 and 42.616 a small diagonal stroke of ink is discernable that could be the right down stroke of *ʿayin*. Allegro's reading is, therefore, plausible.

151 Cf. also Maier, *Texte vom Toten Meer*, 2:137. For the translation here, see the discussion concerning מעורבים in Chapter One.

152 Allegro, DJD 5.91.

153 Scholem, "Physiognomy," 491.21–22.

154 Cf. n. 1 above for the suggestion that גולש could mean "flowing."

Notes and Comments on Readings in 4Q186 3

Upon inspection of the plate with the fragments, I found that 4Q186 3 is not on this plate. One of the curators told me it should be on another plate, but it has not been located yet. The readings, therefore, could not be checked.[155]

4Q186 3 1: Allegro reads מבי·],[156] but Strugnell suggests מבית֯].[157] A small trace of ink appears to the left of *yod*, which might be the bottom stroke of *taw*.[158] As in 4Q186 1 iii 9, this probably refers to the "house of light" or the "house of darkness," again using מן instead of ב.

4Q186 3 2: Allegro reads בכתן,[159] but Strugnell correctly notes that there is a trace of a letter to the right and reads בכת·].[160] It is evident that another letter is attached to the right leg of *taw*,[161] and I suggest this might be *pe*. A plausible reconstruction is בכתפ֯](י)ו ("with his shoulder(s)"). It could be a reference to the shoulder(s) of a described individual. Another possibility is to understand it as analogous to ברגל השור in 4Q186 1 ii 9. In this case it refers to the shoulders of a zodiacal sign. Moreover, if the space to the left of both מבית֯] and בכתפ֯]י represents the column margin, the entire first line could originally have had something like that in 4Q186 1 ii 8.[162] If this were so, בכתפ֯]י refers to a position in a sign of the zodiac in which an individual is said to have been born.

4Q186 3 3: Reading יפי] instead of ו֯פי] as Allegro does,[163] it is possible to translate "beautiful," but due to the fragmentary state it is impossible to determine what was referred to as "beautiful."

Notes and Comments on Readings in 4Q186 4

4Q186 4 3: The text has בהבת֯], but Allegro assumes a scribal error and reads בהמת֯ו (presupposing a join with 4Q186 5) as in 4Q186 1 ii 9.[164] Bergmeier suggests a reconstruction, following Allegro's arrangement, in which 4Q186 2 i is part of the section concerning the zodiacal sign *Taurus*. He therefore reads 4Q186 2 i 9 as [בפרסות השור וזה ה]ואה בהמתו [ש]ו֯[ר] ("in

155 Checked at the scroll laboratory of the IAA on September 22, 2005.

156 Allegro, DJD 5.91.

157 Strugnell, "Notes en marge du volume V," 275.

158 See PAM 43.344; 43.438.

159 Allegro, DJD 5.91.

160 Strugnell, "Notes en marge du volume V," 275.

161 See PAM 43.344; 43.438.

162 4Q186 3 1 would probably be too long if 4Q186 1 ii 8 were exactly copied, but this can easily be solved by leaving הואה in וזה הואה המולד out.

163 Allegro, DJD 5.91.

164 Allegro, DJD 5.91. See also e.g. Vermes, *Complete Dead Sea Scrolls*, 358; García Martínez and Tigchelaar, *DSSSE*, 382-83; Ejrnæs, Holst and Müller, *Dødehavsskrifterne*, 496. The reading of *taw* is, of course, based on the emendation, so we should be cautious about accepting בהמת֯ו as a certain reading.

the hoofs of *Taurus*. And this is his zodiacal sign: *Taurus*").[165] This reconstruction is entirely based on Allegro's arrangement of the fragments. It is impossible to determine its plausibility for the isolated fragment 4Q186 4. In this case only the reading בהמת̊[ו remains. It is not possible to identify the zodiacal sign referred to by בהמת̊[ו in 4Q186 4, assuming that this is what בהמה refers to. Maier, however, does not suppose a scribal error and translates "e]s ist in der Jungf[rau (?)]," probably reading ה]ואה בהבתו[לה], but it is not clear what this might mean.[166] If this were seen as an equivalent to ברגל השור (4Q186 1 ii 9), one would rather expect it to be preceded by a reference to the birth of the individual, such as the words ילוד עליו instead of ה]ואה. Maier's reading is without basis in the text and does not clarify its interpretation. Allegro's reading remains reasonable but needs to be adapted to the new arrangement בהמת̊[ו. It is plausible that 4Q186 4 3 referred to an individual's zodiacal sign.

Although this fragment is very small, it is interesting for two reasons.

First, it shows that the phrase העמוד השני is not followed directly by the words רוח לו as in 4Q186 1 ii 6–7, but a word beginning with *šin*. Equivalent to 4Q186 1 ii 7 one may read והואה מן ה]ע̊מוד השני ש[in 4Q186 4 1. Unfortunately, it is not possible to determine which word might follow the reference to "the second column." But it is evident that the reference to "the second column" in *4QZodiacal Physiognomy* does not occupy a set position in the text.

Second, 4Q186 4 shows several elements known from 4Q186 1 ii to appear near each other but also some differently from 4Q186 1 ii. In 4Q186 4 2 the word מולדו occurs, which can be understood equivalent to המולד in 4Q186 1 ii 8 as a reference to a person's horoscope. But it is also clear that, different from 4Q186 1 ii 6–8, there are fewer lines between the reference to העמוד השני and that to מולדו in 4Q186 4 1–2. On the other hand, the number of lines between מולדו and וזה ה]ואה בהמת̊[ו in 4Q186 4 2–3 is equal to that between המולד and וזה בהמתו in 4Q186 1 ii 8–9. Due to the amount of space available (assuming a regular column width of ca. 8-9 cm), it can almost certainly be ruled out that a reference to the "house of light" and the "house of darkness" stood between both words in 4Q186 4 1–2. At the same time, the fragmentary state does not allow a clear reconstruction. Perhaps in the case of 4Q186 4 2–3 one can assume a reference to an individual's horoscope specified as an ecliptical part of a zodiacal sign similar to 4Q186 1 ii 8–9, but this is far from certain.

[165] Bergmeier, *Glaube als Gabe*, 80 (see Chapter Three). Cf. Albani, "Horoscopes in the Qumran Scrolls," 309.

[166] Maier, *Texte vom Toten Meer*, 2:136. See also Albani, "Horoscopes in the Qumran Scrolls," 284 n. 22.

Comments on 4Q186 5

This fragment appears only in PAM 43.438 where it facilitates Allegro's arrangement of fragments for 4Q186 2 i.[167] It cannot be joined to another extant fragment of *4QZodiacal Physiognomy*. The fragment is too small to provide any meaningful information. Perhaps l.1 originally had הע[מונ]ד, and line 2 [ילונ]ד, but this is far from certain.

Notes and Comments on Readings in 4Q186 6

Allegro joins this small fragment to 4Q186 4,[168] but this join seems incorrect. The two small strokes of ink in 4Q186 6 1 cannot belong to 4Q186 4 3.[169] The legs of *he* in ה[ואה (4Q186 4 3) are too far apart to be the continuation of these strokes. It is, therefore, doubtful whether Allegro's arrangement is correct, and 4Q186 6 is best regarded as a separate fragment.

4Q186 6 2: Allegro reads [·ני זותן].[170] But it is possible to see the remaining stroke of ink to the left as the upper part of a right down stroke of *ʿayin* or *šin*,[171] reading עני, "poor," or שני, in which case it is possible to reconstruct והואה מן העמוד ה[שׁני. Thus, this would be a third occurrence of the phrase העמוד השני in the extant text of *4QZodiacal Physiognomy*. Furthermore, if this reconstruction is accepted, it also demonstrates that the words העמוד השני are not necessarily directly followed by רוח לו as in 4Q186 1 ii 6–7.

4Q186 6 3: Subsequent to a small trace of ink to the left, there is a stroke of ink that might be the down stroke of *gimel* or *nun*, although the latter seems more likely because the trace of ink seems to stand too close for it to be a *gimel*. This letter (probably *nun*) is most likely followed by *he*. It is not possible to determine if *lamed* is part of one word with [·ג̇/נ̇ה̇ן], or if it begins a new word.

4QPhysiognomy ar (4Q561)

Notes and Comments on Readings in 4Q561 1 i

This fragment seems to have preserved the left part of a column. This is indicated by the blank space following נגיד in l.2, and also by the larger size of final *dalet* that is more likely to occur at the end of a line.

167 See also the notes and comments on readings for 4Q186 2 i 7 in this appendix.

168 Allegro, DJD 5.91. See also Maier, *Texte vom Toten Meer*, 2:136; García Martínez and Tigchelaar, *DSSSE*, 382-83.

169 PAM 41.804; 42.616.

170 Allegro, DJD 5.91. See also García Martínez and Tigchelaar, *DSSSE*, 382-83.

171 PAM 41.804; 42.616.

4Q561 1 i 1: Starcky reads ן,[172] whereas Wise has [ו]הי.[173] The state of this fragment, however, is somewhat better in PAM 41.954 than in PAM 43.598. It is, therefore, possible to discern more letters than previous scholars have done. In PAM 43.598 *nun* is no longer visible, but in PAM 41.954 the vertical down stroke and the lower horizontal stroke are clearly discernible. The lower horizontal stroke of *nun* is connected to another vertical stroke that is probably the down stroke of *waw*.[174] Finally, *he* is possible because its upper horizontal stroke is partly visible. Before *he* there is enough space for three or four letters, if indeed בׄין begins l.2.

4Q561 1 i 2: Starcky reads אוׄרין, "brillants,"[175] while Beyer has חׄורין, "weiß."[176] In order to avoid חורין having a masculine plural ending, which conflicts with עינוה[י]ן and the ending of the other adjective לאכומן, Beyer explains it as a feminine plural of חורי.[177] Against these previous readings must be stated that *reš* is not preceded by *waw* but by *ʾalep*. In PAM 41.954 the left top and the attached left leg of *ʾalep* that bends inwards are clearly visible, and perhaps a slight trace even of the right leg. It is unlikely that *ʾalep* is preceded by another letter. The photographs show no evidence of this. Also, there does not seem to be enough space for another letter in the preceding word בׄין. The reading, therefore, of the second word in this line must be אׄרין. Although this reading is strange, following Beyer's reasoning for חורין one might suggest that ארין is feminine plural from *ארי. In addition, one must assume this to be defective for *אורי to make sense of the reading אׄרין in 4Q561 1 i 2.

4Q561 1 i 3: Starcky reads [ו]שפיר, "et beau,"[178] but Wise reads only שׄפיר, although translating "(and) attractive."[179] A slight trace of the left down stroke of *šin* is visible in both PAM 41.954 and 43.598. Whether or not *waw* might be reconstructed depends on where the right margin of the

172 *Preliminary Concordance*, 2359. See also K. Beyer, *Die aramäischen Texte vom Toten Meer: Samt den Inschriften aus Palästina, dem Testament Levis aus der Kairoer Genisa, der Fastenrolle und den alten talmudischen Zitaten: Aramaistische Einleitung, Text, Übersetzung, Deutung, Grammatik/Wörterbuch, Deutsch-aramäische Wortliste, Register* (vol. 2; Göttingen: Vandenhoeck & Ruprecht, 2004), 163, who supplements עלוה[י].

173 R. Eisenman and M. Wise, *Dead Sea Scrolls Uncovered: The First Complete Translation and Interpretation of 50 Key Documents withheld for over 35 Years* (New York: Penguin, 1992), 264. See also García Martínez and Tigchelaar, *DSSSE*, 1116; Wise, "4Q561," 228; Holst and Høgenhaven, "Physiognomy," 36.

174 See PAM 41.954 and cf. also 4Q561 1 i 1: עינוה[י]ן; 1 i 3: ושנוהי.

175 *Preliminary Concordance*, 2010; Starcky, "Un texte messianique araméen," 64. See also Wise, "4Q561," 228-29, where Wise reads אוׄרין, "light."

176 Beyer, *Die aramäischen Texte*, 2:163. See also García Martínez and Tigchelaar, *DSSSE*, 1116-17 (חורין, "pale"); Holst and Høgenhaven, "Physiognomy," 36 (חׄוׄרין).

177 Beyer, *Die aramäischen Texte*, 2:393.

178 *Preliminary Concordance*, 2438; Starcky, "Un texte messianique araméen," 64. See also Beyer, *Die aramäischen Texte*, 2:163 ([ו]שׄפיר); García Martínez and Tigchelaar, *DSSSE*, 1116 ([ו]שפיר); Holst and Høgenhaven, "Physiognomy," 36 ([ו]שׄפיר).

179 Wise, "4Q561," 228-29. But cf. Eisenman and Wise, *Scrolls Uncovered*, 264.

column is set. If בׄין in l.2 represents the right margin, it seems that a reconstructed *waw*-conjunctive would extend beyond this margin.[180] If this is so, it is unlikely that *waw* preceded ש֯פיר. As in 4Q186 1 i 8, however, one might suggest a scribal error and assume that <ו>ש֯פיר is meant.

At the end of this line, following ודקנה, a small dot appears.[181] Perhaps this represents a guide dot placed by the person who manufactured the original scroll?[182] But guide dots usually appear at the beginning and end of sheets, which does not seem to be the case here.

4Q561 1 i 4: Starcky reads ד֯ק, "petite,"[183] whereas others seem certain of the first letter and read דק.[184] The extant manuscript only shows a small trace of ink preceding *qop*, which might be the serif of *dalet*.[185] Paleographically it is also possible to read ר֯ק, "empty, void," but if this is an adjective for ודקנה its sense is not clear. The reading ד֯ק is therefore preferred.

Subsequent to להוׄה, the lacuna provides enough space for לא to have been preceded by *waw*-conjunctive.[186]

Regarding שׂגׄיא, the upper parts of the right and left down strokes of *šin* are visible, as is the upper part of the right down stroke of *gimel*.[187]

4Q561 1 i 5: Starcky reads [מ]מחקי[ן], "glabres."[188] The reading of *yod* before the lacuna, however, is not completely certain.

Just after the first gap, a small trace of a letter is visible. Starcky reads ו[ב֯]ין, "et entre."[189] García Martínez and Tigchelaar do not suggest a reading.[190] Wise reads *mem* instead of *bet* and reconstructs ו[ה]מ֯נו, "and a]r[e."[191] Finally, Holst and Høgenhaven read *yod*, וב[י֯]ן. The trace of the letter con-

[180] Cf. 4Q561 1 i 3 for the amount of space needed for וש in ושנוהי.

[181] See also Holst and Høgenhaven, "Physiognomy," 36.

[182] See PAM 43.598. Perhaps a slight trace of another guide dot appears subsequent to נגיד in l.2. For guide dots, see Tov, *Scribal Practices*, 61-68.

[183] *Preliminary Concordance*, 2133; Starcky, "Un texte messianique araméen," 64. Wise, "4Q561," 228, presumably reads likewise, but the *circellus* is misplaced above *qop*.

[184] See García Martínez and Tigchelaar, *DSSSE*, 1116; Beyer, *Die aramäischen Texte*, 2:163; Holst and Høgenhaven, "Physiognomy," 36.

[185] See PAM 41.954; 43.598.

[186] Cf. also *Preliminary Concordance*, 2262; Starcky, "Un texte messianique araméen," 64; García Martínez and Tigchelaar, *DSSSE*, 1116; Beyer, *Die aramäischen Texte*, 2:163; Wise, "4Q561," 228; Holst and Høgenhaven, "Physiognomy," 36.

[187] See PAM 41.954. Cf. also *Preliminary Concordance*, 2411 (שׂ֯גיא); Starcky, "Un texte messianique araméen," 64; García Martínez and Tigchelaar, *DSSSE*, 1116 (שגיא); Beyer, *Die aramäischen Texte*, 2:163 (שׂגיא); Wise, "4Q561," 228 (שׂ[ג]יא); Holst and Høgenhaven, "Physiognomy," 36 (שגיא).

[188] *Preliminary Concordance*, 2279; Starcky, "Un texte messianique araméen," 64. See also García Martínez and Tigchelaar, *DSSSE*, 1116; Beyer, *Die aramäischen Texte*, 2:163; Wise, "4Q561," 228; Holst and Høgenhaven, "Physiognomy," 36.

[189] *Preliminary Concordance*, 2085; Starcky, "Un texte messianique araméen," 64. See also Beyer, *Die aramäischen Texte*, 2:163.

[190] García Martínez and Tigchelaar, *DSSSE*, 1116.

[191] Wise, "4Q561," 228-29. But cf. Eisenman and Wise, *Scrolls Uncovered*, 264, ובין.

sists of a down stroke to the left of which is attached, first, an upper horizontal stroke, and, second, a lower horizontal stroke. This description excludes *yod*. Reading *bet* is possible, and Starcky's reconstruction is plausible. Paleographically *mem* seems possible too, but the problem with Wise's reconstruction is that his reading of the word following the lacuna is unlikely and, therefore, his suggested reading for the lacuna too. Although I follow Starcky's reading, it must be noted that the amount of space between [מ]מחקי̊ן] and]ב̊ין seems rather large.

Following the second lacuna, Starcky reads מה]דמין, which he apparently translates as "*minces*."[192] Wise, however, initially following Starcky's reading, provides a more apt translation, "mal]formed."[193] Another reconstruction is proposed by Beyer, who reads מג]דמין, "gegliedert."[194] Recently, Wise has offered yet another reading,] דקין, "thin,"[195] but this reading is very unlikely. First, *dalet* is immediately after the lacuna. There is no space between the two. Second, and more important, *qop* is impossible for two reasons. The upper horizontal stroke curves upwards to the right, and there is a diagonal stroke extending to the left.[196] This corresponds to *mem* rather than *qop*.[197] Having said that, Beyer's reading מג]דמין seems preferable to Starcky's מה]דמין, because in the latter case one would expect to see to the right of *dalet* a trace of the upper horizontal stroke of *he* extending to the left.[198]

Finally, Starcky reads לעבי[ן], "épais."[199] The reading of *yod*, however, is not so certain. Wise correctly places a *circellus*.[200] In PAM 41.954 it is clear that the stroke of ink moves from a horizontal to a vertical line, most probably representing one continuous move of the pen. This description fits *reš* or *taw*.[201] This means that a reading לעבר], or, לעבת], is paleographically possible. But לעבי̊[ן] makes more sense in this context.

192 *Preliminary Concordance*, 2137; Starcky, "Un texte messianique araméen," 64.

193 Eisenman and Wise, *Scrolls Uncovered*, 264.

194 K. Beyer, *Die aramäischen Texte vom Toten Meer: Ergänzungsband: Samt den Inschriften aus Palästina, dem Testament Levis aus der Kairoer Genisa, der Fastenrolle und den alten talmudischen Zitaten: Aramaistische Einleitung, Text, Übersetzung, Deutung, Grammatik/Wörterbuch, Deutsch-aramäische Wortliste, Register* (Göttingen: Vandenhoeck & Ruprecht, 1994), 125. More recently, Beyer, *Die aramäischen Texte*, 2:163-64, provides both readings (מג/ה]דמין), but only translates the first. See also García Martínez and Tigchelaar, *DSSSE*, 1116-17, "st]umped"; Holst and Høgenhaven, "Physiognomy," 37, "sh]runken."

195 Wise, "4Q561," 228-29.

196 See PAM 41.954; 43.598.

197 Cf. 4Q561 1 i 1: מערבין; 1 i 2: לאכומן; 1 i 3: ודקנה; 1 ii 8: לקומתה.

198 See PAM 41.954; 43.598.

199 *Preliminary Concordance*, 2326; Starcky, "Un texte messianique araméen," 64. See also García Martínez and Tigchelaar, *DSSSE*, 1116; Beyer, *Die aramäischen Texte*, 2:163; Holst and Høgenhaven, "Physiognomy," 36.

200 Wise, "4Q561," 228. The *circellus*, however, is wrongly placed above *bet* instead of *yod*.

201 Cf. 4Q561 1 i 4: אברוהי.

Notes on Readings in 4Q561 1 ii

4Q561 1 ii 2: Starcky reads ל·מלי as the beginning of this line, but adds a question mark regarding the root מלא.[202] García Martínez and Tigchelaar read ו[מ]תמלי, "[and fi]lled (?)."[203] Wise, finally, has ק֯ל֯ ת֯מלי, "a voice full (?)."[204] First, if the join between the two fragments here is accepted, then it is evident that at least *lamed* has to be read in l.2. The tip of the upper stroke of *lamed* is clearly discernible in the fragment bearing l.1 in PAM 41.954 and 43.598. Second, it is very probable that *mem* is preceded by *taw*. A slight trace of ink is visible at the right edge of the leather, which is far enough from the left leg to represent the right leg of *taw*.[205] It is questionable, however, whether there is enough space between *taw* and *lamed* for *mem* to read ל[מ]תמלי.[206] Perhaps if *mem* is written as a narrower sign. Furthermore, in order for l.2 to have the same right margin as l.1, this line cannot begin with *lamed*; *lamed* must be preceded by another letter. Wise takes *lamed* as the final letter, reads ק֯ל֯, and suggests a blank between ק֯ל֯ and ת֯מלי. It seems, however, that a wide letter such as *qop* would extend beyond the right margin.[207] Rather, *lamed* is probably preceded by a narrower letter such as *waw*. Therefore, I tentatively suggest reading [ו]ל֯[מ]ת֯מלי, *itpe'el* participle of מלא.

Starcky reads יתן, but does not suggest a reconstruction.[208] Wise, however, reads ות֯ן and suggests ות֯[ק]ף, "a voice full (?) and strong."[209] Wise's reading is possible. His reconstruction, however, assumes too much text devoted to the voice (two lines), which is unlikely in *4QPhysiognomy ar* where the descriptions are short and to the point.

4Q561 1 ii 3: Starcky reads ארין, "bright," corresponding to א֯רין in 4Q561 1 i 2.[210] Wise, however, reads אריך, "long."[211] There are no other

[202] *Preliminary Concordance*, 2281. See also Eisenman and Wise, *Scrolls Uncovered*, 264.

[203] García Martínez and Tigchelaar, *DSSSE*, 1116-17. See also Holst and Høgenhaven, "Physiognomy," 36.

[204] Wise, "4Q561," 228-29.

[205] See PAM 42.438; 43.598. Cf. 4Q561 1 ii 8: לקומתה. It seems unlikely to read *nun* preceded by another letter because there would too much space between them, cf. 4Q561 1 ii 6: ואנון.

[206] Cf. the space occupied by *mem* and *taw* in 4Q561 1 ii 8: לקומתה.

[207] Cf. the width of *qop* together with *lamed* in ק֯ל]ה in 4Q561 1 ii 1.

[208] *Preliminary Concordance*, 2281. See also García Martínez and Tigchelaar, *DSSSE*, 1116 (··ן). See also Holst and Høgenhaven, "Physiognomy," 36 (ית֯ן).

[209] Wise, "4Q561," 228-29. This reconstruction is probably due to the occurrence of the root תקף in the composition that Starcky designates *ar S*, see *Preliminary Concordance*, 1. It consists of three fragments. Wise understands these fragments to be part of *4QPhysiognomy ar*.

[210] *Preliminary Concordance*, 2010.

[211] Eisenman and Wise, *Scrolls Uncovered*, 264. See also García Martínez and Tigchelaar, *DSSSE*, 1116-17; Beyer, *Die aramäischen Texte*, 2:164; Wise, "4Q561," 228-29; Holst and Høgenhaven, "Physiognomy," 36.

examples of final *kap* in this manuscript. As for final *nun*, the down stroke following ארי in this line is straight, whereas other examples of final *nun* in the extant text of *4QPhysiognomy ar* curve downwards to the left. This makes *nun* unlikely, and *kap* is, therefore, the most likely reading.

Before this fragment ends at the left, Starcky suggests reading בׄחׄ.[212] García Martínez and Tigchelaar acknowledge the presence of strokes of ink but do not propose a reading.[213] Initially, Wise did not indicate the remaining letter traces,[214] but recently he has read [וה]אׄ], "[and] is [."[215] This latter reading is impossible. First, the two down strokes, connected by a horizontal (slightly diagonal) upper stroke, are too close to each other to be *ʾalep*, and the right down stroke is too vertical.[216] A reading *he*, or *ḥet* is more probable. Second, preceding this letter, two lower horizontal strokes are clearly visible, which might belong to *bet*, *kap*, *mem*, *nun*, or *pe*.[217] However, Wise seems correct in assuming a blank space after the final letter, which means this is the end of a word.[218]

4Q561 1 ii 4: Starcky reads שער.[219] García Martínez and Tigchelaar reconstruct *waw*-conjunctive, [ו]שער.[220] It is possible to discern a tip of the head of *waw*, וׄשער.[221]

Starcky reconstructs the hair of the beard as being thick, שׂגׄ]יא.[222] The manuscript, however, is better preserved in PAM 42.438 than in 43.598. From PAM 42.438 it is evident that a vertical down stroke stands immediately next to the left down stroke of *šin*. This excludes the possibility of *gimel* whose right down stroke is more diagonal so one would expect the upper part to stand further to the left of *šin* than its lower part.[223] Instead, the down stroke might very well be the right leg of *ḥet*. Therefore, I suggest the reading שׄחׄ]ם, "dark, black."

212 *Preliminary Concordance*, 2010.

213 García Martínez and Tigchelaar, *DSSSE*, 1116 (ן···).

214 Eisenman and Wise, *Scrolls Uncovered*, 264. See also Holst and Høgenhaven, "Physiognomy," 36.

215 Wise, "4Q561," 228-29.

216 Cf. 4Q561 1 i 1: שגיא; 1 i 4: שׂגיא; 1 ii 3: אריך.

217 See PAM 42.438; 43.598.

218 The leather has further deteriorated in PAM 43.598, but in PAM 42.438 more leather is visible, supporting the observation of a blank space.

219 *Preliminary Concordance*, 2412. See also Eisenman and Wise, *Scrolls Uncovered*, 264; Beyer, *Die aramäischen Texte*, 2:164.

220 García Martínez and Tigchelaar, *DSSSE*, 1116. See also Wise, "4Q561," 228; Holst and Høgenhaven, "Physiognomy," 36.

221 See PAM 42.438; 43.598.

222 *Preliminary Concordance*, 2411. See also Eisenman and Wise, *Scrolls Uncovered*, 264 (שג]יא); Beyer, *Die aramäischen Texte*, 2:164 (שג]יא); García Martínez and Tigchelaar, *DSSSE*, 1116 (שג]יא); Wise, "4Q561," 228 (שגׄ]יא); Holst and Høgenhaven, "Physiognomy," 36 (שגׄ]יא).

223 Cf. 4Q561 1 i 1: שגיא.

4Q561 1 ii 5: Starcky reads להוון בין עבין ל]דקין.[224]
4Q561 1 ii 6: Starcky reads ואנון,[225] but, because the head of *yod* is more pronounced than that of *waw* in this manuscript, I read the feminine form ואנין.

The second word in the remaining part of this line Starcky reads קטיוׄן, "short."[226] García Martínez and Tigchelaar read קטינ]ן.[227] Beyer, however, reads קטינן, "zierlich."[228] This latter reading seems correct. It can be compared with the pre-final *nun* and final *nun* in מתקנן[in *ar S* 1 2.[229] Also, the feminine form קטינן accords well with ואנין.
4Q561 1 ii 8: Following לקומתה, the next word probably begins with *waw*[230] and not *yod*.[231]
4Q561 1 ii 9: The only indication for this line is the presence of the upper part of *lamed* beneath *he* of לקומתה in the preceding line.

Notes on Readings in 4Q561 2

4Q561 2 1: Wise suggests identifying *bet*, but the leather shows traces of two letters. The lower horizontal stroke is possibly the base of *mem* followed by *waw*, as in לשמקמוקׄי in 4Q561 5 2.[232]
4Q561 2 2: Starcky reads אמׄןׄ, "elbows."[233] Recently, however, Wise has read אפׄהׄ, "whose nose."[234] The leather is very damaged, which makes it difficult to determine the correct reading with certainty. Following *ʾalep* the vertical stroke seems to curve to the left.[235] This description does not seem to fit the other occurrences of *mem* or *pe* in this manuscript[236] but seems

[224] *Preliminary Concordance*, 2326. See also Wise, "4Q561," 228-29.
[225] *Preliminary Concordance*, 2035. See also García Martínez and Tigchelaar, *DSSSE*, 1116; Wise, "4Q561," 228; Holst and Høgenhaven, "Physiognomy," 36.
[226] *Preliminary Concordance*, 2386.
[227] García Martínez and Tigchelaar, *DSSSE*, 1116. See also Wise, "4Q561," 228 (קטינׄ]ן, initially, however, Wise reads קטי<נ>ין]ן, see Eisenman and Wise, *Scrolls Uncovered*, 264); Holst and Høgenhaven, "Physiognomy," 36. Wise, "4Q561," 228, reconstructs the masculine plural ואברוהי[in 4Q561 1 ii 4, and assumes that the extant text of ll.5-6 refers to the limbs, but this is not likely for l.6.
[228] Beyer, *Die aramäischen Texte*, 2:164.
[229] See PAM 41.954, 43.598, and *Preliminary Concordance*, 2450. The reading of Wise, "4Q561," 228, מתקן [, is incorrect, as is that of Holst and Høgenhaven, "Physiognomy," 30, מתקם[.
[230] *Preliminary Concordance*, 2386; Wise, "4Q561," 228.
[231] Eisenman and Wise, *Scrolls Uncovered*, 264; García Martínez and Tigchelaar, *DSSSE*, 1116; Holst and Høgenhaven, "Physiognomy," 36.
[232] See PAM 41.954; 42.438; 43.598.
[233] *Preliminary Concordance*, 2024. See also Eisenman and Wise, *Scrolls Uncovered*, 264; García Martínez and Tigchelaar, *DSSSE*, 1116; Holst and Høgenhaven, "Physiognomy," 37.
[234] Wise, "4Q561," 228-29.
[235] See PAM 41.954; 43.598.
[236] 4Q561 1 i 1: מעׄרבין; 1 i 2: לאכומן, אפה; 1 ii 7: טפרוהׄי; 1 ii 8: לקומתה.

more in accordance with *kap* or *nun*.[237] Regarding the final letter, Wise's suggestion for *he* is highly unlikely. The stroke of ink seems to thicken at the top, and one would expect to see the horizontal upper stroke of *he* that protrudes to the left. Starcky's reading of final *nun* seems better.[238] Finally, between possible *nun* and final *nun*, it seems possible to read *waw* or *yod*.[239] Due to the damaged state of the leather it is not possible to determine whether a ligature occurred. But the traces of ink between *nun* and final *nun* seem to indicate a rather pronounced head, which would fit *yod* better than *waw* in this manuscript. A reading אנין seems, therefore, paleographically the most preferable.

The next word Starcky reads as ברין], "protruding, pointed."[240] Recently, Wise has suggested reading בין, "between,"[241] but this is impossible given the amount of space filled by this word and the fact that two letters stand between the first and the final one.[242] Moreover, reading the first letter as *bet* seems difficult. The letter seems too high, and the remaining traces of ink indicate that the upper part of the down stroke slants slightly to the right.[243] Reading *kap* seems more plausible.[244] The two small down strokes between *kap* and final *nun* can be identified as *waw* and *yod*. Therefore, I suggest reading כוין.

4Q561 2 3: Starcky reads פתין ושקוהין] בין דקין[,[245] but recently Wise has suggested reading פתין שקוהי], "are bro[a]d, whose thighs [."[246] From the available photographs it is not possible to discern a trace of ink to the right of *šin* that indicates *waw*,[247] but without *waw* the space would perhaps be too large. Both Starcky and Wise assume that פתין begins 1.3. But it is not certain that פתין represents the right margin of the column and the beginning of this line because it does not stand in the same vertical line as די of 1.2.

[237] 4Q561 1 i 1: עינוה[י];1 i 2: לאכומן, נגיד; 1 ii 4: דקנה; 1 ii 6: ואנין; 1 ii 7: כעבין.
[238] 4Q561 1 i 1: מערבין; 1 ii 6: ואנין; 1 ii 7: כעבין.
[239] 4Q561 1 ii 6: ואנין; 1 ii 9: לקומתה.
[240] *Preliminary Concordance*, 2097. See also Eisenman and Wise, *Scrolls Uncovered*, 264; García Martínez and Tigchelaar, *DSSSE*, 1116; Holst and Høgenhaven, "Physiognomy," 37.
[241] Wise, "4Q561," 228-29.
[242] See PAM 41.954; 43.598, and cf. 4Q561 1 i 2: בין; 1 ii 5: בין.
[243] See PAM 41.954; 43.598, and cf. 4Q561 1 i 1: מערבין; 1 i 4: אברוהי.
[244] Cf. 4Q561 1 i 2: לאכומן; 1 ii 7: כעבין.
[245] *Preliminary Concordance*, 2439. See also Eisenman and Wise, *Scrolls Uncovered*, 264; García Martínez and Tigchelaar, *DSSSE*, 1116; Beyer, *Die aramäischen Texte*, 2:164; Holst and Høgenhaven, "Physiognomy," 37.
[246] Wise, "4Q561," 228-29. It is not clear why Wise adds brackets in "bro[a]d."
[247] See PAM 41.954; 43.598.

4Q561 2 4: Starcky reads עבי̊ן.[248] In PAM 43.598 a slight trace of ink is discernable that possibly represents the upper part of *nun*.

The second word extant in this line Starcky reads as כ̊ף. Wise initially reconstructs a *waw*-conjunctive [ו]כף,[249] but recently agrees with Starcky's reading.[250] Preceding final *pe*, a stroke of ink is visible that probably represents the serif of a letter. The most sensible reading is *kap*, although this letter does not seem to have such a pronounced serif in this manuscript. Depending on the width of *kap* there is perhaps enough space to reconstruct a *waw*, but this is not necessary.

Finally, Starcky reads רגל̊ו̊[נ]הי,[251] but Beyer seems to reconstruct a singular כף רגלה.[252] It is difficult to determine with certainty which letter the slight trace of ink to the left of *lamed* belongs to. Given the fact that כ̊ף is singular, there is no need to reconstruct a plural.[253] Therefore, I agree with Beyer and read the singular רגלה̊[ן], "(the sole of) his foot."

4Q561 2 5: Starcky reads נ̈[ג]יד.[254] Wise, however, reads שג[י]א, "exce[edingly so."[255] Discernible is, first, the upper part of a down stroke that seems to thicken at the top, and, second, a diagonal down stroke to which is attached a left leg that bends inwards.[256] Apart from the proper identification of the letters these strokes represent, it must be stated that for both readings one would expect to see at least a trace of *yod* to the left of *gimel*.[257] Instead, it seems to be the end of a word followed by a blank space. Furthermore, the amount of space preceding [לה allows for the reconstruction of an entire word instead of just two letters and a blank space. Having said that, the two latter strokes seem to more closely resemble *ʾalep* than *gimel* in this manuscript, because of the pronounced left leg.[258] But the first down stroke seems to move to the right, which would exclude it from being the right down stroke of *ʾalep*. The reading *gimel*, therefore, is more

[248] *Preliminary Concordance*, 2326. See also García Martínez and Tigchelaar, *DSSSE*, 1116; Wise, "4Q561," 228 (לעבי̊ן); Holst and Høgenhaven, "Physiognomy," 37 (לעבין).

[249] Eisenman and Wise, *Scrolls Uncovered*, 264. See also García Martínez and Tigchelaar, *DSSSE*, 1116; Holst and Høgenhaven, "Physiognomy," 37.

[250] Wise, "4Q561," 228.

[251] *Preliminary Concordance*, 2401. See also Eisenman and Wise, *Scrolls Uncovered*, 264 (רגל[ו]הי); García Martínez and Tigchelaar, *DSSSE*, 1116 (רגל[ו]הי); Wise, "4Q561," 228 (רגל̊[ו]הי); Holst and Høgenhaven, "Physiognomy," 37 (רגל[ו]הי).

[252] Beyer, *Die aramäischen Texte*, 2:164.

[253] Cf. the plural in 4Q186 2 i 5: וכפות רגליו, and the singular in *The Book of the Reading of the Hands by an Indian Sage*: בכף רגלו (see Scholem, "Physiognomy," 491.21).

[254] *Preliminary Concordance*, 2309. See also Eisenman and Wise, *Scrolls Uncovered*, 264; García Martínez and Tigchelaar, *DSSSE*, 1116; Holst and Høgenhaven, "Physiognomy," 37.

[255] Wise, "4Q561," 228-29.

[256] See PAM 41.954; 43.598.

[257] Cf. 4Q561 1 i 1: שגיא; 1 i 2: נגיד.

[258] Cf. 4Q561 1 i 1: שגיא; 1 i 2: נגיד; 1 i 4: שׂגיא; 2 4: רגלה̊.

likely. If the down stroke preceding *gimel* is part of *nun*, it must be noted that it seems to stand too close to *gimel*.[259] But if it represents the left down stroke of *šin*, one would expect to see a trace of the middle down stroke, unless the top layer of the leather is missing. Due to the fragmentary state of the manuscript it is not possible to determine the identity of this letter.

Starcky indicates that he sees another trace of ink before [·לה,[260] but this cannot be determined from the photographs.[261] Therefore, I read [לה.[262]

4Q561 2 6: Wise initially reads [לתיפל],[263] but recently Wise has read [לֹתֹ ופקֹ].[264] García Martínez and Tigchelaar read [ת ופל].[265] Wise is correct to read *lamed* before *taw*. Its upper part is clearly visible.[266] To read *taw* next seems difficult. Contrary to the other occurrences of *taw* in this manuscript, the upper part of the left down stroke bends to the left instead of the right.[267] Next, I think *yod* is more probable than *waw* because the tip of the head is more pronounced. Following *pe*, both *lamed* and *qop* seem possible.

4Q561 2 7: Starcky reads [מסך ·], which García Martínez and Tigchelaar translate as a noun, "from the end of."[268] Wise, however, reconstructs the verbal form ל[מסך, "to] come to an end."[269] The reading depends on one's understanding of the text. If מסך is translated as a noun it can be understood to specify a certain part of the body. But if ל[מסך is taken as a verbal form it might be understood as part of a prediction concerning the individual's life on the basis of his bodily features. It then indicates perhaps the number of years after which a person's life will come to an end.[270]

259 Cf. 4Q561 1 i 2: נגיד.

260 *Preliminary Concordance*, 2254. See also Eisenman and Wise, *Scrolls Uncovered*, 264.

261 See PAM 41.954; 43.598.

262 See also García Martínez and Tigchelaar, *DSSSE*, 1116; Wise, "4Q561," 228; Holst and Høgenhaven, "Physiognomy," 37.

263 Eisenman and Wise, *Scrolls Uncovered*, 264.

264 Wise, "4Q561," 228.

265 García Martínez and Tigchelaar, *DSSSE*, 1116. See also Holst and Høgenhaven, "Physiognomy," 37.

266 See PAM 43.598.

267 Cf. 4Q561 2 3: פתין; 6 2: כתפה.

268 *Preliminary Concordance*, 2463. García Martínez and Tigchelaar, *DSSSE*, 1116-17. See also Wise, "Aramaic Horoscope," 568; Holst and Høgenhaven, "Physiognomy," 38.

269 Eisenman and Wise, *Scrolls Uncovered*, 264-65; Wise, "4Q561," 228-29. Cf. also Maier, *Texte vom Toten Meer*, 2:739, "zu]enden."

270 In Babylonian physiognomics, sometimes a concrete amount of time was given for the duration of people's lives, but more often it was couched in general terms ("his days will be long/short"), see Böck, *Die babylonisch-assyrische Morphoskopie*, 30-31. Greco-Roman physiognomics was less predictive, but in Greek *zodiologia* concrete numbers of years were given for people's lives, see e.g. the *zodiologia* in CCAG 12.173–191.

4Q561 2 8: The fragment shows the traces of probably three letters.[271] The second letter probably represents the serif and upper horizontal stroke of *dalet*. This is possibly followed by *waw* or *yod*. The first letter preserves perhaps the upper part of the down stroke and the upper horizontal stroke of *qop*.

Notes on Readings in 4Q561 3

4Q561 3 1: The down stroke curves to the left, which is typical of final *nun* in this manuscript.
4Q561 3 2: Starcky reads ר[וח לה.[272] Starcky's reconstruction is possible but far from certain. Another possibility is perhaps מ[וח לה, "he has a skull," but this is not likely since it does not accord with the way that descriptions of the body are given in the text (noun with suffixed pronoun). Also, one would perhaps expect to see a trace of the lower horizontal stroke of *mem*. See further the section on body and spirit in *4QPhysiognomy ar* in Chapter One.
4Q561 3 3: Starcky reads עקא [, "narrow (?)."[273] From his translation it seems as if Starcky understands עקא as describing a certain part of the body as being narrow. But it is not necessary to take עקא in this sense. It can also be taken as an indication of the individual's present or future situation. In this way it might refer to him being in trouble or distress.
4Q561 3 4: Starcky reads שׁ֯ערן.[274] The left down stroke of *šin* is visible.[275]

The following word is read by Starcky as ע֯ב֯ות.[276] One can discern slight traces of the upper part of the right and left down stroke of *ʿayin*, and the upper horizontal stroke with serif and lower horizontal stroke of *bet*.[277]

Notes on Readings in 4Q561 4

4Q561 4 1: Wise reads ו[ה֯מו, "and th[ey are."[278] But it is very unlikely that the remaining trace of ink to the right of *mem* is *he*.[279] If it were the

[271] See PAM 43.598.

[272] *Preliminary Concordance*, 2402; Starcky, "Un texte messianique araméen," 62. See also Maier, *Texte vom Toten Meer*, 2:739; É. Puech (ed.), *Qumrân Grotte 4.XXII: Textes araméens, Première partie, 4Q529-549* (DJD 31; Oxford: Clarendon, 2001), 143; Wise, "4Q561," 228; Holst and Høgenhaven, "Physiognomy," 37.

[273] *Preliminary Concordance*, 2358. See also Maier, *Texte vom Toten Meer*, 2:739; Wise, "4Q561," 228; Holst and Høgenhaven, "Physiognomy," 37.

[274] *Preliminary Concordance*, 2412. See also Beyer, *Die aramäischen Texte*, 2:164; Wise, "4Q561," 228; Holst and Høgenhaven, "Physiognomy," 37.

[275] See PAM 41.944; 43.598.

[276] *Preliminary Concordance*, 2326. See also Maier, *Texte vom Toten Meer*, 2:739 ("ihre Haare stark"); Wise, "4Q561," 228-29 (ומן די יהוין להן] שׂערן ע֯ב֯ות, "[Anyone whose body] hair [will be] thick"); Holst and Høgenhaven, "Physiognomy," 37 (שׂערן ע֯ב֯ות [, "thick hair").

[277] See PAM 41.944; 43.598.

right end of the upper horizontal stroke one would except to see traces of the middle and left part too. But perhaps the top layer of the leather is missing next to the gap. Even then, *he* is unlikely because of the amount of space between *mem* and the trace of ink. It is conceivable that the trace of ink represents the left part of another letter and that a second letter followed before *mem*. On the other hand, if one letter is to be reconstructed, the trace of ink is perhaps the right upper part of *ṭet*. But in this case one would have to assume the left vertical stroke to be lower than the right one.[280] A clear reading is, unfortunately, not possible.

4Q561 4 2: Starcky reads]ת֯רגל, "dwarf."[281] Wise partially agrees with Starcky and reads [דקנה] תרגל, "whose beard] is curly."[282] Holst and Høgenhaven read]מ֯רגל, "from the foot."[283] It seems more likely to read *taw* instead of *mem*.[284] If the horn of *mem* were intended one would expect it to be more perpendicular on the diagonal stroke.

4Q561 4 3: Wise reads]ו֯לא֯, "but not," while Holst and Høgenhaven have]ע֯ל, "upon."[285] The stroke of ink preceding *lamed* is too fine to be the head of *waw*. It is possibly the upper part of the left down stroke of *ʿayin*. Following *lamed* it is not clear from the photograph whether a trace of ink is discernible or whether it is just shade.[286]

Notes on Readings in 4Q561 5

4Q561 5 1: Although the two traces are not joined, they perhaps belong to *šin*.

4Q561 5 2: Starcky reads]לשמקמיק ·[, "be red."[287] Beyer, however, reads *waw* instead of *yod*: שמקמוק, "rot,"[288] while García Martínez and Tigchelaar read *yod* after *qop*: [... בין ...]לשמקמיקי [, "[... between ...] and reddish [."[289] In this manuscript *waw* is less pronounced than *yod*. Also, it is evident that *yod* is the final letter.[290] I read]לשמקמוקי[.

---

278 Wise, "4Q561," 230-31.

279 See PAM 43.598.

280 Cf. 4Q561 1 ii 6: קטינ֯ן֯; 1 ii 7: טפרוה֯ני.

281 *Preliminary Concordance*, 2451.

282 Wise, "4Q561," 230-31.

283 Holst and Høgenhaven, "Physiognomy," 37-38.

284 See PAM 43.598.

285 Wise, "4Q561," 230-31; Holst and Høgenhaven, "Physiognomy," 37-38.

286 See PAM 43.598.

287 *Preliminary Concordance*, 2325. See also Wise, "4Q561," 228-29.

288 Beyer, *Die aramäischen Texte*, 2:164.

289 García Martínez and Tigchelaar, *DSSSE*, 1118-19. See also Holst and Høgenhaven, "Physiognomy," 37.

290 See PAM 42.438 and 43.598.

In *The Physiognomy of R. Ishmael*, אדום, "red," is used to describe the face (ומי שפניו אדומות), but also מאדמון מעט, "slightly red," for the eyes.[291] It is possible that the complexion of the face is described in 4Q561 5 2.

4Q561 5 3: Starcky reads ב[ררי, from ברא, "external."[292] Initially following Starcky's reading, Wise, however, translates ב[ררי as "cl]ear" (ברר).[293] But recently Wise has suggested כ[דרי, "[whose forehead (?)] will b[e gl]obular."[294] In accordance with the following word, וסגלגל, Wise understands both בררי and כדרי to be singular masculine adjectives. Such an understanding is similar to the suggestion made earlier for ארין in 4Q561 1 i 2, that it is derived from *אורי (see above). It is, therefore, possible to assume an adjective *בררי or *כדרי in *4QPhysiognomy ar*. But Wise's reading of *dalet* is not likely. The letter lacks the typical tick upwards at the right of the horizontal stroke. Therefore, I follow Starcky's reading. If ב[ררי and וסגלגל form one clause, I understand ב[ררי as a masculine singular adjective in conformity with וסגלגל.

It is possible that the text describes one of the eyes as bright and round. In *The Physiognomy of R. Ishmael*, the root ברר is used in two instances to describe a person's eyes or eye as bright.[295] Because of the context of these two examples, together with an Aramaic quote from a *merkabah* tractate in the text *Tanna d'be Eliahu*,[296] Scholem concludes that bright eyes signal a wicked person.[297] However, *The Book of the Reading of the Hands by an Indian Sage* foretells a good fortune for a person whose eyes are clear.[298] In physiognomic contexts the word סגלגל can be used to describe the head and the eyes.[299] There is, however, one grammatical problem with understanding ב[ררי וסגלגל in 4Q561 5 3 as a description of the individual's eye. The word עין is, normally speaking, feminine, whereas סגלגל and בררי are masculine adjectives. Fortunately, the Cairo Genizah text T.-S. K 21.88 provides a solution. Here, one of the person's eyes is singled out and described as being round: ועינו אחד זגלגל.[300] The forms used are masculine.

[291] Scholem, "Physiognomy," 483.47; 486.75; Scholem, "Fragment," 184 §18, 185 §33.

[292] *Preliminary Concordance*, 2096.

[293] Eisenman and Wise, *Scrolls Uncovered*, 264-65. See also García Martínez and Tigchelaar, *DSSSE*, 1118-19. See also Holst and Høgenhaven, "Physiognomy," 37-38.

[294] Wise, "4Q561," 228-29.

[295] Scholem, "Physiognomy," 480.6: ועיניו ברורות; 486.75: ברור; Scholem, "Fragment," 182 §2, 185 §33.

[296] שנו חכמים במרכבה. אם חזית גבר דשביט ודברור ודבררן עיניה. דע דהוא ביש וחטא קדם אלהן:, see M. Friedman, *Seder Eliahu Rabba and Seder Eliahu Zuta (Tanna d'be Eliahu): Pseudo-Seder Eliahu Zuta* (Vienna, 1904; repr., Jerusalem: Wahrmann, 1969), 162.

[297] Scholem, "Fragment," 187 n. 7.

[298] Scholem, "Physiognomy," 491.10: ועיניו זכות [...] יבא לו טוב ויהיה משובח.

[299] See above the comments on 4Q186 1 i 8 for further references.

[300] T.-S. K 21.88 1/ב 22, Gruenwald, "Jewish Physiognomic," 314.

Before the fragment breaks off, Starcky reads לה̊ו̊]נה.[301] Subsequent to *he* a trace of ink is still visible that is possibly *waw*.[302] The position of the letters is similar to other occurrences of להוה in *4QPhysiognomy ar*.[303] There is, therefore, no need to read לה]נוה[304] or לה]ן.[305]

4Q561 5 4: The typical curve of final *nun* is visible.[306] Perhaps one must reconstruct להו[ן̊ with Wise,[307] but this is not certain.

Both the right and left down strokes of *šin*, שׂער, are visible.[308]

Notes on Readings in 4Q561 6

4Q561 6 1: The down stroke curves to the left, which is typical of final *nun* in this manuscript.[309]

4Q561 6 3: Starcky correctly reads על]ן, since the fragment breaks off without any certain indication of a blank space following *lamed*.[310]

4Q561 6 4: Starcky reads רב]ן,[311] but Wise has רב].[312] It cannot be ruled out that a letter followed *bet*.

Notes on Readings in 4Q561 7

4Q561 7 1: Starcky reads ושגיא̊ין].[313] But Wise reads ושגיאן], "and extensive," while Holst and Høgenhaven read ישגיאו], "they will be abundant."[314] From PAM 43.598 it seems as if the final letter originally extended below the line. Therefore, I agree with Wise's reading. After this letter, the photographs show no trace of another letter.[315]

Also, the faint trace of a guide dot to the right of ושגיא̊ן is possibly discernible.[316]

301 *Preliminary Concordance*, 2238. See also Wise, "4Q561," 228.

302 See PAM 42.438; 43.598.

303 See 4Q561 1 ii 1; 6 3. Cf. also 4Q561 1 ii 5: להון.

304 Cf. Eisenman and Wise, *Scrolls Uncovered*, 264; Holst and Høgenhaven, "Physiognomy," 37.

305 Cf. García Martínez and Tigchelaar, *DSSSE*, 1118-19, "for him [."

306 See PAM 42.438; 43.598. Cf. also *Preliminary Concordance*, 2254; García Martínez and Tigchelaar, *DSSSE*, 1118; Holst and Høgenhaven, "Physiognomy," 37.

307 Wise, "4Q561," 228.

308 See PAM 42.438; 43.598. See also *Preliminary Concordance*, 2412; García Martínez and Tigchelaar, *DSSSE*, 1118; Beyer, *Die aramäischen Texte*, 2:164; Wise, "4Q561," 228; Holst and Høgenhaven, "Physiognomy," 37.

309 Cf. e.g. 4Q561 1 i 3: שוין; 1 ii 7: כעבין; 2 3: פתין.

310 *Preliminary Concordance*, 2464. See Eisenman and Wise, *Scrolls Uncovered*, 264; Holst and Høgenhaven, "Physiognomy," 37. But cf. Wise, "4Q561," 228: על]ן.

311 *Preliminary Concordance*, 2398. See also Holst and Høgenhaven, "Physiognomy," 37.

312 Eisenman and Wise, *Scrolls Uncovered*, 264; Wise, "4Q561," 228.

313 *Preliminary Concordance*, 2411.

314 Wise, "4Q561," 230-31; Holst and Høgenhaven, "Physiognomy," 37-38.

315 See PAM 41.944; 43.598.

316 See PAM 43.598.

4Q561 7 2: Starcky reads]יׄתעבדׅן[,[317] an *itpe'el* of עבד, "it will happen." Probably due to the rather long additional stroke of *taw*, Wise understands it as a final letter and reads] עב ת··[.[318] But it is unlikely that there are two separate words here. Holst and Høgenhaven read]ׅויתעבן, "and they will grow thic[k]."[319] Although their reading of *waw* is plausible, it must be noted that the word does not stand on the same margin line as ושגיאׅן in 1.1.

If Starcky's reading is accepted, it yields another instance of a prediction concerning something that will happen to the type of person described. Regarding Holst and Høgenhaven's reading, the question is what its meaning is. It is not likely to be a description of a part of the body because these are expressed by means of adjectives or participles. If it is a prediction, does it intend to describe how a part of the body will develop and look like in the future? This seems rather unlikely. Although it is not possible to discern a trace of a letter following *bet* from the available photographs, I accept Starcky's reading.[320]

[317] *Preliminary Concordance*, 2328.
[318] Wise, "4Q561," 230.
[319] Holst and Høgenhaven, "Physiognomy," 37-38.
[320] See PAM 41.944; 43.598.

APPENDIX II

PHYSIOGNOMIC CONSCIOUSNESS IN THE DEAD SEA SCROLLS AND SECOND TEMPLE PERIOD LITERATURE

INTRODUCTION

More or less simultaneously with the highlighted "physiognomic consciousness" expressed by Ben Sira (see below), there seems to be an emergence of a marked interest in the physical description of people's bodies in Second Temple period Judaism, at least as far as our sources are concerned.[1] The concrete descriptions of the human body in literary texts from the Hellenistic-Early Roman period find, to a certain degree, a remarkable parallel in Greco-Egyptian and Jewish documentary papyri.

The texts briefly discussed here show that physiognomic interest, as conveyed in technical form by the two learned lists from the Dead Sea Scrolls, *4QZodiacal Physiognomy* and *4QPhysiognomy ar*, was not an isolated phenomenon in Second Temple period Judaism. These texts demonstrate in different ways an interest in the appearance of the human body, its descriptions, and what is signified by it.

HEBREW BIBLE

The Hebrew Bible does not contain any physiognomic texts, although later medieval Jewish physiognomic texts elaborated on passages such as Gen 5:1[2] and Isa 3:9,[3] with the latter providing the technical term הכרת פנים for "physiognomics" in Jewish tradition.

[1] Perhaps also indicative of this marked interest are the more elaborate characterizations of David's brothers in 11QPs[a] 28:3–12, see n. 6 below.

[2] Gen 5:1 is quoted in (1) *The Physiognomy of R. Ishmael* §2, see Scholem, "Physiognomy," 477-80; Scholem, "Fragment," 176, 180-182); (2) *The Secret of Physiognomy*, see Scholem, 'Physiognomy," 493.2; (3) T.-S. K 21.95.L 2b/3, see Schäfer, *Geniza Fragmente*, 1984), 137; Schäfer, "Fragment zur Metoposkopie und Chiromantik," 85-86, 91.

[3] Isa 3:9 is quoted in *The Secret of Physiognomy*, see Scholem, "Physiognomy," 492.1–2, 494.9; Scholem, "Fragment," 176.

The only exception is perhaps the *Book of Proverbs* with possible echoes of Babylonian physiognomic omens (e.g. 10:10; 16:30).[4] Even if this were the case, it does not indicate much of a pursuit of physiognomics.[5]

In addition, 1 Sam 16:7 can perhaps be understood as an example of physiognomic consciousness in a negative sense. When Samuel sees the first son of Jesse, he thinks Eliab is the anointed one. God, however, cautions him not to look at his appearance (מראהו) or at his stature (גבה קומתו). These are things humans look on, but God looks on the heart (i.e. the inner person).[6] Physiognomic consciousness in a negative sense, however, works well alongside a positive one. The edge of the statement in 1 Sam 16:7 becomes clear when it is compared with what is said earlier of Saul as the anointed one. In 1 Sam 10:23–24 Saul is said to be head and shoulders taller than all the people (ויגבה מכל־העם משכמו ומעלה), whereupon Samuel asks the people if they see whom God has chosen, for there is no one like him among all the people.[7]

Ptolemaic Palestine: Physical Descriptions of Slaves in a Letter from Tobias to Apollonius

Greco-Egyptian and Jewish papyri demonstrate the identification purposes of physical descriptions. It is a peculiar feature of the Ptolemaic administration in the third century BCE that it introduced physical descriptions of individual citizens for identification purposes in documents such as military enrolment, wills, deeds of sale, contracts, and proclamations for the capture of runaway slaves. The descriptive elements follow a more or less fixed order: age, stature, complexion, other distinguishing marks, and, finally, scars or moles (but there is variation, even within one and the same document).[8]

4 See G.E. Bryce, "Omen-Wisdom in Ancient Israel," *JBL* 94 (1975): 19-37.

5 For more general examples of body imagery in the Hebrew Bible, see e.g. S. Schroer and T. Staubli, *Die Körpersymbolik der Bibel* (Darmstadt: Wissenschaftliche Buchgesellschaft, 1998); S. Gillmayr-Bucher, "Body Images in the Psalms," *JSOT* 28 (2004): 301-26.

6 The LXX has an additional psalm (Ps 151) not in the MT in which David sings of this occasion saying in v.5 that his brothers were handsome (καλοὶ) and tall (μεγάλοι), whereas he was smaller than them (v.1). In 11QPs[a] 28: 9–10 the brothers are handsome of shape (יפי התור), handsome of appearance (ויפי המראה), tall of stature (הגבהים בקומתם), and handsome by their hair (היפים בשערם).

7 See also 1 Sam 9:2. Cf. e.g. T.N.D. Mettinger, *King and Messiah: The Civil and Sacral Legitimation of the Israelite Kings* (ConBOT 8; Lund: CWK Gleerup, 1976), 175-79.

8 See e.g. J. Fürst, "Untersuchungen zur Ephemeris des Diktys von Kreta (Fortsetzung)," *Phil* 61 (1902): 374-440, 593-622; J. Hasebroek, *Das Signalement in den Papyrusurkunden* (PH 3; Berlin: VwV, 1921); K. Jax, "Zur literarischen und amtlichen Personenbeschreibung," *Klio* 29 (1936): 151-63; G. Hübsch, *Die Personalangaben als Identifizierungsvermerke im Recht der gräko-ägyptischen Papyri* (BJA 20; Berlin: Duncker and Humblot, 1968); Evans, *Physiognomics*, 52; W. Clarysse (ed.), *The Petrie Papyri Second Edition (P. Petrie²): Volume*

There is evidence that at least some Jews were familiar with these identification methods of the Ptolemaic administration. In a letter from one of the Zenon papyri, dated May 12, 257 BCE and presumably sent from the Tobiad family estate in Transjordan, Tobias informs Apollonius, the finance minister of Ptolemy II Philadelphus (285–246), that he has sent him four young house-slaves together with a eunuch. Tobias says he has appended the descriptions (τὰς εἰκόνας)[9] of the boys for the information of Apollonius. The descriptions given by Tobias carefully follow the general order of the descriptive elements in Ptolemaic papyri:

Haimos, about 10 dark skin curly hair black eyes rather big jaws with moles on the right jaw uncircumcised	Attikos, about 8 light skin curly hair nose somewhat flat black eyes, scar below the right eye uncircumcised	Audomos, about 10 black eyes curly hair, nose flat protruding lips scar near the right eyebrow circumcised	Okaimos, about 7 round face, nose flat, grey eyes fiery complexion long straight hair scar on forehead above the right eyebrow circumcised[10]

These descriptions of the children show that Jews such as Tobias were familiar with the customs of the Ptolemaic administration for the identification of individual persons in official documents.[11] In addition, this evidence may bear importance for understanding the appearance of an interest in the physical description of people's bodies in Second Temple period

---

1, The Wills (CH 2; Brussels: Koninklijke Academie voor Wetenschappen, Letteren en Schone Kunsten van België, 1991), 49-55; F.A.J. Hoogendijk, "*Parachoresis* von Katökenland," in *Papyri in Honorem Johannis Bingen Octogenarii (P. Bingen)* (eds. R. De Smet, H. Melaerts and C. Saerens; SVBOGLP 5; Leuven: Peeters, 2000), 249-59; Sassi, *Science of Man*, 26-27.

[9] Cf. G. Misener, "Iconistic Portraits," *CP* 19 (1924): 97-123, at 97: "The *εἰκονισμός* is a terse asyndetic description of personal appearance designed either as a means of legal identification or as an ornament of style."

[10] *P.Cair.Zen.* 59076 = *CPJ* 1:125–127, translation from 127.

[11] Due to the formula of greeting at the beginning of the letter ("many thanks to the gods"), scholars assume that Tobias is representative of a Second Temple period Jew open to Hellenistic influences. See e.g. V. Tcherikover, *Hellenistic Civilization and the Jews* (trans. S. Applebaum; Philadelphia: The Jewish Publication Society of America and Jerusalem: Magnes, 1959; repr., Peabody, Massachusetts: Hendrickson, 2004), 71; Hengel, *Judentum und Hellenismus*, 488; L.L. Grabbe, *Judaism from Cyrus to Hadrian* (London: SCM, 1994), 196; L.L. Grabbe, *Judaic Religion in the Second Temple Period: Belief and Practice from the Exile to Yavneh* (London: Routledge, 2000), 39-40. It has not been noted before, but, in addition to the greeting formula, the use of the Ptolemaic identification formula for physical descriptions of individuals may be cited as another example of this influence.

Cf. also *P.Cair.Zen.* 59015 *verso* for a draft of a letter written by Zenon in 259 or 258 BCE to Epikrates, a person residing in or connected with Palestine, asking him to recover some runaway slaves of which he says he gives their names and descriptions (τὰς εἰκόνας, l.20, but these were not provided in the draft). See C.C. Edgar (ed.), *Catalogue général des antiquités égyptiennes du Musée du Caire: Zenon Papyri, Nos 59001-59139* (vol. 1; Cairo: L'Institute Français d'Archéologie Orientale, 1925), 35.

literature. Based on the sources at our disposal, it seems that before the Hellenistic period there is not the same interest for more detailed and concrete physical descriptions and portrayals of individuals. It is possible that Jewish familiarity with the demands of the Ptolemaic administration in the third century BCE regarding physical descriptions for identification purposes functioned as a catalyst for the concern with the physical description of individual human bodies in literary sources. The papyri are suggestive. The distinguishing character of marks and moles appears also in 4Q534 (*4QBirth of Noah*[a]), and the color and type of hair as distinguishing features receive attention in *1 En.* 106.[12]

BEN SIRA

It is not until the Hellenistic period that we come across a Jewish writer who displays some physiognomic awareness. The Jewish sage Ben Sira, active at the beginning of the second century BCE, seems, on the one hand, to agree with the general idea that an individual's character can be known by his outward appearance and behavior, because he says that:

> By appearance a man is known, and an intelligent person is known by (his) face. A man's clothing and hearty laughter, and a person's way of walking announce what he is like.[13]

It seems likely that Ben Sira here gives expression to a notion that was more widely current in Jewish society in his days and already before him, as is indicated by the letter from Tobias to Apollonius. On the other hand, however, Ben Sira seems to express reservations, because the point made is not to judge on appearance:

> Do not praise a person for his good looks, and do not loathe a man for his appearance.[14]

[12] The literary form of the description of Sarai in the *Genesis Apocryphon* falls within an ancient Near Eastern tradition of description songs, but it is far less poetic than Song 4:1–7; 5:10–16; 7:2–10 and other ancient Near Eastern examples. It mainly uses the adjectives "beautiful" or "perfect" in its descriptions. Cf. W. Hermann, "Gedanken zur Geschichte des altorientalischen Beschreibungsliedes," *ZAW* 75 (1963): 176-97. However, the few concrete descriptions of Sarai's hair as soft, skin as fair (lit. white), and fingers as long and slender are very interesting and suggestive. They perhaps betray the poem's indebtedness to the plain physical descriptions known from Greco-Egyptian and Jewish papyri.

[13] Sir 19:29–30. Cf. also the more concrete, practical examples of this statement in Sir 12:11; 13:6–7.11; 14:8–10; 19:26–27; 25:17.21 (cf. *T. Reu.* 3:10; 4:1); 26:9. The Christian pilgrim Palladius, a fifth-century bishop of Helenopolis, wrote about his journeys and in his prologue quoted Sir 19:30, urging his readers to look especially at the physical appearance of the ascetics. See Frank, *Memory of the Eyes*, 138.

[14] Sir 11:2.

Although it is possible that he had an ambivalent attitude on this, both points of advice show that there was a concern with the semiotic value of people's bodies and appearances that Ben Sira felt needed to be addressed.

4QBirth of Noah$^{A, B}$ *ar* (4Q534 and 4Q535)

The Aramaic text *4QBirth of Noah ar* is extant in three copies (4Q534–536).[15] The first two copies preserve descriptions of certain features of the human body that appear in a narrative framework. From the two columns preserved of 4Q534 it seems that the physical appearance of perhaps two persons is depicted, while the third fragment of 4Q535 provides the weight of a newborn baby that is probably identical with the figure described in 4Q534 1 i. The text's content, however, goes beyond simply listing what would be signified by the appearance and weight of the human body.

4Q534 1 i is about a figure identified as "the elect of God" (l.10). He is considered like an intelligent person (כלטיש) in his youth (l.4),[16] but also like a man who does not know anything "until he knows the three books" (ll.4–5).[17] From then on he will be wise (l.6).[18] Counsel and good sense

[15] See Puech, DJD 31.117-70.

[16] Puech, DJD 31.134, 137: "Pendant sa jeunesse, il sera considéré comme un être intelligent." Starcky, "Un texte messianique araméen," at 56, reads כלטיש (passive participle of לטש), "comme aiguisé," as a reference to the acuity of knowledge. Although this reading is paleographically difficult, it is possible and makes sense in the context. Carmignac, "Les Horoscopes," 216-17, reads כליש, "Pendant sa jeunesse il sera comme un lion." Cf. O. Betz, "'Kann denn aus Nazareth etwas Gutes kommen?': Zur Verwendung von Jesaja Kap. 11 in Johannes Kap. 1," in *Wort und Geschichte: Festschrift für Karl Elliger zum 70. Geburtstag* (eds. H. Gese and H.P. Rüger; AOAT 18; Neukirchen-Vluyn: Neukirchener, 1973), 9-16, at 10; J.R. Davila, "4QMess ar (4Q534) and Merkavah Mysticism," *DSD* 5 (1998): 367-81, at 373 n. 25. J.A. Fitzmyer, "The Aramaic 'Elect of God' Text from Qumran Cave IV," in *Essays on the Semitic Background of the New Testament* (J.A. Fitzmyer; London: Geoffrey Chapman, 1971), 127-60, at 146 (originally published in *CBQ* 27 [1965]: 348-72), questions Jean Carmignac's suggestion on orthographical and phonological grounds, and transcribes כלהון instead; F. García Martínez, "*4QMess Ar* and the *Book of Noah*," in *Qumran and Apocalyptic: Studies on the Aramaic Texts from Qumran* (F. García Martínez; STDJ 9; Leiden: E.J. Brill, 1992), 1-44, at 8, and Zimmermann, *Messianische Texte*, 173, criticize previous readings, but do not attempt a reconstruction.

[17] Different suggestions have been made for the character and identity of these three books: eschatological and perhaps astrological (Starcky, "Un texte messianique araméen," 56-57); esoteric (Dupont-Sommer, "Deux documents horoscopiques," 247 n. 2); the *Book of* הגי (1QSa 1:7; CD 10:6; 13:2), the *Rule of the Community*, and the *Damascus Document* (Carmignac, "Les Horoscopes," 212); apocalyptic, heavenly, not specific, real books (Fitzmyer, "The Aramaic 'Elect of God' Text," 147-48); the Enochic *Astronomical Book*, the *Book of the Watchers*, and the *Book of Dreams* (P. Grelot, "Hénoch et ses écritures," *RB* 82 [1975]: 481-500, 492. Cf. also J.T. Milik, "Écrits préesséniens de Qumrân: d'Hénoch à Amram," in *Qumrân*, 91-106, at 94; Milik, *The Books of Enoch*, 66; García Martínez, "*4QMess Ar*," 8-9, 19-20; Puech, DJD 31.124, 137-38; A. Caquot, "Les prodromes du déluge: Légen-

will be with him, he will know the secrets of man, his wisdom shall come to all peoples, and he will know the secrets of all human beings (ll.7–8).[19] However, this individual apparently experiences opposition from others. All their calculations against him will fail, but the opposition of all human beings will be great (l.9).[20] Next, the text reveals that the individual dis-

des araméennes de Qoumrân," *RHPR* 83 [2003]: 41-59, at 58); the traditional threefold corpus of biblical books (Davila, "4QMess ar," 375).

[18] The text possibly attributes to this elect figure some wisdom of sages-seers "coming to himself on his knees" (למאתה לה̇ ע̇ל ארכובתה̇, l.6), which perhaps describes the elect of God as a student at the feet of a teacher. Starcky, "Un texte messianique araméen," 52, reads the final word as ארכובתה, "his knees" (cf. e.g. Carmignac, "Les Horoscopes," 216-17; Fitzmyer, "The Aramaic 'Elect of God' Text," 148; Zimmermann, *Messianische Texte*, 173). But Grelot, "Hénoch et ses écritures," 493-94, reads ארכובת[א] and interprets it as a geographical term used in a mystical sense for a higher celestial sphere; in the Babylonian Talmud ארכובתא denotes the third and uppermost dike layer, cf. *b. B. Meṣiʿa* 103b (see also Milik, *The Books of Enoch*, 41 n. 1; Milik, "Écrits préesséniens de Qumrân," 94; García Martínez, "*4QMess Ar*," 10, 20-21; Davila, "4QMess ar," 376-77). The second interpretation is unlikely, because in this manuscript *ʾalep* is used as the definite article, not *he*; as the final letter here is *he*, it probably functions as a possessive suffix. The final word, therefore, cannot refer to a celestial sphere, and is best understood in the sense of knees, cf. Puech, DJD 31.139 (the *lamed* in לה is understood as an ethical dative, rather than a dative of direction).

[19] Starcky, "Un texte messianique araméen," 57-58, interprets ll.7–8 as references to the wisdom of Solomon as the King-Messiah, possibly to the Servant of the Lord, and thus infers the figure's messianic character. Cf. Carmignac, "Les Horoscopes," 214; M. Hengel, "Jesus als messianischer Lehrer der Weisheit und die Anfänge der Christologie," in *Sagesse et Religion: Colloque de Strasbourg (octobre 1976)* (ed. E. Jacob; Paris: Presses universitaires de France, 1979), 147-88, at 173; Betz, "'Kann denn aus Nazareth,'" 10-12. Fitzmyer, "The Aramaic 'Elect of God' Text," 150, acknowledges the allusion to Solomon's wisdom, but denies that this establishes the messianic identity of the elect figure (cf. García Martínez, "*4QMess Ar*," 11). Grelot, "Hénoch et ses écritures," 495, suggests that knowledge of the secrets of man and of all living things applies to the content of the three books that Enoch transmitted to posterity (alluded to in *Jub.* 4:16–24), and which Noah will bequeath to the people (cf. García Martínez, "*4QMess Ar*," 11, 21-22; Puech, DJD 31.140). A. Caquot, "*4QMess Ar* 1 i 8-11," *RevQ* 15/57-58 (1991): 145-55, at 146-47, argues, firstly, that the secrets are secrets concerning mankind, not secrets held by humanity, secondly, that חייא probably only refers to human beings and not animals, and, finally, that the words וחוכמתה לכול עממיא תהך are a specification of אנשא in the first part, indicating that the elect figure "aura dans l'esprit l'humanité entière" (cf. García Martínez, "*4QMess Ar*," 10-11; Zimmermann, *Messianische Texte*, 179-81; Puech, DJD 31.140). Davila, "4QMess ar," 377, suggests that the elect's "knowledge of the mysteries of others foreshadows the similar knowledge of the Merkavah mystic in the Hekhalot Rabbati."

[20] Since the calculations of the elect figure himself (חשבונוהי) are mentioned in l.10–11, Starcky, "Un texte messianique araméen," 58, assumes an opposition expressed by the adversative use of על (cf. Carmignac, "Les Horoscopes," 216; Dupont-Sommer, "Deux documents horoscopiques esséniens," 248; Fitzmyer, "The Aramaic 'Elect of God' Text," 151; Grelot, "Hénoch et ses écritures," 495-96; Hengel, "Jesus als messianischer Lehrer," 173; Betz, "'Kann denn aus Nazareth,'" 10; García Martínez, "*4QMess Ar*," 4, 11; Davila, "4QMess ar," 374; Zimmermann, *Messianische Texte*, 174, 181-82). Contrary to this interpretation, Caquot, "*4QMess Ar*," 148-49, argues that the possessive suffix in חשבוניהון has objective value, meaning the calculations of which the men are the object, and not their calculations against the elect figure. Consequently, חשבוניהון עלוהי יסופו does not mean "their calculations against him (i.e. the elect one) will fail," but "the calculations concerning them will be accomplished according to him" (cf. Puech, DJD 31.140-41).

cussed in the text is the elect of God (l.10),[21] and there is a reference to his horoscope (מולדה), which, unfortunately, is not further specified.[22] Before the text becomes too fragmentary, it states that the elect's calculations will last forever (l.11).

The words "elect of God" are very intriguing and, understandably, the identification of this figure has given rise to many suggestions by scholars.[23] Although initially scholars favored the elect of God being a future messianic figure, a later consensus emerged to identify this character as Noah, but both identifications are possible and the real identity of the figure remains open.[24]

Regardless of the elect one's true identity, what is of interest here is the description of several features of this figure's body given in ll.1–3.[25] While the greatest part of the text is concerned with describing how the elect of God fares through life, this is preceded by a section describing the newborn that pays special attention to different sorts of marks and moles on his body:

1. on the hand [and] his two k[n]ees,[26] [and colo]rs[27] of a mark (שומה). Red are

---

Different explanations have been proposed for the word מסרת (cf. Zimmermann, *Messianische Texte*, 174-75; Puech, DJD 31.141). Because of the antagonism described in the first half of l.9, I follow Fitzmyer, "The Aramaic 'Elect of God' Text," 151, in understanding מסרת as a form derived from סרר, analogous to מגלת from גלל (cf. Grelot, "Hénoch et ses écritures," 496; García Martínez, "*4QMess Ar*," 4, 10-11).

[21] For several options concerning how this line can be divided into different sentences, see García Martínez, "*4QMess Ar*," 13-14; Puech, DJD 31.141-42).

[22] Cf. Starcky, "Un texte messianique araméen," 61-62; García Martínez, "*4QMess Ar*," 14; Puech, DJD 31.142-43.

[23] Proposed identifications include a future messiah (Starcky, "Un texte messianique araméen." Cf. also Carmignac, "Les Horoscopes," 217; Dupont-Sommer, "Deux documents horoscopiques esséniens," 248-53; Betz, "'Kann denn aus Nazareth,'" 10-11; Zimmermann, *Messianische Texte*, 200-3); Noah (Fitzmyer, "The Aramaic 'Elect of God' Text," 151-53, 158-60. Cf. also Grelot, "Hénoch et ses écritures," 497-98; Milik, "Écrits préesséniens de Qumrân," 94-95; J. Starcky, "Le Maître de Justice et Jésus," *MdB* 4 (1978): 51-55, at 54-55; García Martínez, "*4QMess Ar*," 17-24; Puech, DJD 31.); Melchizedek (J.C. Greenfield, "Prolegomenon," in *3 Enoch or The Hebrew Book of Enoch* [ed. H. Odeberg; New York: Ktav, 1973], xi-xlvii, at xx-xi); *Henoch redivivus* (Caquot, "*4QMess Ar*," 155); similarities with a Merkavah mystic (Davila, "4QMess ar," 367-69).

[24] Cf. D. Dimant, "Review: É. Puech (ed.), *Qumrân Grotte 4.XXII: Textes araméens, Première partie, 4Q529-549* (DJD 31; Oxford: Clarendon, 2001)," *DSD* 10 (2003): 292-304, at 297-98.

[25] 4Q534 1 ii has two scattered remarks on body marks that perhaps belong to another figure, but the text is too fragmentary (4Q534 1 ii+2 2: טלופחא; 5: שׂוׄמׄן בעׄ[נ]ינוהי).

[26] The first words of this line are not the beginning of the text but the continuation of a sentence from a preceding column, which perhaps mentioned marks appearing on the hand and his two knees. As for this last word, most scholars think the manuscript is too fragmentary. Starcky, "Un texte messianique araméen," 55, reads אׄ[ו]כמׄה, "black," but Puech, DJD 31.132-35, reads אׄ[ר]כׄובׄתׄה, "his knees," and connects it with the preceding [ו]תׄרתין.

[27] Cf. Puech, DJD 31.135.

2. [his] hairs (שערֹתְ]ה)[and] moles (טלופחין) on [his body,[28]] *vacat*
3. and small marks on his thighs, but [mole]s (וֹ]שער[יֹן) different from one another.[29]

It is tempting to attribute predictive value to these descriptions,[30] but these references to marks and moles on the body are probably not related to what is said about the elect figure's life in such a direct way.[31] They may signify the special character of this figure and, being distinguishing body marks, function as identity markers.[32] *4QBirth of Noah*^{a} *ar* thus displays physiognomic consciousness by attributing special value to the bodily signs. The description of the body serves to identify the elect of God. The concern with distinguishing body marks for establishing a person's identity is paralleled by Greco-Egyptian and Jewish papyri (see above).

In addition to these observations regarding different marks and moles on the body of the elect of God, the third fragment of 4Q535 gives information concerning the weight of the newborn. It is, however, unlikely that this information has any physiognomic sense; as far as I know there are no references to the concrete weight of people in physiognomic texts:

2. at the fif]th [hour] of the night he is born and comes out heal[thy

[28] Most scholars understand this as a reference to the hair on the head (cf. Puech, DJD 31.135-36, except Fitzmyer, "The Aramaic 'Elect of God' Text," 145), which is described as being red (Puech, DJD 31.134-35, connects שׂמׂקׂמק]י[ן with שומה, but now relates it to שערׂתְ]ה [personal communication 29.09.2005]). It is also possible, I suggest, that it refers to the color of the body hair. Perhaps these hairs are described as being red due to the color of the moles that appear on the whole body, על[גופוה׳] (Puech, DJD 31.136 reconstructs על[אנפוה׳], "on his face," but cf. García Martínez, "*4QMess Ar*," 6-7). In Babylonian physiognomics attention is paid to hairs growing from moles. Also, much consideration is given to the color of the different sorts of moles and whether they appear on the whole body. Cf. Böck, *Die babylonisch-assyrische Morphoskopie*, 210-11, 200-1, 228-29.

[29] Starcky, "Un texte messianique araméen," 56, understands שנין to mean "teeth," and reconstructs [עריכ]יֹן שנין, "well-ordered teeth" (cf. Carmignac, "Les Horoscopes," 216); Dupont-Sommer, "Deux documents horoscopiques esséniens," 246-47, takes שנין as the plural participle of the verb שני, "to be different" (cf. also Grelot, "Hénoch et ses écritures," 491; García Martínez, "*4QMess Ar*," 7; Davila, "4QMess ar," 373; Zimmermann, *Messianische Texte*, 172), and reconstructs [תרת]יֹן, "two", referring to small marks on the thighs that differ from one another; Fitzmyer, "The Aramaic 'Elect of God' Text," 145, interprets שנין as the plural of שנה, "year," and reconstructs [בתר תרת]ין שנין, "after two years"; Grelot, "Hénoch et ses écritures," 491, reads [שער]ין, "barley," in the sense of a mark or mole (cf. also Puech, DJD 31.136). For this sense of שערה cf. Fitzmyer, "The Aramaic 'Elect of God' Text," 145.

[30] Cf., for example, the following Babylonian omens:

> If red *pindû*-moles cover his whole body, his days will be long, he will become old.
> If red *kittabru*-moles cover his whole body, he will become an important person and his name famous, his days will be long. (see Böck, *Die babylonisch-assyrische Morphoskopie*, 200-1, 228-29)

[31] Cf. García Martínez, "*4QMess Ar*," 17-18, 44, who emphasizes the importance of learning the three books as the cause of further things happening in the life of the elect figure.

[32] Cf. Zimmermann, *Messianische Texte*, 176, 192-94.

3.]he weighs three hundred and fi[fty(-one)] shekels[[33]

THE DESCRIPTION OF THE NEWBORN NOAH IN *1 ENOCH* 106

The description of the newborn Noah's miraculous bodily features in *1 En.* 106 has an important narrative function. Noah's appearance leads his father Lamech to think that Noah is not his child but that of an angel:

> And when the child was born, his body was whiter than snow and redder than a rose, his hair was all white and like white wool and curly. Glorious <was his face>. When he opened his eyes, the house shone like the sun. And he stood up from the hands of the midwife, and opened his mouth and praised the Lord of eternity. And Lamech was afraid of him, and he fled and came to Methuselah his father. And he said to him, 'A strange child has been born to me. He is not like men, but (like) the sons of the angels of heaven. His form is strange, not like us. His eyes are like the rays of the sun, and glorious is his face. I think that he is not from me, but from the angels. And I fear him, lest something happen in his days on the earth. I beg you, father, and beseech you, go to Enoch our father and learn the truth from him, for his dwelling is with the angels.'[34]

This description obviously differs from that of the elect figure in 4Q534 (*4QBirth of Noah^{a} ar*), and opinions differ as to the relationship between the two texts.[35] Be that as it may, the bodily description is significant. For the characters in the story, Noah's remarkable appearance identifies his angelic nature. Enoch, however, reveals the true significance of Noah's looks to be indicative of his righteousness and perfection that qualify him as a survivor of the flood and father of a new people. The narrative function of Noah's extraordinary appearance is to enable Enoch to make this revelation known.[36] What is striking is that the description of the human body, however wondrous it might be, is used explicitly to signify, in this case, a revelation to the intended reader.

[33] Cf. Puech, DJD 31.157-58.

[34] *1 En.* 106:2–7. Translation from Nickelsburg, *1 Enoch 1*, 536. Methuselah goes to Enoch and tells him what Lamech told him; the description is recounted a third time. Cf. also the parallels in 1Q19 3; 1QapGen ar 2–5, see J.A. Fitzmyer, *The Genesis Apocryphon of Qumran Cave 1 (1Q20): A Commentary* (BibOr 18/B; Rome: Pontificio Istituto Biblico, 2004), 122-44, 258-60; Nickelsburg, *1 Enoch 1*, 539-50.

[35] García Martínez, "*4QMess Ar*," 23, 27-28; Zimmermann, *Messianische Texte*, 194-95.

[36] Nickelsburg, *1 Enoch 1*, 539, 543-44. For the importance of white hair in *Jub.* 23:25, see A. Caquot, "Les enfants aux cheveux blancs: Réflexions sur un motif," in *Mélanges d'histoire des religions offerts à Henri-Charles Puech* (Paris: Presses Universitaires de France, 1974), 161-72; H. Garcia, "L'enfant vieillard, l'enfant aux cheveux blanc et le Christ polymorphe," *RHPR* 80 (2000): 479-501, at 481-83.

THE DESCRIPTION OF SARAI'S BEAUTY IN THE *GENESIS APOCRYPHON*

Another Aramaic text also evinces a physiognomic interest in a narrative setting. In column 20 of the *Genesis Apocryphon* (1QapGen ar) the beautiful appearance of Sarai is praised in an inserted poem that is extant in ll.2–8.[37] The literary style of praise is clear by the multiple use of the demonstrative adverb כמא ("how great") introducing descriptions of a part of Sarai's body:

2. [ ]how splen[did] and beautiful the form of her face, and how
3. [plea]sant [and] soft the hair of her head; how lovely are her eyes, and how graceful is her nose; all the radiance of
4. her face [ ]; how lovely is her breast, and how beautiful is all her whiteness! Her arms, how beautiful! And her hands, how
5. perfect! And (how) attractive all the appearance of her hands! How lovely (are) her palms, and how long and dainty all the fingers of her hands. Her feet,
6. how beautiful! How perfect are her legs! There are no virgins or brides who enter a bridal chamber more beautiful than she. Indeed, she greatly
7. surpasses in beauty all women; and in her beauty she ranks high above all of them. Yet with all this beauty there is much wisdom in her; and whatever she has
8. is lovely.[38]

The poem describes Sarai's body according to the *a capite ad calcem* principle. It begins with the head and runs down to the feet. This structure is similar to that used in the descriptions of the body in *4QZodiacal Physiognomy* and *4QPhysiognomy ar*.[39] Scholars have related the poem's description technique to the genre known in Arabic as *waṣf*, which was introduced in nineteenth-century literary criticism to characterize the genre of some of the songs in the Song of Songs. *Waṣf* songs are recited during Arab marriage celebrations. Their purpose is to praise the beauty of the wedding couple, especially of the bride. Although the generic name *waṣf* is of Arabic origin, the literary form of the description of Sarai stands within an Ancient Near Eastern tradition of description songs.[40]

Subsequent to the praise of Sarai's bodily features for their magnificence, the poem says that no virgin or bride who enters the bridal chamber

[37] For the poetic character, cf. J.C. VanderKam, "The Poetry of 1QApGen XX, 2-8a," *RevQ* 10/37 (1979): 57-66. In general see Fitzmyer, *Genesis Apocryphon*, 193-97.

[38] Translation from Fitzmyer, *Genesis Apocryphon*, 101.

[39] Cf. also Gordis, "Document in Code," 38; Lange, "Essene Position on Magic," 388.

[40] See e.g. M.H. Goshen-Gottstein, "Philologische Miszellen zu den Qumrantexten," *RevQ* 2/5 (1959): 43-51, at 46-48; Hermann, "Geschichte des altorientalischen Beschreibungsliedes"; Fitzmyer, *Genesis Apocryphon*, 193; D. Bernat, "Biblical *Waṣf*s Beyond Song of Songs," *JSOT* 28 (2004): 327-49.

is more beautiful than her (l.6), and also that with all this beauty there is great wisdom with her, and that everything she does with her hands is perfect (ll.7–8). The praise of her body is meant to imply Sarai's impeccable character as a wife, which is clear from the marital context that is evoked.[41] This physiognomic awareness corresponds to the interest of Babylonian physiognomic omens in women. These physiognomic omens are primarily concerned with women as mothers and wives. It is possible that in Mesopotamia, physiognomics was used to assess a woman's fertility and ability to give birth, as well as to predict her role in her future household and the way she will treat her husband.[42]

4QBarkhi Nafshi$^{A, C}$ (4Q434 and 4Q436), and *4QWiles of the Wicked Woman* (4Q184)

Two passages from the *Barkhi Nafshi* hymns, extant in five copies (4Q434–438), perhaps have a physiognomic interest.[43] It was argued that the references to parts of the human body in 4Q434 1 i and 4Q436 1 i served a physiognomic function in the Qumran community that presumably read these poems. The purpose of these references would have been to reinforce the notion that divine election to the community was reflected in the state of each individual's body.[44] The poems do not actually describe what the body should look like. They describe how parts of the body have been perfected by God so that the poet acts perfectly accordingly.

It has been argued that *4QWiles of the Wicked Woman* contains a short anti-*waṣf* song in ll.2–4.[45] Instead of singing the praises of the physical features, the poem describes the evil, corrupting nature of different parts of the "Wicked Woman's" body. Its function is to warn the reader to keep away from the "Wicked Woman" and her evil ways.[46] Again, the poem does

[41] Cf. also Goshen-Gottstein, "Philologische Miszellen," 48.

[42] Böck, *Die babylonisch-assyrische Morphoskopie*, 36-38, 58-59.

[43] For the texts, see M. Weinfeld and D. Seely, "4QBarkhi Nafshi," in *Qumran Cave 4.XX: Poetical and Liturgical Texts, Part 2* (eds. E. Chazon et al.; DJD 29; Oxford: Clarendon, 1999), 255-334.

[44] G.J. Brooke, "Body Parts in *Barkhi Nafshi* and the Qualifications for Membership of the Worshipping Community," in *Sapiential, Liturgical and Poetical Texts from Qumran: Proceedings of the Third Meeting of the International Organization for Qumran Studies, Oslo 1998* (eds. D.K. Falk, F. García Martínez and E.M. Schuller; STDJ 35; Leiden: Brill, 2000), 79-94.

[45] For the text, see Allegro, DJD 5.82-85; Strugnell, "Notes en marge du volume V," 263-68.

[46] Bernat, "Biblical *Waṣfs*," 346-47. On 4Q184, and its polemical stance, see further e.g. J.M. Allegro, "'The Wiles of the Wicked Woman': A Sapiential Work from Qumran's Fourth Cave," *PEQ* 96 (1964): 53-55; J. Carmignac, "Poème allégorique sur la secte rivale," *RevQ* 5/19 (1965): 361-74; A.M. Gazov, "Double Meaning in a Qumran Work ('The Wiles of the Wicked Woman')," *RevQ* 6/22 (1967): 279-85; H. Burgmann, "'The Wicked

not give any actual descriptions. However, it complements the *Barkhi Nafshi* hymns in stating the corrupting nature of body parts as opposed to those perfected by God. It is possible that people reading both texts may have related them in this way.

BODY AND SOUL IN THE *TESTAMENTS OF THE TWELVE PATRIARCHS*[47]

In some of the testaments from the *Testaments of the Twelve Patriarchs* the notion of sympathy between body and soul/spirit figures prominently, expressing the correspondence and reciprocal influence between body and soul.[48] For example, in the *Testament of Simeon*, Simeon describes to his children the effects of the spirit of envy on the body, concluding that that attitude makes the soul savage and corrupts the body.[49] Talking about his brother, Simeon adds that:

> Therefore, Joseph was attractive in shape and beautiful in appearance, because nothing evil dwelt in him; for the face reveals any trouble of the spirit.[50]

Before Simeon started telling his children about his jealousy of Joseph, he already made a statement concerning the correspondence between body and soul/spirit, saying that the Most High had given courage to men in both soul and body.[51]

Woman': Der Makkabäer Simon?," *RevQ* 8/31 (1974): 323-59; R.D. Moore, "Personification of the Seduction of Evil: 'The Wiles of the Wicked Woman,'" *RevQ* 10/40 (1981): 505-19; J.M. Baumgarten, "On the Nature of the Seductress in *4Q184*," *RevQ* 15/57-58 (1991): 133-43; J. Maier, "Wiles of the Wicked Woman," in *Encyclopedia of the Dead Sea Scrolls*, 976; B.R. Rossing, "City Visions, Feminine Figures and Economic Critique: A Sapiential *Topos* in the Apocalypse," in *Conflicted Boundaries in Wisdom and Apocalypticism*, 181-96, at 186-87.

47 Scholars disagree on the origin of the *Testaments of the Twelve Patriarchs* being Jewish or Christian. In its final form the Greek text is Christian, but it is possible that individual testaments like that of Levi or Naphtali go back to an Aramaic or Hebrew *Vorlage*, fragments of which were discovered at Qumran and the Cairo Genizah. Some scholars, therefore, assume that the other testaments in the *Testaments of the Twelve Patriarchs* also possibly have an Aramaic or Hebrew *Vorlage*, but this is far from certain. It is, therefore, not clear whether the passages from these texts represent Second Temple period Judaism or Late Antique Christianity. See e.g. M. de Jonge, *Pseudepigrapha of the Old Testament as Part of Christian Literature: The Case of the* Testaments of the Twelve Patriarchs *and the Greek* Life of Adam and Eve (SVTP 18; Leiden: Brill, 2003); Drawnel, *New Interpretation of the Levi Document*; J.C. Greenfield, M.E. Stone and E. Eshel, *The Aramaic Levi Document: Edition, Translation, Commentary* (SVTP 19; Leiden: Brill, 2004); J.R. Davila, *The Provenance of the Pseudepigrapha: Jewish, Christian, or Other?* (JSJSup 105; Leiden: Brill, 2005), 5, 232.

48 Cf. also Wis 8:19–20 (I owe this reference to Annemieke ter Brugge); T.-S. K 21.95.L 2a/15–2b/3. See Schäfer, *Geniza Fragmente*, 136-37; Davila, *Descenders to the Chariot*, 183.

49 *T. Sim.* 4:8. Cf. also the influence of anger on soul and body in *T. Dan* 3.

50 *T. Sim.* 5:1.

51 *T. Sim.* 2:5.

The *Testament of Naphtali* provides a theological justification for the correspondence between body and spirit saying that:

> the Lord creates the body in resemblance to the spirit, and puts in the spirit according to the power of the body.[52]

Because of this correspondence between body and spirit created by God, the appearance of the body shows the character of the spirit. This was explicitly said of Joseph in the *Testament of Simeon*. Joseph's lack of an evil attitude within him registers through his shape being attractive and his appearance beautiful.

Apart from the theological basis, the articulated mutual relationship between body and spirit in these passages from the *Testaments of the Twelve Patriarchs* is remarkably similar to the basic premise governing Greco-Roman physiognomics. In the Peripatetic treatise *Physiognomonica* the author states that:

> soul and body react on each other; when the character of the soul changes, it changes also the form of the body, and conversely, when the form of the body changes, it changes the character of the soul.[53]

This reciprocal influence between body and soul implies a correspondence between both. Therefore, people believed that physiognomists could look at the human body and discern people's characters or spirits, because these register on the shape and appearance of the body. The passages taken from the *Testaments of the Twelve Patriarchs* demonstrate familiarity with this basic premise of Greco-Roman physiognomics. The salient difference being that God is believed responsible for the sympathy between body and soul.

Finally, there are two passages in the *Testament of Joseph* that betray physiognomic consciousness. As he was being besieged and threatened by Photimar's wife, Joseph fasted, as a result of which he:

> appeared to the Egyptians as one living in luxury, for those who fast because of God receive beauty of face.[54]

[52] *T. Naph.* 2:2. See M. Kister, "Physical and Metaphysical Measurements Ordained by God in the Literature of the Second Temple Period," in *Reworking the Bible: Apocryphal and Related Texts at Qumran: Proceedings of a Joint Symposium by the Orion Center for the Study of the Dead Sea Scrolls and Associated Literature and the Hebrew University Institute for Advanced Studies Research Group on Qumran, 15-17 January, 2002* (eds. E.G. Chazon, D. Dimant and R.A. Clements; STDJ 58; Leiden: Brill, 2005), 153-76. As to the importance of measuring and weighing people, see also G.A. Anderson, "Two Notes on Measuring Character and Sin at Qumran," in *Things Revealed*, 141-47.

[53] Pseudo-Aristotle, *Physiognomonica* 808b 11–14 (cf. also 805a 1–18).

[54] *T. Jos.* 3:4. Cf. Hollander and de Jonge, *The Testaments of the Twelve Patriarchs*, 376. Asceticism as a more extreme form of fasting resulted in another physiognomy (sunken eyes, emaciated cheeks, and thinning hair), one looking less like a life lived luxuriously, which was recorded in late antiquity by Christian pilgrims. Cf. Frank, *Memory of the Eyes*, 137-44.

And when Joseph tells of his trip to Egypt as a slave, he recounts why one of the slave traders does not believe him when he says that he is a home-born slave, saying:

> You are not a slave, for also your appearance reveals who you are.[55]

[55] *T. Jos.* 11:3. Cf. Hollander and de Jonge, *Testaments of the Twelve Patriarchs*, 395. On the issue of a distinct physiognomy of slaves in antiquity, see e.g. Gleason, *Making Men*, 35-36; Sassi, *Science of Man*, 11-13, 80; Isaac, *Invention of Racism*, 170-94, 211-13. Physiognomic features of a slavish, servile nature can be found in, for example, Pseudo-Aristotle, *Physiognomonica* 812b 36–813a 2; Anonymous Latin, *De physiognomonia liber* §§24, 51, 69; Förster, *Scriptores Physiognomonici*, 1:314, 359.

FIGURE 1

4QZODIACAL PHYSIOGNOMY (4Q186) FRAGMENTS 2, 4 – 7

See DJD 5.Plate XXXI and PAM 43.438.
For fragments 1 – 3, see DJD 5.Plate XXXI

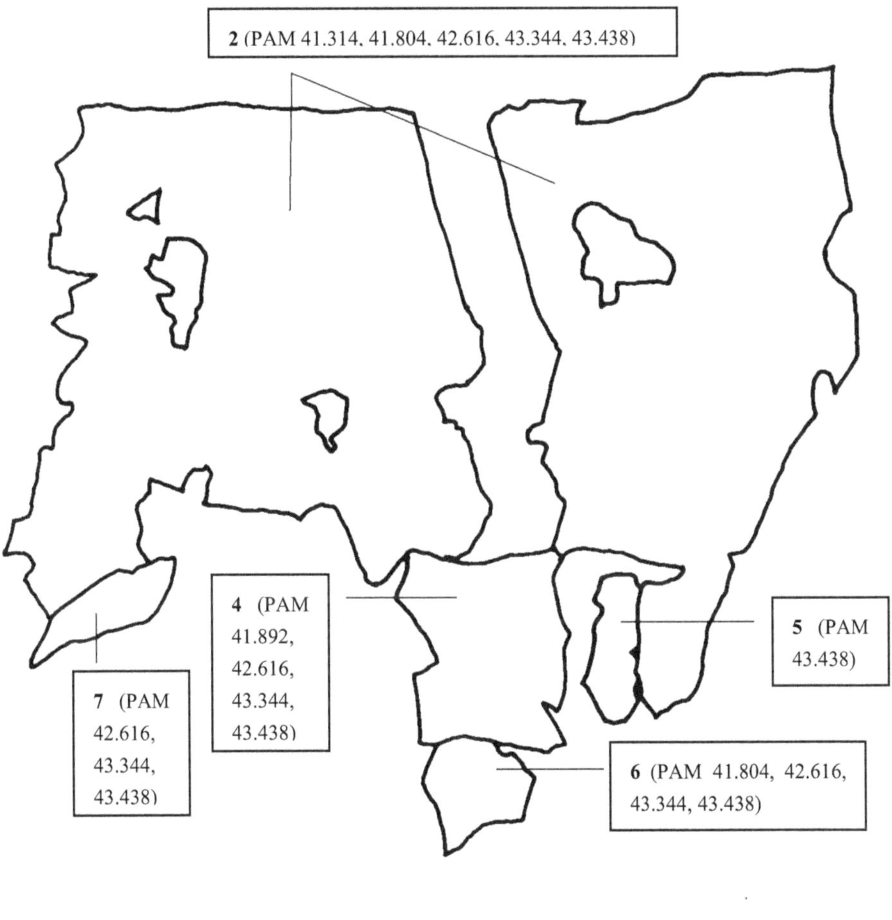

PLATE I

RECENT PHOTOGRAPH (B-45417) OF IAA #109 CONTAINING
4QZODIACAL PHYSIOGNOMY (4Q186)

Photo Clara Amit, Courtesy of the Israel Antiquities Authority

BIBLIOGRAPHY

Abry, J.-H., "Les *anaphorai* des signes du zodiaque dans les écrits astrologiques," in *Sciences exactes et sciences appliquées à Alexandrie (IIIe siècle av. J.-C. – Ier siècle ap. J.-C.): Actes du Colloque International de Saint-Étienne (6-8 juin 1996)* (eds. G. Argoud and J.-Y. Guillaumin; Mémoires 16; Saint-Étienne: Université de Saint-Étienne, 1998), 305-24

Albani, M., "Der Zodiakos in 4Q318 und die Henoch-Astronomie," *MBFJTFL* 7 (1993): 3-42

———, *Astronomie und Schöpfungsglaube: Untersuchungen zum astronomischen Henochbuch* (WMANT 68; Neukirchen-Vluyn: Neukirchener, 1994)

———, "Horoscopes in the Qumran Scrolls," in *The Dead Sea Scrolls After Fifty Years: A Comprehensive Assessment* (eds. P.W. Flint and J.C. VanderKam; vol. 2; Leiden: E.J. Brill, 1999), 279-330

———, "Horoscopes," in *Encyclopedia of the Dead Sea Scrolls* (eds. L.H. Schiffman and J.C. VanderKam; New York: Oxford University Press, 2000), 370-73

———, "'Kannst du die *Sternbilder* hervortreten lassen zur rechten Zeit …?' (Hi 38,32): Gott und Gestirne im Alten Testament und im Alten Orient," in *Das biblische Weltbild und seine altorientalische Kontexte* (eds. B. Janowski and B. Ego; FAT 32; Tübingen: J.C.B. Mohr [Paul Siebeck], 2001), 181-226

Albani, M. and U. Gleßmer, "Un instrument de mesures astronomiques à Qumrân," *RB* 104 (1997): 88-115

Alexander, L., "The Living Voice: Scepticism towards the Written Word in Early Christian and in Graeco-Roman Texts," in *The Bible in Three Dimensions: Essays in Celebration of Forty Years of Biblical Studies in the University of Sheffield* (eds. D.J.A. Clines, S.E. Fowl and S.E. Porter; JSOTSup 87; Sheffield: Sheffield Academic Press, 1990), 221-47

Alexander, P.S., "The Historical Setting of the Hebrew Book of Enoch," *JJS* 28 (1977): 156-80

———, "3 (Hebrew Apocalypse of) Enoch," *OTP* 1:223-315

———, "Incantations and Books of Magic," in *Emil Schürer: The History of the Jewish People in the Age of Jesus Christ (175 B.C.-A.D. 135)* (vol. 3.1; eds. G. Vermes, F. Millar and M. Goodman; Edinburgh: T&T Clark, 1986), 342-79

———, "Physiognonomy [sic!], Initiation, and Rank in the Qumran Community," in *Geschichte – Tradition – Reflexion: Festschrift für Martin Hengel zum 70. Geburtstag, Band I Judentum* (eds. H. Cancik, H. Lichtenberger and P. Schäfer; Tübingen: J.C.B. Mohr [Paul Siebeck], 1996), 385-94

———, "The Redaction-History of *Serekh ha-Yaḥad*: A Proposal," *RevQ* 17/65-68 (1996): 437-56

———, "'Wrestling Against Wickedness in High Places': Magic in the Worldview of the Qumran Community," in *The Scrolls and the Scriptures: Qumran Fifty Years After* (eds. S.E. Porter and C.A. Evans; JSPSup 26; Sheffield: Sheffield Academic Press, 1997), 318-37

———, "Jewish Elements in Gnosticism and Magic *c.* CE 70–*c.* CE 270," in *The Cambridge History of Judaism, Volume Three: The Early Roman Period* (eds. W. Horbury, W.D. Davies and J. Sturdy; Cambridge: Cambridge University Press, 1999), 1052-78

———, "The Demonology of the Dead Sea Scrolls," in *The Dead Sea Scrolls After Fifty Years: A Comprehensive Assessment* (eds. P.W. Flint and J.C. VanderKam; vol. 2; Leiden: E.J. Brill, 1999), 331-53

———, "Enoch and the Beginnings of Jewish Interest in Natural Science," in *The Wisdom Texts from Qumran and the Development of Sapiential Thought* (eds. C. Hempel, A. Lange and H. Lichtenberger; BETL 159; Leuven: Peeters, 2002), 223-43

———, "Contextualizing the Demonology of the Testament of Solomon," in *Demons: The Demonology of Israelite-Jewish and Early Christian Literature in Context of their Environment* (eds. A. Lange, H. Lichtenberger and K.F.D. Römheld; Tübingen: Mohr Siebeck, 2003), 613-35

———, "Literacy among Jews in Second Temple Palestine: Reflections on the Evidence from Qumran," in *Hamlet on a Hill: Semitic and Greek Studies Presented to Professor*

T. Muraoka on the Occasion of his Sixty-Fifth Birthday (eds. M.F.J. Baasten and W.Th. van Peursen; OLA 118; Leuven: Peeters, 2003), 3-24

———, "Predestination and Free Will in the Theology of the Dead Sea Scrolls," paper read at a seminar on *Divine and Human Agency in Paul and his Cultural Environment* at the University of Aberdeen, August 18-21, 2004 (http://www.abdn.ac.uk/divinity/Gathercole/paper-alexander.htm)

Alexander, P.S. and G. Vermes (eds.), *Qumran Cave 4.XIX: Serekh ha-Yaḥad and Two Related Texts* (DJD 26; Oxford: Clarendon, 1998)

Allegro, J.M., "Le travail d'édition des fragments manuscrits de Qumrân: Communication de J.M. Allegro," *RB* 63 (1956): 62-64

———, *The Dead Sea Scrolls: A Reappraisal* (2d ed.; Harmondsworth, Middlesex: Penguin, 1964)

———, "An Astrological Cryptic Document from Qumran," *JSS* 9 (1964): 291-94

———, "'The Wiles of the Wicked Woman': A Sapiential Work from Qumran's Fourth Cave," *PEQ* 96 (1964): 53-55

——— (ed. with the collaboration of A.A. Anderson), *Qumrân Cave 4.I (4Q158-4Q186)* (DJD 5; Oxford: Clarendon, 1968)

Allison, Jr., D.C., *Testament of Abraham* (CEJL; Berlin: Walter de Gruyter, 2003)

Anderson, G.A., "Two Notes on Measuring Character and Sin at Qumran," in *Things Revealed: Studies in Early Jewish and Christian Literature in Honor of Michael E. Stone* (eds. E.G. Chazon, D. Satran and R.E. Clements; JSJSup 89; Leiden: Brill, 2004), 141-47

André, J., *Anonyme Latin: Traité de physiognomonie* (Budé; Paris: Les Belles Lettres, 1981)

Argall, R.A., 1 Enoch *and Sirach: A Comparative Literary and Conceptual Analysis of the Themes of Revelation, Creation and Judgment* (SBLEJL 8; Atlanta: Scholars Press, 1995)

Armstrong, A.MacC., "The Methods of the Greek Physiognomists," *GR* 5 (1958): 52-56

Aune, D.E., *Revelation 1-5* (WBC 52A; Dallas, Texas: Word Books, 1997)

———, *Revelation 17-22* (WBC 52C; Dallas, Texas: Word Books, 1998)

Baccani, D., *Oroscopi Greci: Documentazione papirologica* (RC 1; Messina: Sicania, 1992)

Balme, D.M., "Ἄνθρωπος ἄνθρωπον γεννᾷ: Human is Generated by Human," in *The Human Embryo: Aristotle and the Arabic and European Traditions* (ed. G.R. Dunstan; Exeter: University of Exeter Press, 1990), 20-31

Bar Ilan, M., "Astrology in Ancient Judaism," in *The Encyclopaedia of Judaism: Volume V, Supplement Two* (eds. J. Neusner, A.J. Avery-Peck and W.S. Green; Leiden: Brill, 2004), 2031-37

Bara, J.-F. (ed.), *Vettius Valens d'Antioche: Anthologies, Livre I* (EPRO 111; Leiden: E.J. Brill, 1989)

Barton, T., *Ancient Astrology* (London: Routledge, 1994)

———, *Power and Knowledge: Astrology, Physiognomics, and Medicine under the Roman Empire* (Ann Arbor, Michigan: University of Michigan Press, 1994)

Baumgarten, A.I., *The Flourishing of Jewish Sects in the Maccabean Era: An Interpretation* (JSJSup 55; Leiden: Brill, 1997), 114-36

Baumgarten, J.M., "The 4Q Zadokite Fragments on Skin Disease," *JJS* 41 (1990): 153-65

———, "On the Nature of the Seductress in *4Q184*," *RevQ* 15/57-58 (1991): 133-43

———, "מינוח קומראני ואסטרולוגי ב׳מוסר למבין׳" (Some Qumranic and Astrological Terms in *Mûsār lĕ Mēvîn*)," *Tarbiz* 72 (2003): 321-28

Bautch, K. Coblentz, *A Study of the Geography of 1 Enoch 17-19: "No One Has Seen What I Have Seen"* (JSJSup 81; Leiden: Brill, 2003)

Beaulieu, P.-A., "New Light on Secret Knowledge in Late Babylonian Culture," *ZA* 82 (1992): 98-111

Ben-Dov, J., "The Initial Stages of Lunar Theory at Qumran," *JJS* 54 (2003): 125-38

———, *Astronomy and Calendars at Qumran – Sources and Trends* (Diss. Jerusalem, 2005 [in Hebrew])

———, "Mesopotamian Science in West Semitic Sources: The Case of Qumran," in *The Interpretations of Ancient Astral Science* (eds. H. Falk and D. Brown), *forthcoming*

Ben-Dov, J. and W. Horowitz, "The Babylonian Lunar Three in Calendrical Scrolls from Qumran," *ZA* 95 (2005): 104-20
Ben-Shammai, H., "Saadia's Introduction to Daniel: Prophetic Calculation of the End of Days vs. Astrological and Magical Speculation," *Aleph* 4 (2004): 11-87
Ben Yehuda, E., *Dictionary and Thesaurus of the Hebrew Language* (vol. 5; New York: Thomas Yoseloff, 1960)
Bergland, K.J.H., "Reading Character in the Face: Lavater, Socrates, and Physiognomy," *Word and Image* 9 (1993): 252-69
Bergmeier, R., *Glaube als Gabe nach Johannes: Religions- und theologiegeschichtliche Studien zum prädestinatianischen Dualismus im vierten Evangelium* (BWANT 112; Stuttgart: W. Kohlhammer, 1980)
Bernat, D., "Biblical *Waṣfs* Beyond Song of Songs," *JSOT* 28 (2004): 327-49
Bernstein, M. et al., *Qumran Cave 4.XXVIII: Miscellanea, Part 2* (DJD 28; Oxford: Clarendon, 2001)
Betz, H.D., *The Greek Magical Papyri in Translation, Including the Demotic Spells* (Chicago: University of Chicago Press, 1996)
Betz, O., "'Kann denn aus Nazareth etwas Gutes kommen?': Zur Verwendung von Jesaja Kap. 11 in Johannes Kap. 1," in *Wort und Geschichte: Festschrift für Karl Elliger zum 70. Geburtstag* (eds. H. Gese and H.P. Rüger; AOAT 18; Neukirchen-Vluyn: Neukirchener, 1973), 9-16
Beyer, K., *Die aramäischen Texte vom Toten Meer: Ergänzungsband: Samt den Inschriften aus Palästina, dem Testament Levis aus der Kairoer Genisa, der Fastenrolle und den alten talmudischen Zitaten: Aramaistische Einleitung, Text, Übersetzung, Deutung, Grammatik/Wörterbuch, Deutsch-aramäische Wortliste, Register* (Göttingen: Vandenhoeck & Ruprecht, 1994)
———, *Die aramäischen Texte vom Toten Meer: Samt den Inschriften aus Palästina, dem Testament Levis aus der Kairoer Genisa, der Fastenrolle und den alten talmudischen Zitaten: Aramaistische Einleitung, Text, Übersetzung, Deutung, Grammatik/Wörterbuch, Deutsch-aramäische Wortliste, Register* (vol. 2; Göttingen: Vandenhoeck & Ruprecht, 2004)
Bianchi, U. and Y. Stoyanov, "Dualism," *ER* 4 (2005), 2504-17
Bietenhard, H., *Die himmlische Welt im Urchristentum und Spätjudentum* (WUNT 2; Tübingen: J.C.B. Mohr [Paul Siebeck], 1951)
Biggs, R.D., "An Esoteric Babylonian Commentary," *RA* 62 (1968): 51-58
Blankenburg, M., "Physiognomik, Physiognomie," HWP 7 (1989), 955-63
Blockmuehl, M., "Redaction and Ideology in the *Rule of the Community*," *RevQ* 18/72 (1998): 541-60
Böcher, O., "Engel IV. Neues Testament," *TRE* 9:596-99
Böck, B., "Babylonische Divination und Magie als Ausdruck der Denkstrukturen des altmesopotamischen Menschen," in *Babylon: Focus mesopotamischer Geschichte, Wiege früher Gelehrsamkeit, Mythos in der Moderne: 2. Internationales Colloquium der Deutschen Orient-Gesellschaft, 24.-26. März 1998 in Berlin* (ed. J. Renger; CDOG 2; Saarbrücken: SDV, 1999), 409-25
———, *Die babylonisch-assyrische Morphoskopie* (AfO.B 27; Vienna: Institut für Orientalistik der Universität Wien, 2000)
———, "'An Esoteric Babylonian Commentary' Revisited," *JAOS* 120 (2000): 615-20
———, "Physiognomie und Schicksal? Oder wie der altmesopotamische Mensch mit einem durch ein physiognomisches Omen angekundigtes Unheil umgegangen sein mag," *Sefarad* 62 (2002): 241-57
Boiy, T., *Late Achaemenid and Hellenistic Babylon* (OLA 136; Leuven: Peeters, 2004)
Boll, F., *Sphaera: Neue griechische Texte und Untersuchungen zur Geschichte der Sternbilder* (Leipzig: B.G. Teubner, 1903; repr., Hildesheim: Georg Olms, 1967)
Boll, F., C. Bezold, and W. Gundel, *Sternglaube und Sterndeutung: Die Geschichte und das Wesen der Astrologie* (5d ed.; Darmstadt: Wissenschaftliche Buchgesellschaft, 1966)
Bollók, J., "The Description of Paul in the Acta Pauli," in *The Apocryphal Acts of Paul and Thecla* (ed. J.N. Bremmer; SAAA 2; Kampen: Kok Pharos, 1996), 1-15

Bonner, S.F., *Education in Ancient Rome: From the Elder Cato to the Younger Pliny* (London: Methuen, 1977)
Borger, R., "Geheimwissen," *RlA* 3 (1957-1971), 188-91
Bottéro, J., "Symptômes, signes, écritures: en Mésopotamie ancienne," in *Divination et Rationalité* (eds. J.P. Vernant et al.; Paris: Éditions du Seuil, 1974), 70-197
———, *Mesopotamia: Writing, Reasoning, and the Gods* (Chicago: The University of Chicago Press, 1992) (first published as *Mésopotamie: L'écriture, la raison et les dieux* [Paris: Gallimard, 1987])
Böttrich, C., "Astrologie in der Henochtradition," *ZAW* 109 (1997): 222-45
Bouché-Leclercq, A., *L'histoire de la divination dans l'antiquité* (vol.1; Paris: Ernest Leroux, 1879)
———, *L'astrologie grecque* (Paris: Ernest Leroux, 1899)
Bourdieu, P., *Distinction: A Social Critique of the Judgement of Taste* (trans. R. Nice; London: Routledge & Kegan Paul, 1984)
———, "The Forms of Capital," in *Handbook of Theory and Research for the Sociology of Education* (New York: Greenwood, 1986), 241-58
Brach, J.-P. and W.J. Hanegraaff, "Correspondences," in *Dictionary of Gnosis and Western Esotericism* (ed. W.J. Hanegraaff et al.; vol. 1; Leiden: Brill, 2005), 275-79
Brack-Bernsen, L. and H. Hunger, "The Babylonian Zodiac: Speculations on its Invention and Significance," *Centaurus* 41 (1999): 280-92
Brandt, E.A., "On Secrecy and the Control of Knowledge: Taos Pueblo," in *Secrecy: A Cross-Cultural Perspective* (ed. S.T. Tefft; New York: Human Sciences, 1980), 123-46
Braude, W.G., *Pesikta Rabbati: Discourses for Feasts, Fasts, and Special Sabbaths* (2 vols; YJS 18; New Haven: Yale University Press, 1968)
Bremmer, J.N., "Magic, Martyrdom and Women's Liberation in the Acts of Paul and Thecla," in *The Apocryphal Acts of Paul and Thecla* (ed. J.N. Bremmer; SAAA 2; Kampen: Kok Pharos, 1996), 36-59
———, "Foolish Egyptians: Apion and Anoubion in the *Pseudo-Clementines*," in *The Wisdom of Egypt: Jewish, Early Christian, and Gnostic Essays in Honour of Gerard P. Luttikhuizen* (eds. A. Hilhorst and G.H. van Kooten; AJEC 59; Leiden: Brill, 2005), 311-29
Brooke, G.J., "Body Parts in *Barkhi Nafshi* and the Qualifications for Membership of the Worshipping Community," in *Sapiential, Liturgical and Poetical Texts from Qumran: Proceedings of the Third Meeting of the International Organization for Qumran Studies, Oslo 1998* (eds. D.K. Falk, F. García Martínez and E.M. Schuller; STDJ 35; Leiden: Brill, 2000), 79-94
Broshi, M., "Predestination in the Bible and the Dead Sea Scrolls," in *Bread, Wine, Walls, and Scrolls* (M. Broshi; JSPSup 36; Sheffield; Sheffield Academic Press, 2001), 238-51
Brown, D., *Mesopotamian Planetary Astronomy-Astrology* (CM 18; Groningen: STYX, 2000)
———, "Review: B. Pongratz-Leisten, *Herrschaftswissen in Mesopotamien: Formen der Kommunikation zwischen Gott und König im 2. und 1. Jahrtausend v. Chr.* (SAAS 10; Helsinki: The Neo-Assyrian Text Corpus Project, 1999)," *ZA* 94 (2004): 112-21
Brown, J.A., *John Marco Allegro: The Maverick of the Dead Sea Scrolls* (Grand Rapids, Michigan: Eerdmans, 2005)
Brownlee, W.H., *The Dead Sea Manual of Discipline: Translation and Notes* (BASORSup 10-12; New Haven: American Schools of Oriental Research, 1951)
Bryce, G.E., "Omen-Wisdom in Ancient Israel," *JBL* 94 (1975): 19-37
Burgmann, H., "'The Wicked Woman': Der Makkabäer Simon?," *RevQ* 8/31 (1974): 323-59
Burkert, W., *The Orientalizing Revolution: Near Eastern Influence on Greek Culture in the Early Archaic Age* (RevAnt 5; Cambridge, Massachusetts: Harvard University Press, 1992)
———, *Babylon, Memphis, Persepolis: Eastern Contexts of Greek Culture* (Cambridge, Massachusetts: Harvard University Press, 2004)
Burnett, C.S.F., "The Planets and the Development of the Embryo," in *The Human Embryo: Aristotle and the Arabic and European Traditions* (ed. G.R. Dunstan; Exeter: University of Exeter Press, 1990), 95-112

Burnett, C., K. Yamamoto and M. Yano (eds.), *Abū Maʿšar: The Abbreviation of the Introduction to Astrology, Together with the Medieval Latin Translation by Adelard of Bath* (IPTS 15; Leiden: E.J. Brill, 1994)

Busch, P., *Das Testament Salomos: Die älteste christliche Dämonologie, kommentiert und in deutscher Erstübersetzung* (TUGAL 153; Berlin: Walter de Gruyter, 2006)

Campe, R. and M. Schneider (eds.), *Geschichten der Physiognomik: Text, Bild, Wissen* (RWRL 36; Freiburg im Bresgau: Rombach, 1996)

Caquot, A., "Les enfants aux cheveux blancs: Réflexions sur un motif," in *Mélanges d'histoire des religions offerts à Henri-Charles Puech* (Paris: Presses Universitaires de France, 1974), 161-72

———, "*4QMess Ar* 1 i 8-11," *RevQ* 15/57-58 (1991): 145-55

———, "Les textes de sagesse de Qoumrân (Aperçu préliminaire)," *RHPR* 76 (1996): 1-34

———, "Les prodromes du déluge: Légendes araméennes de Qoumrân," *RHPR* 83 (2003): 41-59

Carmignac, J., "Les Horoscopes de Qumrân," *RevQ* 5/18 (1965): 199-217

———, "Poème allégorique sur la secte rivale," *RevQ* 5/19 (1965): 361-74

Carr, D.M., *Writing on the Tablet of the Heart: Origins of Scripture and Literature* (New York: Oxford University Press, 2005)

Catalogus codicum astrologorum Graecorum (12 vols.; Brussels: Henri Lamertin, 1898-1953)

Cavigneaux, A., "Lexikalische Listen," *RlA* 6 (1980-83), 609-41

Charlesworth, J.H., "A Critical Comparison of the Dualism in 1QS 3:13-4:26 and the 'Dualism' Contained in the Gospel of John," in *John and the Dead Sea Scrolls* (ed. J.H. Charlesworth; New York: Crossroads, 1991), 76-106 (originally published in *NTS* 15 [1968-69]: 389-418)

———, "Treatise of Shem," *OTP* 1:473-86

———, "Jewish Interest in Astrology during the Hellenistic and Roman Period," *ANRW* II 20.2 (1987), 926-50

———, *Die Schrift des Sem* (JSHRZ-NF 2.9; Gütersloh: Gütersloher Verlaghaus, 2005)

Chazon, E.G., "Is *Divrei ha-me'orot* a Sectarian Prayer?," in *The Dead Sea Scrolls: Forty Years of Research* (eds. D. Dimant and U. Rappaport; STDJ 10; Leiden: E.J. Brill and Jerusalem: Magnes, 1992), 3-17

Clarke, A.G., "Metoposcopy: An Art to Find the Mind's Construction in the Forehead," in *Astrology, Science, and Society: Historical Essays* (ed. P. Curry; Woodbridge, Suffolk: Boydell, 1987), 171-95

Clarke, M.L., *Higher Education in the Ancient World* (London: Routledge & Kegan Paul, 1971)

Clarysse, W. (ed.), *The Petrie Papyri Second Edition (P. Petrie²): Volume 1, The Wills* (CH 2; Brussels: Koninklijke Academie voor Wetenschappen, Letteren en Schone Kunsten van België, 1991)

Clines, D.J.A., *Job 1-20* (WBC 17; Dallas, Texas: Word Books, 1989)

Collins, J.J., *Daniel: A Commentary on the Book of Daniel* (Hermeneia; Minneapolis: Fortress, 1993)

———, *Apocalypticism in the Dead Sea Scrolls* (London: Routledge, 1997)

———, "Journeys to the World Beyond in Ancient Judaism," in *Apocalyptic and Eschatological Heritage: The Middle East and Celtic Realms* (ed. M. McNamara; Dublin: Four Courts, 2003), 20-36

Colson, F.H., *Philo: VI* (LCL 289; Cambridge, Massachusetts: Harvard University Press, 1984)

Conybeare, F.C., *Philostratus: The Life of Apollonius of Tyana, the Epistles of Apollonius and the Treatise of Eusebius* (LCL 16; Cambridge, Massachusetts: Harvard University Press, 1989)

Cosgrove, D., "Global Illumination and Enlightenment in the Geographies of Vincenzo Coronelli and Athanasius Kirchner," in *Geography and Enlightenment* (eds. D.N. Livingstone and C.W.J. Withers; Chicago: University of Chicago Press, 1999)

———, *Apollo's Eye: A Cartographic Genealogy of the Earth in the Western Imagination* (Baltimore: Johns Hopkins University, 2001)

Couissin, J., "Suétone physiognomiste dans les *Vies des XII Césars*," *REL* 31 (1953): 234-56
Cowling, M., *The Artist as Anthropologist: The Representation of Type and Character in Victorian Art* (Cambridge: Cambridge University Press, 1989)
Cramer, F.H., *Astrology in Roman Law and Politics* (MAPS 37; Philadelphia: American Philosophical Society, 1954)
Crenshaw, J.L., "The Primacy of Listening in Ben Sira's Pedagogy," in *Wisdom, You Are My Sister: Studies in Honor of Roland E. Murphy, O.Carm., on the Occasion of His Eightieth Birthday* (ed. M.L. Barré; CBQMS 29; Washington, DC: Catholic Biblical Association of America, 1997), 172-87
———, *Education in Ancient Israel: Across the Deadening Silence* (New York: Doubleday, 1998)
Cribiore, R., *Gymnastics of the Mind: Greek Education in Hellenistic and Roman Egypt* (Princeton: Princeton University Press, 2001)
Cross, Jr., F.M., "The Development of the Jewish Script," in *The Bible and the Ancient Near East: Essays in Honor of William Foxwell Albright* (ed. G.E. Wright; Garden City, New York: Doubleday, 1961), 133-202
Cryer, F.H., *Divination in Ancient Israel and its Near Eastern Environment: A Socio-Historical Investigation* (JSOTSup 142; Sheffield: Sheffield Academic Press, 1994)
Cumont, F., "La plus ancienne géographie astrologique," *Klio* 9 (1909) 263-73
———, *L'Égypte des astrologues* (Brussels: Fondation Égyptologique Reine Élisabeth, 1937)
———, *Lux Perpetua* (Paris: Paul Geuthner, 1949)
Currie, H.MacL., "Aristotle and Quintilian: Physiognomical Reflections," in *Aristotle on Nature and Living Things: Philosophical and Historical Studies Presented to David M. Balme on his Seventieth Birthday* (ed. A. Gotthelf; Pittsburgh, Pennsylvania: Mathesis and Bristol: Bristol Classical, 1985), 359-66
Dalley, S. et al., *The Legacy of Mesopotamia* (Oxford: Oxford University Press, 1998)
Daniel, R., "The Testament of Solomon XVIII 27-28, 33-40," in *Festschrift zum 100-jährigen Bestehen der Papyrussammlung der österreichischen Nationalbibliothek: Papryus Erzherzog Rainer* (Vienna: Brüder Hollinek, 1983), 294-304
Daniélou, J., "Un source de la spiritualité chrétienne dans les manuscrits de la Mer Morte: la doctrine des deux esprits," *DV* 25 (1953): 127-36
———, *Philon d'Alexandrie* (Paris: Arthème Fayard, 1958)
Davidson, M.J., *Angels at Qumran: A Comparative Study of 1 Enoch 1-36, 72-108 and Sectarian Writings from Qumran* (JSPSup 11; Sheffield: Sheffield Academic Press, 1992)
Davies, P.R., "Sects from Texts: On the Problems of Doing a Sociology of the Qumran Literature," in *New Directions in Qumran Studies: Proceedings of the Bristol Colloquium on the Dead Sea Scrolls, 8–10 September 2003* (eds. J.G. Campbell, W.J. Lyons and L.K. Pietersen; LSTS 52; London: T&T Clark, 2005), 69-82
Davila, J.R., "4QMess ar (4Q534) and Merkavah Mysticism," *DSD* 5 (1998): 367-81
———, "The Dead Sea Scrolls and Merkavah Mysticism," in *The Dead Sea Scrolls in their Historical Context* (eds. T.H. Lim et al.; Edinburgh: T&T Clark, 2000), 249-64
———, *Descenders to the Chariot: The People behind the Hekhalot Literature* (JSJSup 70; Leiden: Brill, 2001)
———, *The Provenance of the Pseudepigrapha: Jewish, Christian, or Other?* (JSJSup 105; Leiden: Brill, 2005)
De Breucker, G., "Berossos and the Mesopotamian Temple as Centre of Knowledge during the Hellenistic Period," in *Learned Antiquity: Scholarship and Society in the Near-East, the Greco-Roman World, and the Early Medieval West* (eds. A.A. MacDonald, M.W. Twomey and G.J. Reinink; GSCC 5; Leuven: Peeters, 2003), 13-23
De Falco, V., M. Krause and O. Neugebauer, *Hypsikles: Die Aufgangszeiten der Gestirne* (AAWGPHK 3/62; Göttingen: Vandenhoeck & Ruprecht, 1966)
Degkwitz, A., *Die pseudoaristotelischen 'Physiognomonica' Traktat A: Übersetzung und Kommentar* (Diss. Heidelberg, 1988)
———, "Die pseudoaristotelischen 'Physiognomonica,'" in *Geschichten der Physiognomik: Text, Bild, Wissen* (eds. R. Campe and M. Schneider; RWRL 36; Freiburg im Bresgau: Rombach, 1996), 23-44

Delcor, M., "Recherches sur un horoscope en langue hébraïque provenant de Qumrân," in *Religion d'Israël et Proche Orient Ancien: Des Phéniciens aux Esséniens* (M. Delcor; Leiden: E.J. Brill, 1976), 298-319 (originally published in *RevQ* 5/20 [1966]: 521-42)

De Smet, D., "L'alphabet secret des Ismaéliens ou la force magique de l'écriture," in *Charmes et sortilèges: Magie et magiciens* (ed. R. Gyselen; ResOr 14; Bures-sur-Yvette: Groupe pour l'Étude de la Civilisation du Moyen-Orient, 2002), 51-60

Didi-Huberman, G., *Invention of Hysteria: Charcot and the Photographic Iconography of the Salpêtrière* (trans. A. Hartz; Cambridge, Massachusetts: MIT Press, 2003)

Dietrich, M., "Babylonische Sklaven auf der Schreiberschule," in *Veenhof Anniversary Volume: Studies Presented to Klaas R. Veenhof on the Occasion of his Sixty-Fifth Birthday* (eds. W.H. van Soldt et al.; UNHAII 89; Leiden: Nederlands Instituut voor het Nabije Oosten, 2001), 67-81

Dijk, J. van, "Die Inschriftenfunde," in UVB 18 (Berlin: Gebr. Mann, 1962), 39-62

Dillon, J., "Plotinus on Whether the Stars are Causes," in *La science des cieux: Sages, mages, astrologues* (ed. R. Gyselen; ResOr 12; Bures-sur-Yvette: Groupe pour l'Étude de la Civilisation du Moyen-Orient, 1999), 87-92

Dimant, D., "Qumran Sectarian Literature," in *Jewish Writings of the Second Temple Period: Apocrypha, Pseudepigrapha, Qumran Sectarian Writings, Philo, Josephus* (ed. M.E. Stone; CRINT 2.2; Assen: Van Gorcum and Philadelphia: Fortress Press, 1984), 483-550

———, "The Qumran Manuscripts: Contents and Significance," in *Time to Prepare the Way in the Wilderness: Papers on the Qumran Scrolls by Fellows of the Institute for Advanced Studies of the Hebrew University, Jerusalem, 1989-1990* (eds. D. Dimant and L.H. Schiffman; STDJ 16; Leiden: E.J. Brill, 1995), 23-58

———, "Dualism at Qumran: New Perspectives," in *Caves of Enlightenment: Proceedings of the American Schools of Oriental Research Dead Sea Scrolls Jubilee Symposium (1947-1997)* (ed. J.H. Charlesworth; North Richland Hills, Texas: Bibal, 1998), 55-73

———, "Review: É. Puech (ed.), *Qumrân Grotte 4.XXII: Textes araméens, Première partie, 4Q529-549* (DJD 31; Oxford: Clarendon, 2001)," *DSD* 10 (2003): 292-304

———, "Two 'Scientific' Fictions: The So-Called *Book of Noah* and the Alleged Quotation of *Jubilees* in CD 16:3-4," in *Studies in the Hebrew Bible, Qumran, and the Septuagint Presented to Eugene Ulrich* (eds. P.W. Flint, E. Tov and J.C. VanderKam; VTSup 101; Leiden: Brill, 2006), 230–49

Doran, R., "The High Cost of a Good Education," in *Hellenism in the Land of Israel* (eds. J.J. Collins and G.E. Sterling; CJAS 13; Notre Dame, Indiana: University of Notre Dame, 2001), 94-115

Douglas, M. and G. Mars, "Terrorism: A Positive Feedback Game," *Human Relations* 56/7 (2003): 763-86

Drawnel, H., *An Aramaic Wisdom Text from Qumran: A New Interpretation of the Levi Document* (JSJSup 86; Leiden: Brill, 2004)

Drijvers, H.J.W., *The Book of the Laws of Countries: Dialogue on Fate of Bardaiṣan of Edessa* (STT 3; Assen: Van Gorcum, 1965)

———, *Bardaiṣan of Edessa* (Assen: Van Gorcum, 1966)

Drijvers, J.W., "'Hij was een uitzonderlijke schoonheid...': Keizerlijke lichamen in beeld en geschrift," in *Machtige lichamen: Het vingertje van Luns en andere politieke wapens* (eds. C. Santing, H. te Velde and M. Wilke; Amsterdam: Wereldbibliotheek, 2005), 191-207

Drower, E.S., *The Book of the Zodiac* (OTF 36; London: Royal Asiatic Society, 1949)

Dubs, J.-C., "4Q317 et le rôle de l'observation de la Pleine Lune pour la détermination du temps à Qoumrân," in *Le Temps et les Temps dans les littératures juives et chrétiennes au tournant de notre ère* (eds. C. Grappe and J.-C. Ingelaere; JSJSup 112; Leiden: Brill, 2006), 37-54

Duhaime, J., "Les voies des deux esprits (*1QS* iv 2-14): Une analyse structurelle," *RevQ* 19/75 (2000): 349-67

———, "Dualism," in *Encyclopedia of the Dead Sea Scrolls* (eds. L.H. Schiffman and J.C. VanderKam; Oxford: Oxford University Press, 2000), 215-20

———, "Cohérence structurelle et tensions internes dans l'Instruction sur les Deux Esprits (1QS III 13 – IV 26)," in *Wisdom and Apocalypticism in the Dead Sea Scrolls and in the Biblical Tradition* (ed. F. García Martínez; BETL 168; Leuven: Peeters, 2003), 103-31

Duling, D.C., "Testament of Solomon," *OTP* 1:935-87

———, "The Testament of Solomon: Retrospect and Prospect," *JSP* 2 (1988) 87-112

Dupont-Sommer, A., "L'instruction sur les deux Esprits dans le 'Manuel de Discipline,'" *RHR* 142 (1952): 5-35

———, "Le Problème des influences étrangères sur la secte juive de Qoumrân," *RHPR* 35 (1955): 75-94

———, "Deux documents horoscopiques esséniens découverts à Qoumrân, près de la Mer Morte," CRAI (1965), 239-53

Edgar, C.C. (ed.), *Catalogue général des antiquités égyptiennes du Musée du Caire: Zenon Papyri, Nos 59001-59139* (vol. 1; Cairo: L'Institute Français d'Archéologie Orientale, 1925)

Eijk, P.J. van der, "Aristotle's Psycho-Physiological Account of the Soul-Body Relationship," in *Psyche and Soma: Physicians and Metaphysicians on the Mind-Body Problem from Antiquity to Enlightenment* (eds. J.P. Wright and P. Potter; Oxford: Clarendon, 2000), 57-77

Eisenman, R. and M. Wise, *Dead Sea Scrolls Uncovered: The First Complete Translation and Interpretation of 50 Key Documents withheld for over 35 Years* (New York: Penguin, 1992)

Ejrnæs, B., S. Holst and M. Müller (eds.), *Dødehavsskrifterne og de antikke kilder om essæerne* (2d ed.; Copenhagen: ANIS, 2003)

Eppel, R., "Les tables de la Loi et les tables célestes," *RHPR* 17 (1937): 401-12

Eshel,, E., H. Eshel and A. Yardeni, "4QApocryphal Psalm and Prayer," in *Qumran Cave 4.VI: Poetical and Liturgical Texts, Part 1* (eds. E. Eshel et al.; DJD 11; Oxford: Clarendon, 1998), 403-25

Evans, E.C., "Roman Descriptions of Personal Appearance in History and Biography," *HSCP* 46 (1935): 43-84

———, "Galen the Physician as Physiognomist," *TAPA* 76 (1945): 287-98

———, "Literary Portraiture in Ancient Epic: A Study of the Descriptions of Physical Appearance in Classical Epic," *HSCP* 58-59 (1948): 189-217

———, "A Stoic Aspect of Senecan Drama: Portraiture," *TAPA* 81 (1950): 169-84

———, *Physiognomics in the Ancient World* (TAPS 59/5; Philadelphia: American Philosophical Society, 1969)

Evans, J., *The History and Practice of Ancient Astronomy* (New York: Oxford University Press, 1998)

Fahd, T., *La divination arabe: Études religieuses, sociologiques et folkloriques sur le milieu natif de l'Islam* (Leiden: E.J. Brill, 1966)

Falconer, W.A., *Cicero: On Old Age, On Friendship, On Divination* (LCL 154; Cambridge, Massachusetts: Harvard University Press, 2001)

Fernández Marcos, N., "Rewritten Bible or *Imitatio*?: The Vestments of the High-Priest," in *Studies in the Hebrew Bible, Qumran, and the Septuagint Presented to Eugene Ulrich* (eds. P.W. Flint, E. Tov and J.C. VanderKam; VTSup 101; Leiden: Brill, 2006), 321-36

Festugière, A.-J., *La révélation d'Hermès Trismégiste, I: L'astrologie et les sciences occultes* (2d ed.; CEA.EtB; 1950; repr., Paris: Les Belles Lettres, 1983)

Finkel, I.L. (ed. with the collaboration of M. Civil), *The Series SIG_7.ALAN = Nabnītu* (MSL 16; Rome: Pontificium Institutum Biblicum, 1982)

———, "Adad-apla-iddina, Esagil-kīn-apli, and the Series SA.GIG," in *A Scientific Humanist: Studies in Memory of Abraham Sachs* (eds. E. Leichty, M. deJong Ellis and P. Gerardi; OPSNKF 9; Philadelphia: The University Museum, 1988), 143-59

Fitzmyer, J.A., "The Aramaic 'Elect of God' Text from Qumran Cave IV," in *Essays on the Semitic Background of the New Testament* (J.A. Fitzmyer; London: Geoffrey Chapman, 1971), 127-60 (originally published in *CBQ* 27 [1965]: 348-72)

———, "Aramaic," in *Encyclopedia of the Dead Sea Scrolls* (eds. L.H. Schiffman and J.C. VanderKam; Oxford: Oxford University Press, 2000), 48-51

———, *Tobit* (CEJL; Berlin: Walter de Gruyter, 2003)
———, *The Genesis Apocryphon of Qumran Cave 1 (1Q20): A Commentary* (BibOr 18/B; Rome: Pontificio Istituto Biblico, 2004)
Flannery-Dailey, F., *Dreamers, Scribes, and Priests: Jewish Dreams in the Hellenistic and Roman Eras* (JSJSup 90; Leiden: Brill, 2004)
Flinterman, J.-J., *Power,* Paideia, *and Pythagoreanism: Greek Identity, Conceptions of the Relationship between Philosophers and Monarchs, and Political Ideas in Philostratus'* Life of Apollonius (DMAHA 13; Amsterdam: J.C. Gieben, 1995)
Förster, R. *Die Physiognomik der Griechen* (Kiel: Schmidt & Klaunig, 1884)
——— (ed.), *Scriptores Physiognomonici Graeci et Latini* (2 vols.; Leipzig: B.G. Teubner, 1893)
Frank, G., *The Memory of the Eyes: Pilgrims to Living Saints in Christian Late Antiquity* (Berkeley: University of California Press, 2000)
Freedman, D.N., B.E. Willoughby and H.-J. Fabry, "עַמּוּד *'ammûd*," *TWAT* 7:204-9
Freedman, S.M., *If a City is Set on a Height: The Akkadian Omen Series Šumma Alu ina Mēlê šakin: Volume 1: Tablets 1-21* (OPSNKF 17; Philadelphia: The University of Pennsylvania Museum, 1998)
Frey, J., "Different Patterns of Dualistic Thought in the Qumran Library," in *Legal Texts and Legal Issues: Proceedings of the Second Meeting of the International Organization for Qumran Studies Cambridge 1995: Published in Honour of Joseph M. Baumgarten* (eds. M. Bernstein, F. García Martínez and J. Kampen; STDJ 23; Leiden: E.J. Brill, 1997), 275-335
Friedman, M. (ed.), *Pesikta Rabbati: Midrash für den Fest-Cyclus und die ausgezeichneten Sabbathe* (Vienna, 1880)
———, *Seder Eliahu Rabba and Seder Eliahu Zuta (Tanna d'be Eliahu): Pseudo-Seder Eliahu Zuta* (Vienna, 1904; repr., Jerusalem: Wahrmann, 1969)
Frommhold, K., *Bedeutung und Berechnung der Empfängnis in der Astrologie der Antike* (OrA 38; Münster: Aschendorff, 2004)
Fürst, J., "Untersuchungen zur Ephemeris des Diktys von Kreta (Fortsetzung)," *Phil* 61 (1902): 374-440, 593-622
Gager, J.G. (ed.), *Curse Tablets and Binding Spells from the Ancient World* (New York: Oxford University Press, 1992)
Gagnon, R.A.J., "How Did the Rule of the Community Obtain its Final Shape? A Review of Scholarly Research," *JSP* 10 (1992): 61-72
Gandz, S., "The Problem of the Molad," *Studies in Hebrew Astronomy and Mathematics* (S. Gandz; New York: Ktav, 1970), 120-49
Garcia, H., "L'enfant vieillard, l'enfant aux cheveux blanc et le Christ polymorphe," *RHPR* 80 (2000): 479-501
García Martínez, F., "Qumran Origins and Early History: A Groningen Hypothesis," *FO* 25 (1988): 113-36
———, "*4QMess Ar* and the *Book of Noah*," in *Qumran and Apocalyptic: Studies on the Aramaic Texts from Qumran* (F. García Martínez; STDJ 9; Leiden: E.J. Brill, 1992), 1-44
———, "The Prayer of Nabonidus: A New Synthesis," in *Qumran and Apocalyptic: Studies on the Aramaic Texts from Qumran* (F. García Martínez; STDJ 9; Leiden: E.J. Brill, 1992), 116-36
———, *The Dead Sea Scrolls Translated: The Qumran Texts in English* (2d ed.; trans. W.G.E. Watson; Leiden: E.J. Brill and Grand Rapids, Michigan: Eerdmans, 1996)
———, "The Heavenly Tablets in the Book of Jubilees," in *Studies in the Book of Jubilees* (eds. M. Albani, J. Frey and A. Lange; TSAJ 65; Tübingen: Mohr Siebeck, 1997), 243-60
———, "Magic in the Dead Sea Scrolls," in *The Metamorphosis of Magic from Late Antiquity to the Early Modern Period* (eds. J.N. Bremmer and J.R. Veenstra; GSCC 1, Leuven: Peeters, 2002), 13-33
———, "Iranian Influences in Qumran?," in *Apocalyptic and Eschatological Heritage: The Middle East and Celtic Realms* (ed. M. McNamara; Dublin: Four Courts, 2003), 37-49

García Martínez, F. and E.J.C. Tigchelaar, *The Dead Sea Scrolls Study Edition: Volume 1 (1Q1-4Q273)* (Leiden: E.J. Brill and Grand Rapids, Michigan: Eerdmans, 2000)
———, *The Dead Sea Scrolls Study Edition: Volume 2 (4Q274-11Q31)* (Leiden: E.J. Brill and Grand Rapids, Michigan: Eerdmans, 2000)
———, *Fragmenten uit de woestijn: De Dode-Zeerollen opnieuw bekeken* (Zoetermeer: Meinema, 2003)
García Martínez, F. and A.S. van der Woude, "A 'Groningen' Hypothesis of Qumran Origins and Early History," *RevQ* 14/56 (1990): 521-41
———, *De rollen van de Dode Zee: Ingeleid en in het Nederlands vertaald: Deel 2* (Kampen: Kok, 1995)
Gardner, I., The Kephalaia of the Teacher*: The Edited Coptic Manichaean Texts in Translation with Commentary* (NHMS 37; Leiden: E.J. Brill, 1995)
Garnet, P., "Cave 4 ms Parallels to 1QS 5.1-7: Towards a *Serek* Text History," *JSP* 15 (1997): 67-78
Gazov, A.M., "Double Meaning in a Qumran Work ('The Wiles of the Wicked Woman')," *RevQ* 6/22 (1967): 279-85
Geller, M.J., "Astronomy and Authorship," *BSOAS* 53 (1990): 209-13
———, "The Last Wedge," *ZA* 87 (1997): 43-95
———, "New Documents from the Dead Sea: Babylonian Science in Aramaic," in *Boundaries of the Ancient Near Eastern World: A Tribute to Cyrus H. Gordon* (eds. M. Lubetski, C. Gottlieb and S. Keller; JSOTSup 273; Sheffield: Sheffield Academic Press, 1998), 224-29
———, "Graeco-Babyloniaca in Babylon," in *Babylon: Focus mesopotamischer Geschichte, Wiege früher Gelehrsamkeit, Mythos in der Moderne: 2. Internationales Colloquium der Deutschen Orient-Gesellschaft, 24.-26. März 1998 in Berlin* (ed. J. Renger; CDOG 2; Saarbrücken: SDV, 1999), 377-83
———, "The Survival of Babylonian Wissenschaft in Later Tradition," in *The Heirs of Assyria: Proceedings of the Opening Symposium of the Assyrian and Babylonian Intellectual Heritage Project Held in Tvärminne, Finland, October 8-11, 1998* (eds. S. Aro and R.M. Whiting; MS 1; Helsinki: The Neo-Assyrian Text Corpus Project, 2000), 1-6
———, "An Akkadian Vademecum in the Babylonian Talmud," in *From Athens to Jerusalem: Medicine in Hellenized Jewish Lore and in Early Christian Literature* (eds. S. Kottek et al.; Pantaleon 33; Rotterdam: Erasmus, 2000), 13-32
———, "Incipits and Rubrics," in *Wisdom, Gods and Literature: Studies in Assyriology in Honour of W.G. Lambert* (eds. A.R. George and I.L. Finkel; Winona Lake, Indiana: Eisenbrauns, 2000), 225-58
———, "West Meets East: Early Greek and Babylonian Diagnosis," *AfO* 48/49 (2001/2002): 50-75
———, "Akkadian Healing Therapies in the Babylonian Talmud," *Max-Planck-Institut für Wissenschaftsgeschichte Preprint* 259 (2004): 1-60
Gibbs, S.L., *Greek and Roman Sundials* (YSHSM 11; New Haven: Yale University Press, 1976)
Gibson, M., *Born to Crime: Cesare Lombroso and the Origins of Biological Criminology* (Westport, Connecticut: Praeger, 2002)
Gillmayr-Bucher, S., "Body Images in the Psalms," *JSOT* 28 (2004): 301-26
Ginzburg, C., "Morelli, Freud, and Sherlock Holmes: Clues and Scientific Method," in *The Sign of Three: Dupin, Holmes, Peirce* (eds. U. Eco and T.A. Sebeok; Bloomington: Indiana University Press, 1983), 81-118
Gleason, M.W., *Making Men: Sophists and Self-Presentation in Ancient Rome* (Princeton: Princeton University Press, 1995)
Gleßmer, U. and M. Albani, "An Astronomical Measuring Instrument from Qumran," in *The Provo International Conference on the Dead Sea Scrolls: Technological Innovations, New Texts, and Reformulated Issues* (eds. D.W. Parry and E. Ulrich; STDJ 30; Leiden: Brill, 1999), 407-42

Goitein, S.D., *A Mediterranean Society: The Jewish Communities of the Arab World as Portrayed in the Documents of the Cairo Geniza* (vol. 5; Berkeley: University of California Press, 1988)

Goldstein, B.R. and D. Pingree, "Horoscopes from the Cairo Geniza," *JNES* 36 (1977): 113-44

———, "More Horoscopes from the Cairo Geniza," *PAPS* 125 (1981): 155-89

Goold, G.P., *Manilius: Astronomica* (LCL 469; Cambridge, Massachusetts: Harvard University Press, 1997)

Gordis, R., "A Document in Code from Qumran – Some Observations," *JSS* 11 (1966): 37-39

Gordon, N., "4Q186 (4QHoroscope)," in *Additional Genres and Unclassified Texts* (eds. D.W. Parry and E. Tov; DSSR 6; Leiden: Brill, 2005), 220-23

Gordon, R., "*Quaedam veritatis umbrae*: Hellenistic Magic and Astrology," in *Conventional Values of the Hellenistic Greeks* (eds. P. Bilde et al.; SHC 7; Aarhus: Aarhus University Press, 1997), 128-58

Goshen-Gottstein, M.H., "Philologische Miszellen zu den Qumrantexten," *RevQ* 2/5 (1959): 43-51

Gourevitch, D., "La gynécologie et l'obstétrique," *ANRW* II 37.3 (1996), 2098-2103

Grabbe, L.L., *Judaism from Cyrus to Hadrian* (London: SCM, 1994)

———, *Judaic Religion in the Second Temple Period: Belief and Practice from the Exile to Yavneh* (London: Routledge, 2000)

Graf, F., "Gestures and Conventions: The Gestures of Roman Actors and Orators," in *A Cultural History of Gesture: From Antiquity to the Present Day* (eds. J.N. Bremmer and H. Roodenburg; Cambridge: Polity, 1991), 36-58

Greenfield, J.C., "Prolegomenon," in *3 Enoch or The Hebrew Book of Enoch* (ed. H. Odeberg; New York: Ktav, 1973), xi-xlvii

Greenfield, J.C. and M. Sokoloff, "Astrological and Related Omen Texts in Jewish Palestinian Aramaic," *JNES* 48 (1989): 201-14

Greenfield, J.C. et al., "4QZodiology and Brontology ar," in *Qumran Cave 4.XXVI: Cryptic Texts and* Miscellanea, *Part 1* (eds. P. Alexander et al.; DJD 36; Oxford: Clarendon, 2000), 259-74

Greenfield, J.C., M.E. Stone and E. Eshel, *The Aramaic Levi Document: Edition, Translation, Commentary* (SVTP 19; Leiden: Brill, 2004)

Grelot, P., "Hénoch et ses écritures," *RB* 82 (1975): 481-500

Groningen, B.A. van, *Short Manual of Greek Palaeography* (Leiden: A.W. Sijthoff, 1940)

Grözinger, K.-E., *Ich bin der Herr, dein Gott: Eine rabbinische Homilie zum Ersten Gebot (PesR 20)* (FJS 2; Frankfurt: Peter Lang, 1976)

———, "Engel III. Judentum," *TRE* 9:586-96

Gruenwald, I., "קטעים חדשים מספרות הכרת־פנים וסדרי־שרטוטין (Further Jewish Physiognomic and Chiromantic Fragments)," *Tarbiz* 40 (1970-1971): 301-19

———, *Apocalyptic and Merkavah Mysticism* (AGJU 14; Leiden: E.J. Brill, 1980)

Guinan, A.K., "Divination," in *The Context of Scripture: Volume I Canonical Compositions from the Biblical World* (eds. W.W. Hallo et al.; Leiden: Brill, 1997), 421-26

———, "A Severed Head Laughed: Stories of Divinatory Interpretation," in *Magic and Divination in the Ancient World* (eds. L. Ciraolo and J. Seidel; AMD 2, Leiden: Brill/STYX, 2002), 7-40

Gundel, H.G., *Weltbild und Astrologie in den griechischen Zauberpapyri* (MBPF 53; Munich: C.H. Beck, 1968)

———, *Zodiakos, Tierkreisbilder im Altertum: Kosmische Bezüge und Jenseitsvorstellungen im antike Alltagsleben* (KAW 54; Mainz: Philipp von Zabern, 1992)

Gundel, H. and R. Böker, "Zodiakos: Der Tierkreis in der Antike," *RE* 2.19 (1972), 462-709

Gundel, H.G. and A. Kehl, "Horoskop," *RAC* 16 (1994), 597-662

Gundel, W., *Sterne und Sternbilder im Glauben des Altertums und der Neuzeit* (Bonn: Kurt Schroeder, 1922; repr., Hildesheim: Georg Olms, 1981)

———, "Individualschicksal, Menschentypen und Berufe in der antiken Astrologie," *JdC* 4 (1927): 135-93

———, *Dekane und Dekansternbilder: Ein Beitrag zur Geschichte der Sternbilder der Kulturvölker* (SBW 19; Glückstadt: J.J. Augustin, 1936)
———, *Neue astrologische Texte des Hermes Trismegistos: Funde und Forschungen auf dem Gebiet der antiken Astronomie und Astrologie* (ABAWPHA 12; Munich: Bayerischen Akademie der Wissenschaften, 1936)
———, "Dekane," *RE*, Supp. 7 (1940), 116-24
Gundel, W. and H.G. Gundel, *Astrologumena: Die astrologische Literatur in der Antike und ihre Geschichte* (SA.B 6; Wiesbaden: Franz Steiner, 1966)
Gundert, B., "Soma and Psyche in Hippocratic Medicine," in *Psyche and Soma: Physicians and Metaphysicians on the Mind-Body Problem from Antiquity to Enlightenment* (eds. J.P. Wright and P. Potter; Oxford: Clarendon, 2000), 13-35
Gunneweg, J. and M. Balla, "Neutron Activation Analysis: Scroll Jars and Common Wares," in *Khirbet Qumrân et 'Aïn Feshkha II: Études d'anthropologie, de physique et de chimie, Studies of Anthropology, Physics and Chemistry* (eds. J.-B. Humbert and J. Gunneweg; NTOA.SA 3; Fribourg: Academic Press Fribourg and Göttingen: Vandenhoeck & Ruprecht, 2003), 3-53
Hachlili, R., *Ancient Jewish Art and Archaeology in the Land of Israel* (HdO: Siebente Abteilung: Kunst und Archäologie 1/2B4; Leiden: E.J. Brill, 1988)
Haines-Eitzen, K., *Guardians of Letters: Literacy, Power, and the Transmitters of Early Christian Literature* (New York: Oxford University Press, 2000)
Hanson, R.S., "Paleography," in *The Paleo-Hebrew Leviticus Scroll (11QpaleoLev)* (D.N. Freedman and K.A. Mathews; Winona Lake, Indiana: American Schools of Oriental Research, 1985)
Harding, J. and L. Alexander, "Dating the Testament of Solomon," posted May 28, 1999, at http://www.st-andrews.ac.uk/%7Ewww_sd/date_tsol.html
Harril, J.A., "Invective against Paul (2 Cor 10:10), the Physiognomics of the Ancient Slave Body, and the Greco-Roman Rhetoric of Manhood," in *Antiquity and Humanity: Essays on Ancient Religion and Philosophy Presented to Hans Dieter Betz on His 70th Birthday* (eds. A. Yarbro Collins and M.M. Mitchell; Tübingen: Mohr Siebeck, 2001), 191-213
Harris, W.V., *Ancient Literacy* (Cambridge, Massachusetts: Harvard University Press, 1989)
Hasebroek, J., *Das Signalement in den Papyrusurkunden* (PH 3; Berlin: VwV, 1921)
Hassin, R. and Y. Trope, "Facing Faces: Studies on the Cognitive Aspects of Physiognomy," *JPSP* 78 (2000): 837-52
Heeßel, N.P., *Babylonisch-assyrische Diagnostik* (AOAT 43; Münster: Ugarit, 2000)
———, "Stein, Pflanze und Holz: Ein neuer Text zur 'medizinischen Astrologie,'" *Or* 74 (2005): 1-22
Heilen, S., "Zur Datierung des Löwenhoroskops auf dem Nemrud Dağ," *forthcoming*
Hempel, C., *The Laws of the Damascus Document: Sources, Tradition and Redaction* (STDJ 29; Leiden: Brill, 1998)
———, "The Qumran Sapiential Texts and the Rule Books," in *The Wisdom Texts from Qumran and the Development of Sapiential Thought* (eds. C. Hempel, A. Lange and H. Lichtenberger; BETL 159; Leuven: Peeters/Leuven University Press, 2002), 277-95
Hengel, M., "Qumrān und der Hellenismus," in *Qumrân: Sa piété, sa théologie et son milieu* (ed. M. Delcor; BETL 46; Paris-Gembloux: Duculot and Leuven: University Press, 1978), 333-72
———, "Jesus als messianischer Lehrer der Weisheit und die Anfänge der Christologie," in *Sagesse et Religion: Colloque de Strasbourg (octobre 1976)* (ed. E. Jacob; Paris: Presses universitaires de France, 1979), 147-88
———, *Judentum und Hellenismus: Studien zu ihrer Begegnung unter besonderer Berücksichtigung Palästinas bis zur Mitte des 2. Jh.s v. Chr.* (3d ed.; WUNT 10; Tübingen: J.C.B. Mohr [Paul Siebeck], 1988)
———, "Qumran and Hellenism," in *Religion in the Dead Sea Scrolls* (eds. J.J. Collins and R.A. Kugler; SDSSRL; Grand Rapids, Michigan: Eerdmans, 2000), 46-56
Hermann, W., "Gedanken zur Geschichte des altorientalischen Beschreibungsliedes," *ZAW* 75 (1963): 176-97

Hett, W.S., *Aristotle XIV: Minor Works* (LCL 307; Cambridge, Massachusetts: Harvard University Press, 1980)
Hezser, C., *Jewish Literacy in Roman Palestine* (TSAJ 81; Tübingen: Mohr Siebeck, 2001)
Himmelfarb, M., *Ascent to Heaven in Jewish and Christian Apocalypses* (New York: Oxford University Press, 1993)
———, "Torah, Testimony, and Heavenly Tablets: The Claim to Authority of the *Book of Jubilees*," in *A Multiform Heritage: Studies on Early Judaism and Christianity in Honor of Robert A. Kraft* (ed. B.G. Wright; HS 24; Atlanta, Georgia: Scholars Press, 1999), 19-29
Hirschfeld, Y., *Qumran in Context: Reassessing the Archaeological Evidence* (Peabody, Massachusetts: Hendrickson, 2004)
Hock, R.F., "The Educational Curriculum in Chariton's *Callirhoe*," in *Ancient Fiction: The Matrix of Early Christian and Jewish Narrative* (eds. J.A. Brant, C.W. Hedrick and C. Shea; SBLSymS 32; Leiden: Brill, 2005), 15-36
Hollander, H.W. and M. de Jonge, *The Testaments of the Twelve Patriarchs: A Commentary* (SVTP 8; Leiden: E.J. Brill, 1985)
Hollenback, G.M., "The Qumran Roundel: An Equatorial Sundial?," *DSD* 7 (2000): 123-29
———, "More on the Qumran Roundel as an Equatorial Sundial," *DSD* 11 (2004): 289-92
Holst, S. and J. Høgenhaven, "Physiognomy and Eschatology: Some More Fragments of 4Q561," *JJS* 57 (2006): 26-43
Honigmann, E., *Die sieben Klimata und die ΠΟΛΕΙΣ ΕΠΙΣΗΜΟΙ: Eine Untersuchung zur Geschichte der Geographie und Astrologie im Altertum und Mittelalter* (Heidelberg: Carl Winter, 1929)
Hoogendijk, F.A.J., "*Parachoresis* von Katökenland," in *Papyri in Honorem Johannis Bingen Octogenarii (P. Bingen)* (eds. R. De Smet, H. Melaerts and C. Saerens; SVBOGLP 5; Leuven: Peeters, 2000), 249-59
Horn, D.G., *The Criminal Body: Lombroso and the Anatomy of Deviance* (New York: Routledge, 2003)
Horsley, R.A., "The Politics of Cultural Production in Second Temple Judea: Historical Context and Political-Religious Relations of the Scribes Who Produced *1 Enoch*, Sirach, and Daniel," in *Conflicted Boundaries in Wisdom and Apocalypticism* (eds. B.G. Wright III and L.M. Wills; SBLSymS 35; Atlanta: Society of Biblical Literature, 2005), 123-45
Horst, P.W. van der, "Seven Months' Children in Jewish and Christian Literature from Antiquity," in *Essays on the Jewish World of Early Christianity* (P.W. van der Horst; NTOA 14; Freiburg, Switzerland: Universitätsverlag and Göttingen: Vandenhoeck & Ruprecht, 1990), 233-47
———, "Sarah's Seminal Emission: Hebrews 11:11 in the Light of Ancient Embryology," in *Hellenism-Judaism-Christianity: Essays on Their Interaction* (P.W. van der Horst; CBET 8; Kampen: Kok Pharos, 1994), 203-23
Hübner, W., *Die Eigenschaften der Tierkreiszeichen in der Antike: Ihre Darstellung und Verwendung unter besonderer Berücksichtigung des Manilius* (SA.B 22; Wiesbaden: Franz Steiner, 1982)
———, *Zodiacus Christianus: Jüdisch-christliche Adaptationen des Tierkreises von der Antike bis zur Gegenwart* (BKP 144; Meisenheim: Anton Hain, 1983)
———, "Manilius als Astrologe und Dichter," *ANRW* II 32.1 (1984), 126-320
———, *Die Begriffe 'Astrologie' und 'Astronomie' in der Antike: Wortgeschichte und Wissenschaftsthematik, mit einer Hypothese zum Terminus 'Quadrivium'* (AWLMAGSK 7; Wiesbaden: Franz Steiner, 1989)
———, *Grade und Gradbezirke der Tierkreiszeichen: Der Anonyme Traktat De Stellis Fixis, In Quibus Gradibus Oriuntur Signorum: I Quellenkritische Edition, II Kommentar* (SWC; Stuttgart: B.G. Teubner, 1995)
———, *Claudius Ptolemaeus: ΑΠΟΤΕΛΕΣΜΑΤΙΚΑ* (vol. 3.1; Teubner; Stuttgart: B.G. Teubner, 1998)
———, "The Ptolemaic View of the Universe," *GRBS* 41 (2000): 59-93
———, "Zur Verwendung und Umschreibung des Terminus ὡροσκόπος in der astrologischen Lehrdichtung der Antike," *Mene* 1 (2001): 219-38

———, "Pleiaden," *DNP* 9 (2002), 1127-28
———, "Teukros [4]," *DNP* 12/1 (2002), 205
———, "Der *descensus* als ordnendes Prinzip in der 'Naturalis historia' des Plinius," in *Die Enzyklopädie im Wandel vom Hochmittelalter bis zur frühen Neuzeit* (ed. C. Meier; MMS 78; Munich: Wilhelm Fink, 2002), 25-41
———, "Der Tierkreis in der Psalmenmagie," in *Ritual und Poesie: Formen und Orte religiöser Dichtung im Alten Orient, im Judentum und im Christentum* (ed. E. Zenger; HBS 36; Freiburg: Herder, 2003), 293-319
———, "Sulla's Horoscope? (Firm. math. 6,31,1)," in *Horoscopes and Public Spheres: Essays on the History of Astrology* (eds. G. Oestmann, H.D. Rutkin and K. von Stuckrad; RelSoc 42; Berlin: Walter de Gruyter, 2005), 13-35
———, "Δωδεκατημόριον," in *Corona Coronaria: Festschrift für Hans-Otto Kröner zum 75. Geburtstag* (eds. S. Harwardt and J. Schwind; Spudasmata 102; Hildesheim: Georg Holms, 2005), 189-217
Hübsch, G., *Die Personalangaben als Identifizierungsvermerke im Recht der gräko-ägyptischen Papyri* (BJA 20; Berlin: Duncker and Humblot, 1968)
Humbert, J.B. and A. Chambon (eds.), *Fouilles de Khirbet Qumrân et de Aïn Feshkha I: Album de photographies, Répertoire du fonds photographique, Synthèse des notes de chantier du Père Roland de Vaux OP* (NTOA.SA 1, Fribourg, Switzerland: Éditions Universitaires and Göttingen: Vandenhoeck & Ruprecht, 1994)
Hunger, H., *Astrological Reports to Assyrian Kings* (SAA 8; Helsinki: Helsinki University Press, 1992)
Hunger, H. and D. Pingree (eds.), *MUL.APIN: An Astronomical Compendium in Cuneiform* (AfO.B 24; Horn, Austria: Ferdinand Berger, 1989)
———, *Astral Sciences in Mesopotamia* (HdO I/44; Leiden: Brill, 1999)
Ilan, T., *Lexicon of Jewish Names in Late Antiquity: Part I, Palestine 330 BCE – 200 CE* (TSAJ 91; Tübingen: Mohr Siebeck, 2002)
Irby-Massie, G.L. and P.T. Keyser, *Greek Science of the Hellenistic Era: A Sourcebook* (London: Routledge, 2002)
Isaac, B., *The Invention of Racism in Classical Antiquity* (Princeton: Princeton University Press, 2004)
Jackson, H.M., "Notes on the Testament of Solomon," *JSJ* 19 (1988) 19-60
Jax, K., "Zur literarischen und amtlichen Personenbeschreibung," *Klio* 29 (1936): 151-63
Jeyes, U., *Old Babylonian Extispicy: Omen Texts in the British Museum* (UNHAII 64; Istanbul: Nederlands Historisch-Archeologisch Instituut, 1989)
———, "Assurbanipal's *bārûtu*," in *Assyrien im Wandel der Zeiten: XXXIX*[e] *Rencontre Assyriologique Internationale Heidelberg 6.-10. Juli 1992* (eds. H. Waetzoldt and H. Hauptman; HSAO 6; Heidelberg: Heidelberger Orientverlag, 1997), 61-65
Johnston, S.I., "The *Testament of Solomon* from Late Antiquity to the Renaissance," in *The Metamorphosis of Magic from Late Antiquity to the Early Modern Period* (eds. J.N. Bremmer and J.R. Veenstra; GSCC 1, Leuven: Peeters, 2002), 35-49
Jokiranta, J.M., "'Sectarianism' of the Qumran 'Sect': Sociological Notes," *RevQ* 78/20 (2001): 223-39
———, *Identity on a Continuum: Constructing and Expressing Sectarian Social Identity in Qumran* Serakhim *and* Persharim (Diss. Helsinki, 2005)
Jones, A., "Evidence for Babylonian Arithmetical Schemes in Greek Astronomy," in *Die Rolle der Astronomie in den Kulturen Mesopotamiens: Beiträge zum 3. Grazer Morgenländischen Symposion (23.-27. September 1991)* (ed. H.D. Galter; GMS 3; Graz: GrazKult, 1993), 77-94
———, "Studies in the Astronomy of the Roman Period II: Tables for Solar Longitude," *Centaurus* 39 (1997): 211-29
———, *Astronomical Papyri from Oxyrhynchus (P. Oxy. 4133-4300a)* (2 vols; MAPS 233; Philadelphia: American Philosophical Society, 1999)
Jonge, M. de, *Pseudepigrapha of the Old Testament as Part of Christian Literature: The Case of the* Testaments of the Twelve Patriarchs *and the Greek* Life of Adam and Eve (SVTP 18; Leiden: Brill, 2003)

Jouanna, J., *Hippocrates* (trans. M.B. DeBevoise; Medicine and Culture; Baltimore, Maryland: Johns Hopkins University Press, 1999)
Kamlah, E., *Die Form der katalogischen Paränese im Neuen Testament* (WUNT 7; Tübingen: J.C.B. Mohr [Paul Siebeck], 1964)
Keulen, W., "*Ad amussim congruentia*: Measuring the Intellectual in Apuleius," in *Lectiones Scrupulosae: Essays on the Text and Interpretation of Apuleius'* Metamorphoses (eds. W. Keulen, R. Nauta and S. Panayotakis; ANSupp 6; Groningen: Barkhuis, 2006), 168-202
Kidd, D. (ed.), *Aratus: Phaenomena* (CCTC 34; Cambridge: Cambridge University Press, 1997)
King, H., "Making a Man: Becoming Human in Early Greek Medicine," in *The Human Embryo: Aristotle and the Arabic and European Traditions* (ed. G.R. Dunstan; Exeter: University of Exeter Press, 1990), 10-19
———, "Geburt: Medizinisch," *DNP* 4 (1998), 835
Kinnier Wilson, J.V., "Two Medical Texts from Nimrud," *Iraq* 18 (1956): 130-46
———, "The Nimrud Catalogue of Medical and Physiognomic Omina," *Iraq* 24 (1962): 52-62
Kippenberg, H.G. and G.G. Stroumsa (eds.), *Secrecy and Concealment: Studies in the History of Mediterranean and Near Eastern Religions* (SHR 65; Leiden: E.J. Brill, 1995)
Kister, M., "Three Unknown Hebrew Words in Newly-Published Texts from Qumran," *Lešonénu* 63 (2000-2001): 35-40
———, "Wisdom Literature and Its Relation to Other Genres: From Ben Sira to *Mysteries*," in *Sapiential Perspectives: Wisdom Literature in Light of the Dead Sea Scrolls: Proceedings of the Sixth International Symposium of the Orion Center for the Study of the Dead Sea Scrolls and Associated Literature, 20-22 May, 2001* (eds. J.J. Collins and G.E. Sterling; STDJ 51; Leiden: Brill, 2004), 13-47
———, "Physical and Metaphysical Measurements Ordained by God in the Literature of the Second Temple Period," in *Reworking the Bible: Apocryphal and Related Texts at Qumran: Proceedings of a Joint Symposium by the Orion Center for the Study of the Dead Sea Scrolls and Associated Literature and the Hebrew University Institute for Advanced Studies Research Group on Qumran, 15-17 January, 2002* (eds. E.G. Chazon, D. Dimant and R.A. Clements; STDJ 58; Leiden: Brill, 2005), 153-76
Kleinknecht, H., "πνεῦμα im Griechischen," *TWNT* 6:333-57
Klibansky, R., E. Panofsky and F. Saxl, *Saturn and Melancholy: Studies in the History of Natural Philosophy, Religion, and Art* (London: Thomas Nelson, 1964; repr., Nendeln, Liechtenstein: Kraus, 1979)
Klutz, T.E., "The Archer and the Cross: Chorographic Astrology and Literary Design in the *Testament of Solomon*," in *Magic in the Biblical World: From the Rod of Aaron to the Ring of Solomon* (ed. T.E. Klutz; JSNTSup 245; London: T&T Clark, 2003), 219-44
———, *Rewriting the* Testament of Solomon*: Tradition, Conflict and Identity in a Late Antique Pseudepigraphon* (LSTS 53; London T&T Clark, 2005)
Koch, J., *Neue Untersuchungen zur Topographie des babylonischen Fixsternhimmels* (Wiesbaden: Otto Harrassowitz, 1989)
Koch-Westenholz, U., *Mesopotamian Astrology: An Introduction to Babylonian and Assyrian Celestial Divination* (CNIP 19; Copenhagen: Museum Tusculanum, 1995)
———, "Old Babylonian Extispicy Reports," in *Mining the Archives: Festschrift for Christopher Walker on the Occasion of His 60th Birthday* (ed. C. Wunsch; BA 1; Dresden: ISLET, 2002), 131-45
Koep, L., *Das himmlische Buch in Antike und Christentum: Eine religionsgeschichtliche Untersuchung zur altchristlichen Bildersprache* (Theophaneia 8; Bonn: Peter Hanstein, 1952)
Kollmann, B., "Göttliche Offenbarung magisch-pharmakologischer Heilkunst im Buch Tobit," *ZAW* 106 (1994): 289-99
Konstan, D., "Conventional Values of the Hellenistic Greeks: The Evidence from Astrology," in *Conventional Values of the Hellenistic Greeks* (eds. P. Bilde et al.; SHC 7; Aarhus: Aarhus University Press, 1997), 159-76

Kooten, G.H. van, "Enoch, the 'Watchers,' Seth's Descendants and Abraham as Astronomers: Jewish Applications of the Greek Motif of the First Inventor (300 BCE-CE 100)," in *Recycling Biblical Figures: Papers Read at a NOSTER Colloquium in Amsterdam, 12-13 May, 1997* (eds. A. Brenner and J.W. van Henten; STAR 1; Leiden: Deo, 1999), 292-316

Kraus, F.R., *Die physiognomischen Omina der Babylonier* (MVAG 40/2; Leipzig: J.C. Hinrichs, 1935)

———, "Ein Sittenkanon in Omenform," *ZA* 43 (1936): 77-113

———, "Babylonische Omina mit Ausdeutung der Begleiterscheinungen des Sprechens," *AfO* 11 (1936-1937): 219-30

———, *Texte zur babylonischen Physiognomatik* (AfO.B 3; 1939; repr., Osnabrück: Biblio, 1967)

Kroll, W., F. Skutsch and K. Ziegler (eds.), *Firmicus Maternus: Matheseos Libri VIII* (vol. 2; Teubner; Leipzig: B.G. Teubner, 1913)

Kuhn, K.G., "Die Sektenschrift und die iranische Religion," *ZThK* 49 (1952): 296-316

Kuhrt, A. and S. Sherwin-White (eds.), *Hellenism in the East: The Interaction of Greek and Non-Greek Civilizations from Syria to Central Asia after Alexander* (Berkeley: University of California Press, 1987)

Kvalvaag, R.V., "The Spirit in Human-Beings in Some Qumran Non-Biblical Texts," in *Qumran Between the Old and New Testaments* (eds. F.H. Cryer and T.L. Thompson; JSOTSup 290; Sheffield: Sheffield Academic Press, 1998), 159-80

Kvanvig, H.S., *Roots of Apocalyptic: The Mesopotamian Background of the Enoch Figure and the Son of Man* (WMANT 61; Neukirchen-Vluyn: Neukirchener Verlag, 1988)

Lambert, W.G., "A Catalogue of Texts and Authors," *JCS* 16 (1962): 59-77

———, "Enmeduranki and Related Matters," *JCS* 21 (1967): 126-38

———, "The Qualifications of Babylonian Diviners," in *Festschrift für Rykle Borger zu seinem 65. Geburtstag am 24. Mai 1994: tikip santakki mala bašmu...* (ed. S.M. Maul; CM 10; Groningen: Styx, 1998), 141-58

Landsberger, B. and M. Civil (eds.), *The Series ḪAR-ra = ḫubullu: Tablet XV and Related Texts* (MSL 9; Rome: Pontificium Institutum Biblicum, 1967)

Langdon, S., "Miscellanea Assyriaca IV," *Babyloniaca* 7 (1913-1923): 230-36

Lange, A., *Weisheit und Prädestination: Weisheitliche Urordnung und Prädestination in den Textfunden von Qumran* (STDJ 18; Leiden: E.J. Brill, 1995)

———, "Wisdom and Predestination in the Dead Sea Scrolls," *DSD* 2 (1995): 340-54

———, "Kognitives *lqḥ* in Sap A, im Tᵉnak und Sir," *ZAH* 9 (1996): 190-95

———, "The Essene Position on Magic and Divination," in *Legal Texts and Legal Issues: Proceedings of the Second Meeting of the International Organization for Qumran Studies Cambridge 1995: Published in Honour of Joseph M. Baumgarten* (eds. M. Bernstein, F. García Martínez and J. Kampen; STDJ 23, Leiden: E.J. Brill, 1997), 377-435

Lansing Hicks, R., "*Delet* and *Megillah*: A Fresh Approach to Jeremiah xxxvi," *VT* 33 (1983): 46-66

Lavater, J.C., *Physiognomische Fragmente zur Beförderung der Menschenkenntniß und Menschenliebe: Erster Versuch* (Leipzig und Winterthur: M.G. Weidmanns Erben und Reich, und Heinrich Steiner und Compagnie, 1775)

Le Boeuffle, A., *Les noms latins d'astres et de constellations* (Paris: Les Belles Lettres, 1977)

———, *Astronomie, astrologie: lexique latin* (Paris: Picard, 1987)

Lee Too, Y. (ed.), *Education in Greek and Roman Antiquity* (Leiden: Brill, 2001)

Lehmann, M.R., "New Light on Astrology in Qumran and the Talmud," *RevQ* 8/32 (1975): 599-602

Lesky, E. and J.H. Waszink, "Embryologie," *RAC* 4 (1959), 1236-40

———, "Empfängnis," *RAC* 4 (1959), 1245-55

Levison, J.R., "The Angelic Spirit in Early Judaism," in *Society of Biblical Literature 1995 Seminar Papers* (ed. E.H. Lovering, Jr.; *SBLSP* 34; Atlanta, Georgia: Scholars Press, 1995), 464-93

Licht, J., "An Analysis of the Treatise on the Two Spirits in DSD," in *Aspects of the Dead Sea Scrolls* (eds. C. Rabin and Y. Yadin; ScrHier 4; Jerusalem: Magnes, 1958), 88-100

———, "שוקיים סימן לבחירה (Legs as Signs of Election)," *Tarbiz* 35 (1965-1966): 18-26
Lichtenberger, H., *Studien zum Menschenbild in Texten aus der Qumrangemeinde* (SUNT 15; Göttingen: Vandenhoeck & Ruprecht, 1980)
Lieberman, S.J., "Canonical and Official Cuneiform Texts: Towards an Understanding of Assurbanipal's Personal Tablet Collection," in *Lingering over Words: Studies in Ancient Near Eastern Literature in Honor of William L. Moran* (eds. T. Abusch, J. Huehnergard and P. Steinkeller; HSS 37; Atlanta, Georgia: Scholars Press, 1990), 305-36
Lim, T.H. (ed. in consultation with P.S. Alexander), *The Dead Sea Scrolls Electronic Reference Library: Volume I* (Oxford: Oxford University Press and Leiden: Brill, 1997)
Lisān al-ʿArab (Beirut: Dār Iḥyā' al-Turāṯ al-ʿArabī, 1988)
Livingstone, A., *Mystical and Mythological Explanatory Works of Assyrian and Babylonian Scholars* (Oxford: Clarendon, 1986)
Lloyd, G.E.R., "The Hot, the Cold, the Dry and the Wet in Greek Philosophy," *JHS* 84 (1964): 92-106
———, *Ancient Worlds, Modern Reflections: Philosophical Perspectives on Greek and Chinese Science and Culture* (Oxford: Clarendon, 2004)
Long, A.A., "Astrology: Arguments Pro and Contra," in *Science and Speculation: Studies in Hellenistic Theory and Practice* (eds. J. Barnes et al.; Cambridge: Cambridge University Press, 1982), 165-92
Long, P.O., *Openness, Secrecy, Authorship: Technical Arts and the Culture of Knowledge from Antiquity to the Renaissance* (Baltimore: Johns Hopkins University Press, 2001)
Loveday, T. and E.S. Forster, "Physiognomonics," in *The Complete Works of Aristotle: The Revised Oxford Translation* (ed. J. Barnes; vol. 1; Princeton: Princeton University Press, 1995), 1237-50
Luttikhuizen, G.P., *The Revelation of Elchasai: Investigations into the Evidence for a Mesopotamian Jewish Apocalypse of the Second Century and its Reception by Judeo-Christian Propagandists* (TSAJ 8; Tübingen: J.C.B. Mohr [Paul Siebeck], 1985)
Mach, M., *Entwicklungsstadien des jüdischen Engelglaubens in vorrabbinischer Zeit* (TSAJ 34; Tübingen: J.C.B. Mohr [Paul Siebeck], 1992)
———, "Angels," in *Encyclopedia of the Dead Sea Scrolls* (eds. L.H. Schiffman and J.C. VanderKam; New York: Oxford University Press, 2000), 24-27
Magdalino, P., "The Byzantine Reception of Classical Astrology," in *Literacy, Education and Manuscript Transmission in Byzantium and Beyond* (eds. C. Holmes and J. Waring; MMed 42; Leiden: Brill, 2002), 33-57
Maier, G., *Mensch und freier Wille: Nach den jüdischen Religionsparteien zwischen Ben Sira und Paulus* (WUNT 12; Tübingen: J.C.B. Mohr [Paul Siebeck], 1971)
Maier, J., *Die Qumran-Essener: Die Texte vom Toten Meer: Band II: Die Texte der Höhle 4* (vol. 2; Munich: Ernst Reinhardt, 1995)
———, "Wiles of the Wicked Woman," in *Encyclopedia of the Dead Sea Scrolls* (eds. L.H. Schiffman and J.C. VanderKam; New York: Oxford University Press, 2000), 976
Malherbe, A.J., "A Physical Description of Paul," in *Paul and the Popular Philosophers* (A.J. Malherbe; Minneapolis: Fortress, 1989), 170-75 (originally published in *HTR* 79 [1986]: 170-75)
Malina, B.J. and J.H. Neyrey, *Portraits of Paul: An Archaeology of Ancient Personality* (Louisville, Kentucky: Westminster John Knox, 1996)
Mansfeld, J., "Galen, Papias, and Others on Teaching and Being Taught," in *Things Revealed: Studies in Early Jewish and Christian Literature in Honor of Michael E. Stone* (eds E.G. Chazon, D. Satran and R.E. Clements; JSJSup 89; Leiden: Brill, 2004), 317-29
Marrou, H.-I., *Histoire de l'éducation dans l'Antiquité* (2 vols; Paris: Seuil, 1948; repr., 1981)
Martin, D.B., *The Corinthian Body* (New Haven: Yale University Press, 1995)
Massyngbaerde Ford, J., "The Physical Features of the Antichrist," *JSP* (1996): 23-41
Maul, S.M., *Zukunftsbewältigung: Eine Untersuchung altorientalischen Denkens anhand der babylonisch-assyrischen Löserituale (Namburbi)* (BaF 18; Mainz: Philipp von Zabern, 1994)
———, "Omina und Orakel. A. Mesopotamien," *RlA* 10.1/2 (2003), 45-88

May, H.G., "Cosmological Reference in the Qumran Doctrine of the Two Spirits and in Old Testament Imagery," *JBL* 82 (1963): 1-14
McCown, C.C. (ed.), *The Testament of Solomon* (UNT 9; Leipzig: J.C. Hinrichs, 1922)
McEwan, G.J.P., *Priest and Temple in Hellenistic Babylonia* (FAS 4; Wiesbaden: Franz Steiner, 1981)
McEwen, I.K., *Vitruvius: Writing the Body of Architecture* (Cambridge, Massachusetts: MIT Press, 2003)
Mébarki, F. and É. Puech (eds.), *Les Manuscrits de la mer Morte* (Rodez: Rouergue, 2002)
Mendels, D., "Hellenistic Utopia and the Essenes," in *Identity, Religion and Historiography: Studies in Hellenistic History* (D. Mendels; JSPSup 24; Sheffield: Sheffield Academic Press, 1998), 420-39 (originally published in *HTR* 72 [1979]: 207-22)
Mengozzi, A., *Trattato di Sem e altri testi astrologici* (TVOa 7, LSc 1; Brescia: Paideia, 1997)
Mesk, J., "Die Beispiele in Polemons Physiognomonik," *WS* 50 (1932): 51-67
Metso, S., *The Textual Development of the Qumran Community Rule* (STDJ 21; Leiden: Brill, 1997)
Mettinger, T.N.D., *King and Messiah: The Civil and Sacral Legitimation of the Israelite Kings* (ConBOT 8; Lund: CWK Gleerup, 1976)
Michaud, H., "Un mythe zervanite dans un des manuscrits de Qumrân," *VT* 5 (1955): 137-47
Milik, J.T., *Ten Years of Discovery in the Wilderness of Judaea* (2d ed.; SBT 26; trans. J. Strugnell; London: SCM Press, 1963)
———, "Problèmes de la littérature Hénochique à la lumière des fragments araméens de Qumrân," *HTR* 64 (1971): 333-78
———, *The Books of Enoch: Aramaic Fragments of Qumrân Cave 4* (Oxford: Clarendon, 1976)
———, "Écrits préesséniens de Qumrân: d'Hénoch à Amram," in *Qumrân: Sa piété, sa théologie et son milieu* (ed. M. Delcor; BETL 46; Paris-Gembloux: Duculot and Leuven: Leuven University Press, 1978), 91-106
Millard, A., *Reading and Writing in the Time of Jesus* (BibSem 69; Sheffield; Sheffield Academic Press, 2000)
Misener, G., "Loxus, Physician and Physiognomist," *CP* 18 (1923): 1-22
———, "Iconistic Portraits," *CP* 19 (1924): 97-123
Momigliano, A., *Alien Wisdom: The Limits of Hellenization* (Cambridge: Cambridge University Press, 1975)
Monat, P. (ed.), *Firmicus Maternus: Mathesis* (vol. 3; Budé; Paris: Les Belles Lettres, 1997)
Moore, R.D., "Personification of the Seduction of Evil: 'The Wiles of the Wicked Woman,'" *RevQ* 10/40 (1981): 505-19
Moormann, E.M. and M.J. Versluys, "The Nemrud Dağ Project: First Interim Report," *BABesch* 77 (2002): 73-111
Morgenstern, M., "The Meaning of בית מולדים in the Qumran Wisdom Texts," *JJS* 51 (2000): 141-44
Müller, I.W., *Humoralmedizin: Physiologische, pathologische und therapeutische Grundlagen der galenistischen Heilkunst* (Heidelberg: Karl F. Haug, 1993)
Müller-Kessler, C., "Aramäische Beschwörungen und astronomische Omina in nachbabylonischer Zeit: Das Fortleben mesopotamischer Kultur im Vorderen Orient," in *Babylon: Focus mesopotamischer Geschichte, Wiege früher Gelehrsamkeit, Mythos in der Moderne: 2. Internationales Colloquium der Deutschen Orient-Gesellschaft, 24.-26. März 1998 in Berlin* (ed. J. Renger; CDOG 2; Saarbrücken: SDV, 1999), 427-43
———, "A Charm Against Demons of Time," in *Mining the Archives: Festschrift for Christopher Walker on the Occasion of His 60th Birthday* (ed. C. Wunsch; BA 1; Dresden: ISLET, 2002), 183-89
Murphy-O'Connor, J., "La genèse littéraire de la *Règle de la Communauté*," *RB* 76 (1969): 528-49
Naveh, J., "Lamp Inscriptions and Inverted Writing," *IEJ* 38 (1988): 36-43
———, "Fragments of an Aramaic Magic Book from Qumran," *IEJ* 48 (1998): 252-61
———, "Hebrew Graffiti from the First Temple Period," *IEJ* 51 (2001): 194-207
Nebe, G.W., "אבר in 4Q186," *RevQ* 8/30 (1973): 265-66

Ness, L., *Astrology and Judaism in Late Antiquity* (Diss. Oxford, Ohio, 1990), http://www.smoe.org/arcana/diss4.html
Neugebauer, O., "The Rising Times in Babylonian Astronomy," *JCS* 7 (1953): 100-2
———, "Melothesia and Dodecatemoria," AnBib 12 (SBO 3, *OrAnt*; Rome: Pontificio Istituto Biblico, 1959), 270-75
———, "*Decem Tulerunt Fastidia Menses*," *AJP* 84 (1963): 64-65
———, "Notes on Ethiopic Astronomy," *Or* 33 (1964): 50-61
———, *The Exact Sciences in Antiquity* (2d ed.; New York: Dover Publications, 1969)
———, *A History of Ancient Mathematical Astronomy* (3 vols; SHMPS 1; Berlin: Springer, 1975)
———, *Ethiopic Astronomy and Computus* (ÖAWPH.SB 347; Vienna: Österreichischen Akademie der Wissenschaften, 1979)
———, "The 'Astronomical' Chapters of the Ethiopic Book of Enoch (72 to 82)," in The Book of Enoch *or* 1 Enoch*: A New English Edition* (M. Black; SVTP 7; Leiden: E.J. Brill, 1985), 386-414
Neugebauer, O. and H.B. van Hoesen, *Greek Horoscopes* (MAPS 48; Philadelphia: American Philosophical Society, 1959)
Neugebauer, O. and A. Sachs, "The 'Dodekatemoria' in Babylonian Astrology," *AfO* 16 (1952-53): 65-66
Newsom, C.A., "'Sectually Explicit' Literature from Qumran," in *The Hebrew Bible and its Interpreters* (eds. W.H. Propp, B. Halpern and D.N. Freedman; BJSUCSD 1; Winona Lake, Indiana: Eisenbrauns, 1990), 167-87
———, "Angels: Old Testament," *ABD* 1:248-53
———, "4QShirot 'Olat HaShabbath[d]," in *Qumran Cave 4.VI: Poetical and Liturgical Texts, Part 1* (eds. E. Eshel et al.; DJD 11; Oxford: Clarendon, 1998), 253-92
———, *The Self as Symbolic Space: Constructing Identity and Community at Qumran* (STDJ 52; Leiden: Brill, 2004)
Nickelsburg, G.W.E., *1 Enoch 1: A Commentary on the Book of 1 Enoch, Chapters 1-36; 81-108* (Hermeneia; Minneapolis: Fortress, 2001)
Nickelsburg, G.W.E. and J.C. VanderKam, *1 Enoch: A New Translation* (Minneapolis: Fortress, 2004)
Nilsson, M.P., *Geschichte der griechischen Religion: Zweiter Band, Die hellenistische und römische Zeit* (2d ed.; vol. 2; HdA 5/2; Munich: C.H. Beck, 1961)
North, J.D., *Horoscopes and History* (WIST 13; London: The Warburg Institute University of London, 1986)
Nötscher, F., "Himmlische Bücher und Schicksalsglaube in Qumran," *RevQ* 1/3 (1959): 405-11
Nougayrol, J., "La divination babylonienne," in *La divination: Tome Premier* (eds. A. Caquot and M. Leibovici; Paris: Presses Universitaires de France, 1968), 25-81
Nutton, V., "Adamantios," *DNP* 1 (1996), 103
———, "Säftelehre," *DNP* 10 (2001), 1208-10
———, *Ancient Medicine* (London: Routledge, 2004)
Oelsner, J., *Materialien zur babylonischen Gesellschaft und Kultur in Hellenistischer Zeit* (Assyriologia 7; Budapest: Eötvös University, 1986)
Oppenheim, A.L., "Divination and Celestial Observation in the Last Assyrian Empire," *Centaurus* 14 (1969): 97-135
———, "A Babylonian Diviner's Manual," *JNES* 33 (1974): 197-220
———, *Ancient Mesopotamia: Portrait of a Dead Civilization* (Revised Edition Completed by E. Reiner; Chicago: University of Chicago Press, 1977)
Orth, W., "Astrologie und Öffentlichkeit in der frühen römischen Kaiserzeit," in *Kommunikation in politischen und kultischen Gemeinschaften: Stätten und Formen der Kommunikation im Altertum V* (eds. G. Binder and K. Ehlich; BAC 24; Trier: Wissenschaftlicher Verlag Trier, 1996), 99-132
Osten-Sacken, P. von der, *Gott und Belial: Traditionsgeschichtliche Untersuchungen zum Dualismus in den Texten aus Qumran* (SUNT 6; Göttingen: Vandenhoeck & Ruprecht, 1969)

Parpola, S., "Mesopotamian Astrology and Astronomy as Domains of the Mesopotamian 'Wisdom,'" in *Die Rolle der Astronomie in den Kulturen Mesopotamiens: Beiträge zum 3. Grazer Morgenländischen Symposion (23.-27. September 1991)* (ed. H.D. Galter; GMS 3; Graz: GrazKult, 1993), 47-59

———, *Letters from Assyrian and Babylonian Scholars* (SAA 10; Helsinki: Helsinki University Press, 1993)

Parsons, M.C., "'Short in Stature': Luke's Physical Description of Zacchaeus," *NTS* 47 (2001): 50-57

Paul, S.M., "Heavenly Books and the Book of Life," *JANESCU* 5 (1973): 345-53

Pedersén, O., *Archives and Libraries in the Ancient Near East 1500-300 B.C.* (Bethesda, Maryland: CDL, 1998)

Peck, A.L., *Aristotle: Generation of Animals* (LCL 366; Cambridge, Massachusetts: Harvard University Press, 1963)

Penny, D.L. and M.O. Wise, "By the Power of Beelzebub: An Aramaic Incantation Formula from Qumran," *JBL* 113 (1994): 627-50

Perrin, B., *Plutarch's Lives I: Theseus and Romulus, Lycurgus and Numa, Solon and Publicola* (LCL 46; Cambridge, Massachusetts: Harvard University Press, 1967)

Pestalozzi, K. and H. Weigelt (eds.), *Das Antlitz Gottes im Antlitz des Menschen: Zugänge zu Johann Kaspar Lavater* (AGP 31; Göttingen: Vandenhoeck & Ruprecht, 1994)

Pfann, S., "4Qpap cryptA Midrash Sefer Moshe," in *Qumran Cave 4.XXV: Halakhic Texts* (eds. J. Baumgarten et al.; DJD 35; Oxford: Clarendon Press, 1999), 1-24

———, "The Writings in Esoteric Script from Qumran," in *The Dead Sea Scrolls Fifty Years After Their Discovery, 1947-1997* (eds. L.H. Schiffman, E. Tov and J.C. VanderKam; Jerusalem: Israel Exploration Society, 2000), 177-90

———, "Introduction," in *Qumran Cave 4.XXVI: Cryptic Texts and* Miscellanea, *Part 1* (eds. P. Alexander et al.; DJD 36; Oxford: Clarendon Press, 2000), 515-46

Pfann, S.J. and M. Kister, "4QcryptA Words of the Maskil to All Sons of Dawn," in *Qumran Cave 4.XV: Sapiential Texts, Part 1* (eds. T. Elgvin et al.; DJD 20; Oxford: Clarendon Press, 1997), 1-30

Philonenko, M., "Deux horoscopes Qoumrâniens: Identification des personnages," *RHPR* 65/1 (1985): 61-66

———, "La doctrine qoumrânienne des deux Esprits: Ses origines iraniennes et ses prolongements dans le judaïsme essénien et le christianisme antique," in *Apocalyptique iranienne et dualisme qoumrânien* (eds. G. Widengren, A. Hultgård and M. Philonenko; RI 2; Paris: Adrien Maisonneuve, 1995), 163-211

Pingree, D., "Astrology," in *Dictionary of the History of Ideas: Studies of Selected Pivotal Ideas* (vol. 1; ed. P.P. Wiener; New York: Charles Scribner's Sons, 1973), 118-26

———, *Hephaestio Thebanus: Apotelesmatica* (vol.1; Teubner; Leipzig: B.G. Teubner, 1973)

———, *The Yavanajātaka of Sphujidhvaja* (vol. 2; HOS 48; Cambridge, Massachusetts: Harvard University Press, 1978)

———, *Vettius Valens: Anthologiae* (Teubner; Leipzig: B.G. Teubner, 1986)

———, *From Astral Omens to Astrology: From Babylon to Bīkāner* (SOR 78; Rome: Istituto Italiano per l'Africa e l'Oriente, 1997)

Pongratz-Leisten, B., *Herrschaftswissen in Mesopotamien: Formen der Kommunikation zwischen Gott und König im 2. und 1. Jahrtausend v. Chr.* (SAAS 10; Helsinki: The Neo-Assyrian Text Corpus Project, 1999)

Popović, M., "De archeologie van Qumran," in *Fragmenten uit de woestijn: De Dode-Zeerollen opnieuw bekeken* (eds. F. García Martínez and E. Tigchelaar; Zoetermeer: Meinema, 2003), 48-77

———, "A Note on the Reading of שמונה and עמוד השני in *4Q186 2 i 7*," *RevQ* 21/84 (2004): 635-41

———, "Physiognomic Knowledge in Qumran and Babylonia: Form, Interdisciplinarity, and Secrecy," *DSD* 13 (2006): 150-76

———, "The Human Body, Astrology, and Writing in Code: Physiognomic Divination in the Dead Sea Scrolls," paper read at the CNRS Symposium *Divination et écriture, écriture*

de la divination, Paris, December 9, 2005 (the proceedings are to be published by J.-J. Glassner)
Pouilly, J., *La Règle de la Communauté de Qumrân: Son évolution littéraire* (CahRB 17; Paris: Gabalda, 1976)
Preisendanz, K., "Ein Wiener Papyrusfragment zum Testamentum Salomonis," *Eos* 48/3 (1956): 161-67
A Preliminary Concordance to the Hebrew and Aramaic Fragments from Qumrân Caves II-X, Including Especially the Unpublished Material from Cave IV, Printed from a Card Index Prepared by R.E. Brown, S.S., J.A. Fitzmyer, S.J., W.G. Oxtoby, and J. Teixidor, Prepared and Arranged for Printing by H.-P. Richter (5 vols.; *editorum in usum*; Göttingen: Privately Printed, 1988)
Puech, É. (ed.), *Qumrân Grotte 4.XVIII: Textes Hébreux (4Q521-4Q528, 4Q576-4Q579)* (DJD 25; Oxford: Clarendon, 1998)
———, *Qumrân Grotte 4.XXII: Textes araméens, Première partie, 4Q529-549* (DJD 31; Oxford: Clarendon, 2001)
Rackham, H., *Pliny: Natural History* (vol. 1; LCL 330; Cambridge, Massachusetts: Harvard University Press, 1949)
Rau, E., *Kosmologie, Eschatologie und die Lehrautorität Henochs: Traditions- und formgeschichtliche Untersuchungen zum äth. Henochbuch und zu verwandten Schriften* (Diss. Hamburg, 1974)
Reardon, B.P., *Courant littéraires grecs des IIe et IIIe siècles après J.-C.* (Paris: Les Belles Lettres, 1971)
Reed, A.Y., "Abraham as Chaldean Scientist and Father of the Jews: Josephus, *Ant.* 1.154-168, and the Greco-Roman Discourse about Astronomy/Astrology," *JSJ* 35 (2004): 119-58
———, "Heavenly Ascent, Angelic Descent, and the Transmission of Knowledge in 1 Enoch 6–16," in *Heavenly Realms and Earthly Realities in Late Antique Religions* (eds. R.S. Boustan and A.Y. Reed; Cambridge: Cambridge University Press, 2004), 47-66
———, *Fallen Angels and the History of Judaism and Christianity: The Reception of Enochic Literature* (Cambridge: Cambridge University Press, 2005)
Regev, E., "Comparing Sectarian Practice and Organization: The Qumran Sects in Light of the Regulations of the Shakers, Hutterites, Mennonites and Amish," *Numen* 51 (2004): 146-81
Reiner, E., "The Uses of Astrology," *JAOS* 105 (1985): 589-95
———, "Two Babylonian Precursors of Astrology," *NABU* (1993): 21-22
———, *Astral Magic in Babylonia* (TAPS 85/4; Philadelphia: American Philosophical Society, 1995)
———, "Babylonian Celestial Divination," in *Ancient Astronomy and Celestial Divination* (ed. N.M. Swerdlow; Cambridge, Massachusetts: MIT Press, 1999), 21-37
———, "Early Zodiologia and Related Matters," in *Wisdom, Gods and Literature: Studies in Assyriology in Honour of W.G. Lambert* (eds. A.R. George and I.L. Finkel; Winona Lake, Indiana: Eisenbrauns, 2000), 421-27
Reiner, E. and D. Pingree, *Babylonian Planetary Omens Part Three* (CM 11; Groningen: STYX, 1998)
Riedweg, C., *Pythagoras: Leben, Lehre, Nachwirkung; Eine Einführung* (Munich: C.H. Beck, 2002)
Ritter, E.K., "Magical-Expert (=*āšipu*) and Physician (=*asû*): Notes on Two Complementary Professions in Babylonian Medicine," in *Studies in Honor of Benno Landsberger on his Seventy-Fifth Birthday, April 21, 1965* (eds. H.G. Güterbock and T. Jacobsen; AS 16; Chicago: University of Chicago Press, 1965), 299-321
Robbins, F.E., *Ptolemy: Tetrabiblos* (LCL 435; Cambridge, Massachusetts: Harvard University Press, 2001)
Rocca-Serra, G., *Censorinus: Le jour natal* (HDAC 5; Paris: Librairie Philosophique J. Vrin, 1980)
Rochberg, F., *Babylonian Horoscopes* (TAPS 88/1; Philadelphia: American Philosophical Society, 1998)

———, "The Babylonian Origins of the Mandaean Book of the Zodiac," *ARAM* 11-12 (1999-2000): 237-47
———, "The Cultural Locus of Astronomy in Late Babylonia," in *Die Rolle der Astronomie in den Kulturen Mesopotamiens: Beiträge zum 3. Grazer Morgenländischen Symposion (23.-27. September 1991)* (ed. H.D. Galter; GMS 3; Graz: GrazKult, 1993), 31-45
———, "A Babylonian Rising-Times Scheme in Non-Tabular Astronomical Texts," in *Studies in the History of the Exact Sciences in Honour of David Pingree* (eds. C. Burnett et al.; IPTS 54; Leiden: Brill, 2004), 56-94
———, *The Heavenly Writing: Divination, Horoscopy, and Astronomy in Mesopotamian Culture* (Cambridge: Cambridge University Press, 2004)
Rochberg-Halton, F., "The Assumed 29th *aḫû*-Tablet of *Enūma Anu Enlil*," in *Language, Literature, and History: Philological and Historical Studies Presented to Erica Reiner* (ed. F. Rochberg-Halton; AOS 67; New Haven, Connecticut: American Oriental Society, 1987), 327-50
———, "Elements of the Babylonian Contribution to Hellenistic Astrology," *JAOS* 108 (1988): 51-62
———, "Babylonian Horoscopes and Their Sources," *Or* 58 (1989): 102-23
Rodler, L., *Il corpo specchio dell'anima: Teoria e storia della fisiognomica* (Milan: Bruno Mondadori, 2000)
Rohr, R.R.J., *Sundials: History, Theory, and Practice* (trans. G. Godin; New York: Dover Publications, 1996)
Rolfe, J.C., *The Attic Nights of Aulus Gellius* (LCL 195; Cambridge, Massachusetts: Harvard University Press, 1954)
———, *The Attic Nights of Aulus Gellius* (LCL 200; Cambridge, Massachusetts: Harvard University Press, 1968)
Rossing, B.R., "City Visions, Feminine Figures and Economic Critique: A Sapiential *Topos* in the Apocalypse," in *Conflicted Boundaries in Wisdom and Apocalypticism* (eds. B.G. Wright III and L.M. Wills; SBLSymS 35; Atlanta: Society of Biblical Literature, 2005), 181-96
Rowland, I.D. and T.N. Howe (eds.), *Vitruvius: Ten Books on Architecture* (Cambridge: Cambridge University Press, 1999)
Runia, D.T., *Philo of Alexandria: On the Creation of the Cosmos according to Moses* (PACS 1; Leiden: Brill, 2001)
Sachs, A., "Babylonian Horoscopes," *JCS* 6 (1952): 49-75
———, *Late Babylonian Astronomical and Related Texts* (BUS 18; Providence, Rhode Island: Brown University Press, 1955)
Sallmann, K. (ed.), *Censorinus: De die natalie liber* (Teubner; Leipzig: B.G. Teubner, 1983)
Sassi, M.M., "Physiognomy," *OCD* 1181
———, *The Science of Man in Ancient Greece* (trans. P. Tucker; Chicago: University of Chicago Press, 2001)
Scarborough, J., "The Pharmacology of Sacred Plants, Herbs, and Roots," in *Magika Hiera: Ancient Greek Magic and Religion* (eds. C.A. Faraone and D. Obbink; New York: Oxford University Press, 1991), 138-74
Schäfer, P., *Geniza Fragmente zur Hekhalot-Literatur* (TSAJ 6; Tübingen: J.C.B. Mohr [Paul Siebeck], 1984)
———, "Ein neues Fragment zur Metoposkopie und Chiromantik," in *Hekhalot-Studien* (P. Schäfer; TSAJ 19; Tübingen: J.C.B. Mohr [Paul Siebeck], 1988), 84-95 (originally published in *FJB* 13 [1985]: 61-82)
Schäfer, P. et al. (eds.), *Synopse zur Hekhalot-Literatur* (TSAJ 2; Tübingen: J.C.B. Mohr [Paul Siebeck], 1981)
———, *Übersetzung der Hekhalot-Literatur II §§81-334* (TSAJ 17; Tübingen: J.C.B. Mohr [Paul Siebeck], 1987)
Schams, C., *Jewish Scribes in the Second Temple Period* (JSOTSup 291; Sheffield: Sheffield Academic Press, 1998)
Schiaparelli, G., *Die Astronomie im Alten Testament* (Gießen: J. Ricker [Alfred Töpelmann], 1904)

Schiffman, L.H., *Reclaiming the Dead Sea Scrolls: The History of Judaism, the Background of Christianity, the Lost Library of Qumran* (Philadelphia: The Jewish Publication Society, 1994)

———, "4QMysteries[a]," in *Qumran Cave 4.XV: Sapiential Texts, Part 1* (eds. T. Elgvin et al.; DJD 20; Oxford: Clarendon, 1997), 33-97

Schiffman, L.H. and M.D. Swartz, *Hebrew and Aramaic Incantation Texts from the Cairo Genizah: Selected Texts from Taylor-Schechter Box K1* (STS 1; Sheffield: Sheffield Academic Press, 1992)

Schmid, A., *Augustus und die Macht der Sterne: Antike Astrologie und die Etablierung der Monarchie in Rom* (Cologne: Böhlau, 2005)

Schmidt, F., "Astrologie juive ancienne: Essai d'interprétation de *4QCryptique (4Q186)*," *RevQ* 18/69 (1997): 125-41

———, "Astrology and Predestination in Qumran," *Qadmoniot* 30 (1997): 115-18 (in Hebrew)

———, "Ancient Jewish Astrology: An Attempt to Interpret 4QCryptic (4Q186)," in *Biblical Perspectives: Early Use and Interpretation of the Bible in Light of the Dead Sea Scrolls: Proceedings of the First International Symposium of the Orion Center for the Study of the Dead Sea Scrolls and Associated Literature, 12-14 May, 1996* (eds. M.E. Stone and E.G. Chazon; STDJ 28; Leiden: E.J. Brill, 1998), 189-205

———, "'Recherche son thème de géniture dans le Mystère de ce qui doit être:' Astrologie et prédestination à Qoumrân," in *Qumrân et le judaïsme du tournant de notre ère* (eds. A. Lemaire and S. Mimouni; CREJ 39; Leuven: Peeters, 2006), *forthcoming*

Schmidt, J., "Physiognomik," *RE* 39 (1941), 1064-74

Scholem, G., "הכרת פנים וסדרי שרטוטין ('Physiognomy and Chiromancy)," in *ספר אסף: קובץ מאמרי מחקר* (eds. M.D. Cassuto, J. Klausner and J. Gutmann; Jerusalem: Mossad Harav Kook, 1953), 459-95

———, "Ein Fragment zur Physiognomik und Chiromantik aus der spätantiken jüdischen Esoterik," in *Liber Amicorum: Studies in Honour of Professor Dr. C.J. Bleeker* (SHR 17; Leiden: E.J. Brill, 1969), 175-93

Schroer, S. and T. Staubli, *Die Körpersymbolik der Bibel* (Darmstadt: Wissenschaftliche Buchgesellschaft, 1998)

Schüngel-Straumann, H., *Tobit* (HThKAT; Freiburg: Herder, 2000)

Scott, A., *Origen and the Life of the Stars: A History of an Idea* (Oxford: Clarendon, 1991)

Scurlock, J., "Review: B. Böck, *Die babylonisch-assyrische Morphoskopie* (AfO.B 27; Vienna: Institut für Orientalistik der Universität Wien, 2000)," *JAOS* 123 (2003): 395-99

Scurlock, J. and B.R. Andersen, *Diagnoses in Assyrian and Babylonian Medicine: Ancient Sources, Translations, and Modern Medical Analyses* (Urbana: University of Illinois Press, 2005)

Segal, M.H., *A Grammar of Mishnaic Hebrew* (Oxford: Clarendon, 1927; repr., 1978)

Segert, S., "Die Sprachenfrage in der Qumrāngemeinschaft," in *Qumran-Probleme: Vorträge des Leipziger Symposions über Qumran-Probleme vom 9. bis 14. Oktober 1961* (ed. H. Bardtke; DAWBSSA 42; Berlin: Akademie Verlag, 1963), 315-39

Seitz, O.J.F., "Two Spirits in Man: An Essay in Biblical Exegesis," *NTS* 6 (1959): 82-95

Sekki, A.E., *The Meaning of* Ruaḥ *at Qumran* (SBLDS 110; Atlanta, Georgia: Scholars Press, 1989)

Shaked, S., "Qumran and Iran: Further Considerations," *IOS* 2 (1972): 433-46

Shookman, E. (ed.), *The Faces of Physiognomy: Interdisciplinary Approaches to Johann Caspar Lavater* (Columbia, South Carolina: Camden House, 1993)

Siegrist, C., "'Letters of the Divine Alphabet' – Lavater's Concept of Physiognomy," in *The Faces of Physiognomy: Interdisciplinary Approaches to Johann Caspar Lavater* (ed. E. Shookman; Columbia, South Carolina: Camden House, 1993), 25-39

Singer, P.N., *Galen: Selected Works* (Oxford: Oxford University Press, 1997)

Sjöberg, E., "רוּחַ im palästinischen Judentum," *TWNT* 6:373-87

Smith, R., *Aristotle: Prior Analytics: Translated, with Introduction, Notes, and Commentary* (Indianapolis: Hackett, 1989)

Smith, W.D., *Hippocrates: Volume VII* (LCL 477; Cambridge, Massachusetts: Harvard University Press, 1994)
Soubiran, J. (ed.), *Vitruve: De l'architecture, Livre 9* (Budé; Paris: Les Belles Lettres, 1969)
Spinelli, E., "Sesto Empirico e l'astrologia," in *Traditions of Theology: Studies in Hellenistic Theology, Its Background and Aftermath* (eds. D. Frede and A. Laks; PhA 89; Leiden: Brill, 2002), 239-73
Staden, H. von, "Body, Soul, and Nerves: Epicurus, Herophilus, Erasistratus, the Stoics, and Galen," in *Psyche and Soma: Physicians and Metaphysicians on the Mind-Body Problem from Antiquity to Enlightenment* (eds. J.P. Wright and P. Potter; Oxford: Clarendon, 2000), 79-116
Starcky, J., "Le travail d'édition des fragments manuscrits de Qumrân: Communication de J. Starcky," *RB* 63 (1956): 66-67
———, "Les quatre étapes du messianisme à Qumrân," *RB* 70 (1963): 481-505
———, "Un texte messianique araméen de la grotte 4 de Qumrân," in *Mémorial du cinquantenaire 1914-1964, École des Langues Orientales Anciennes de l'Institut Catholique de Paris* (TICP 10; Paris: Bloud & Gay, 1964), 51-66
———, "Le Maître de Justice et Jésus," *MdB* 4 (1978): 51-55
Starr, I., *Queries to the Sungod: Divination and Politics in Sargonid Assyria* (SAA 4; Helsinki: Helsinki University Press, 1990)
Stegemann, H., "Zu Textbestand und Grundgedanken von 1QS III, 13-IV, 26," *RevQ* 13/49-52 (1988): 95-131
———, "Methods for the Reconstruction of Scrolls from Scattered Fragments," in *Archaeology and History in the Dead Sea Scrolls: The New York University Conference in Memory of Yigael Yadin* (ed. L.H. Schiffman; JSPSup 8; Sheffield: JSOT Press, 1990), 189-220
Steinkeller, P., "Of Stars and Men: The Conceptual and Mythological Setup of Babylonian Extispicy," in *Biblical and Oriental Essays in Memory of William L. Moran* (ed. A. Gianto; BibOr 48; Rome: Pontificio Istituto Biblico, 2005), 11-47
Stok, F., "La fisiognomica fra teoria e pratica," in *Sciences exactes et sciences appliquées à Alexandrie (IIIe siècle av. J.-C. – Ier siècle ap. J.-C.): Actes du Colloque International de Saint-Étienne (6-8 juin 1996)* (eds. G. Argoud and J.-Y. Guillaumin; Mémoires 16; Saint-Étienne: Publications de l'Université de Saint-Étienne, 1998), 173-87
Stol, M., *Birth in Babylonia and the Bible: Its Mediterranean Setting* (CM 14; STYX: Groningen, 2000)
Stone, M.E., "Lists of Revealed Things in the Apocalyptic Literature," in *Magnalia Dei: The Mighty Acts of God: Essays on the Bible and Archaeology in Memory of G. Ernest Wright* (eds. F.M. Cross, Jr., W.E. Lemke and P.D. Miller; Garden City, New York: Doubleday, 1976), 414-52
———, "The Book of Enoch and Judaism in the Third Century B.C.E.," *CBQ* 40 (1978): 479-92
———, *Scriptures, Sects and Visions: A Profile of Judaism from Ezra to the Jewish Revolts* (Cleveland: Collins, 1980)
———, "Enoch, Aramaic Levi and Sectarian Origins," *JSJ* 19 (1988): 159-70
Strauß, H., *Hiob: 2. Teilband, 19,1-42,17* (BKAT 16/2; Neukirchen-Vluyn: Neukirchener, 2000)
Strugnell, J., "Notes en marge du volume V des 'Discoveries in the Judaean Desert of Jordan,'" *RevQ* 7/26 (1970): 163-276
Strugnell, J., D.J. Harrington and T. Elgvin (eds.), *Qumran Cave 4.XXIV: Sapiential Texts, Part 2: 4QInstruction (Mûsār lĕ Mēvîn): 4Q415 ff. with a Re-edition of 1Q26 by John Strugnell and Daniel J. Harrington, S.J. and an Edition of 4Q423 by Torleif Elgvin, in Consultation with Joseph A. Fitzmyer, S.J.* (DJD 34; Oxford: Clarendon, 1999)
Stuckrad, K. von, *Frömmigkeit und Wissenschaft: Astrologie in Tanach, Qumran und frührabbinischer Literatur* (EH 23/572; Frankfurt: Peter Lang, 1996)
———, *Das Ringen um die Astrologie: Jüdische und christliche Beiträge zum antiken Zeitverständnis* (RVV 49; Berlin: Walter de Gruyter, 2000)

———, "Jewish and Christian Astrology in Late Antiquity – A New Approach," *Numen* 47 (2000): 1-40

Sullivan, K.P., *Wrestling with Angels: A Study of the Relationship between Angels and Humans in Ancient Jewish Literature and the New Testament* (AGJU 55; Leiden: Brill, 2004)

Svenberg, E., *Lunaria et zodiologia Latina* (SGLG 16; Gothenburg: University of Gothenburg Press, 1963)

Swartz, M.D., "The Dead Sea Scrolls and Later Jewish Magic and Mysticism," *DSD* 8 (2001): 182-93

Swerdlow, N.M., *The Babylonian Theory of the Planets* (Princeton: Princeton University Press, 1998)

Tarabout, G., "Pratiques astrologiques, double autorité textuelle (monde indien)," paper read at the CNRS Symposium *Divination et écriture, écriture de la divination*, Paris, December 10, 2005 (the proceedings are to be published by J.-J. Glassner)

Tcherikover, V., *Hellenistic Civilization and the Jews* (trans. S. Applebaum; Philadelphia: The Jewish Publication Society of America and Jerusalem: Magnes, 1959; repr., Peabody, Massachusetts: Hendrickson, 2004)

Temkin, O., "Greek Medicine as Science and Craft," *Isis* 44 (1953): 213-25

Tester, S.J., *A History of Western Astrology* (Woodbridge: Boydell, 1987)

Thackeray, H.St.J., *Josephus: The Jewish War, Books I-II* (LCL 203; Cambridge, Massachusetts: Harvard University Press, 1997)

Thiering, B., "The Qumran Sundial as an Odometer Using Fixed Lengths of Hours," *DSD* 9 (2002): 347-63

Thompson, E.M., *An Introduction to Greek and Latin Palaeography* (Oxford: Oxford University Press, 1912; repr., New York: Burt Franklin, 1964)

Tigchelaar, E.J.C., "A Newly Identified 11QSerekh ha-Yaḥad Fragment (11Q29)?," in *The Dead Sea Scrolls: Fifty Years After Their Discovery: Proceedings of the Jerusalem Congress, July 20-25, 1997* (eds. H. Schiffman, E. Tov and J.C. VanderKam; Jerusalem: Israel Exploration Society in cooperation with The Shrine of the Book, Israel Museum, 2000), 285-92

———, *To Increase Learning for the Understanding Ones: Reading and Reconstructing the Fragmentary Early Jewish Sapiential Text 4QInstruction* (STDJ 44; Leiden: Brill, 2001)

———, "Your Wisdom and Your Folly: The Case of 1-4QMysteries," in *Wisdom and Apocalypticism in the Dead Sea Scrolls and in the Biblical Tradition* (ed. F. García Martínez; BETL 168; Leuven: Peeters, 2003), 67-88

———, "Astronomie en kalenders; horoscopen en astrologie," in *Fragmenten uit de woestijn: De Dode-Zeerollen opnieuw bekeken* (eds. F. García Martínez and E. Tigchelaar; Zoetermeer: Meinema, 2003), 167-83

———, "'These Are the Names of the Spirits of…': A Preliminary Edition of *4QCatalogue of Spirits (4Q230)* and New Manuscript Evidence for the *Two Spirits Treatise (4Q257* and *1Q29a)*," *RevQ* 21/84 (2004): 529-47

Toepel, A., "Planetary Demons in Early Jewish Literature," *JSP* 14 (2005) 231-38

Torijano, P.A., *Solomon the Esoteric King: From King to Magus, Development of a Tradition* (JSJSup 73; Leiden: Brill, 2002)

Touwaide, A., "Physiognomik," *DNP* 9 (2000), 997-98

Tov, E., *Scribal Practices and Approaches Reflected in the Texts Found in the Judean Desert* (STDJ 54; Leiden: Brill, 2004)

Tov, E. et al. (eds.), *The Texts from the Judaean Desert: Indices and an Introduction to the* Discoveries in the Judaean Desert *Series* (DJD 39; Oxford: Clarendon, 2002)

Tredennick, H., *Aristotle: Prior Analytics* (LCL 325; Cambridge, Massachusetts: Harvard University Press, 1962)

Ungnad, A., "Besprechungskunst und Astrologie in Babylonien," *AfO* 14 (1941-44): 251-84

Urban, H.B., "Elitism and Esotericism: Strategies of Secrecy and Power in South Indian Tantra and French Masonry," *Numen* 44 (1997): 1-38

———, "The Torment of Secrecy: Ethical and Epistemological Problems in the Study of Esoteric Traditions," *HR* 37 (1998): 209-48

VanderKam, J.C., "The Poetry of 1QApGen XX, 2-8a," *RevQ* 10/37 (1979): 57-66

———, *Enoch and the Growth of an Apocalyptic Tradition* (CBQMS 16; Washington, DC: Catholic Biblical Association of America, 1984)
VanderKam, J.C. and P.W. Flint, *The Meaning of the Dead Sea Scrolls: Their Significance for Understanding the Bible, Judaism, Jesus, and Christianity* (London: T&T Clark, 2005)
Vanstiphout, H.L.J., "*Miḫiltum*, or the Image of Cuneiform Writing," in *The Image in Writing* (VisRel 6; Leiden: E.J. Brill, 1988), 152-68
———, "The nth Degree of Writing at Nineveh," *Iraq* 66 (2004): 51-54
Vanstiphout, H.L.J. and N. Veldhuis, "Ṭuppi ilāni takāltu pirišti šamê u erṣetim," *AIUON* 55 (1995): 30-32
Vázquez Allegue, J., *La "Regla de la comunidad" de Qumrán* (BEBM 8; Salamanca: Sígueme, 2006)
Veldhuis, N., *Elementary Education at Nippur: The Lists of Trees and Wooden Objects* (Diss. Groningen, 1997)
———, "Reading the Signs," in *All Those Nations...: Cultural Encounters within and with the Near East: Studies Presented to Han Drijvers at the Occasion of his Sixtyfifth Birthday by Colleagues and Students* (eds. H.L.J. Vanstiphout et al.; COMERS/ICOG Communications 2; Groningen: STYX, 1999), 161-74
Vermes, G., "The Leadership of the Qumran Community: Sons of Zadok-Priests-Congregation," in *Geschichte – Tradition – Reflexion: Festschrift für Martin Hengel zum 70. Geburtstag, Band I Judentum* (eds. H. Cancik, H. Lichtenberger and P. Schäfer; Tübingen: J.C.B. Mohr [Paul Siebeck], 1996), 375-84
———, *The Complete Dead Sea Scrolls in English* (London: Allen Lane The Penguin Press, 1997)
Vince, J.H., *Demosthenes III* (LCL 299; Cambridge, Massachusetts: Harvard University Press, 1986)
Vogt, S., *Aristoteles: Opuscula VI: Physiognomonica* (Aristoteles, Werke in deutscher Übersetzung 18.6; Berlin: Akademie Verlag, 1999)
Wächter, L., "Astrologie und Schicksalsglaube im rabbinischen Judentum," *Kairos* 11 (1969): 181-200
Waerden, B.L. van der, "History of the Zodiac," *AfO* 16 (1952-1953): 216-30
———, "Die Zahlen der Texte VAT 7815 und 7816," in *Gestirn-Darstellungen auf babylonische Tontafeln* (E. Weidner; ÖAWPH.SB 254/2; Vienna: Hermann Böhlaus, 1967), 50-52
———, *Die Astronomie der Griechen: Eine Einführung* (Darmstadt: Wissenschaftliche Buchgesellschaft, 1988)
Waltke, B.K. and M. O'Connor, *An Introduction to Biblical Hebrew Syntax* (Winona Lake, Indiana: Eisenbrauns, 1990)
Weinfeld, M. and D. Seely, "4QBarkhi Nafshi," in *Qumran Cave 4.XX: Poetical and Liturgical Texts, Part 2* (eds. E. Chazon et al.; DJD 29; Oxford: Clarendon, 1999), 255-334
Wernberg-Møller, P., *The Manual of Discipline: Translated and Annotated with an Introduction* (STDJ 1; Leiden: E.J. Brill, 1957)
———, "A Reconsideration of the Two Spirits in the Rule of the Community (1QSerek III,13-IV,26)," *RevQ* 3/11 (1961): 413-41
Werrett, I., *Ritual Purity and the Dead Sea Scrolls* (Diss. St Andrews, 2006)
Wertheimer, S.A. and A.J. Wertheimer (eds.), *Batei Midrashot: Twenty-Five Midrashim Published for the First Time from Manuscripts Discovered in the Genizoth of Jerusalem and Egypt* (2d enl. and am. ed.; vol. 2; Jerusalem: Ktab Wasepher, 1968)
Wilcke, C., *Wer las und schrieb in Babylonien und Assyrien: Überlegungen zur Literalität im Alten Zweistromenland* (BAWPHK.S 6; Munich: C.H. Beck, 2000)
Wilde, A. de, *Das Buch Hiob: Eingeleitet, übersetzt und erläutert* (OTS 22; Leiden: E.J. Brill, 1981)
Wischmeyer, O., *Die Kultur des Buches Jesus Sirach* (BZNW 77; Berlin: Walter de Gruyter, 1995)
Wise, M.O., "Thunder in Gemini: An Aramaic Brontologion (4Q318) from Qumran," in *Thunder in Gemini And Other Essays on the History, Language and Literature of Second*

Temple Palestine (M.O. Wise; JSPSup 15; Sheffield: Sheffield Academic Press, 1994), 13-50

———, "A Horoscope Written in Code (4Q186)," in *The Dead Sea Scrolls: A New Translation* (eds. M.O Wise, M.G. Abegg, Jr. and E.M. Cook; New York: HarperCollins, 2005), 275-78

———, "An Aramaic Horoscope (4Q561)," in *The Dead Sea Scrolls: A New Translation* (eds. M.O Wise, M.G. Abegg, Jr. and E.M. Cook; New York: HarperCollins, 2005), 567-68

———, "4Q561 (4QPhysiognomy/Horoscope ar)," in *Additional Genres and Unclassified Texts* (eds. D.W. Parry and E. Tov; DSSR 6; Leiden: Brill, 2005), 228-31

Wiseman, D.J. and J.A. Black, *Literary Texts from the Temple of Nabû* (CTN 4; British School of Archaeology in Iraq, 1996)

Wlosok, A., *Laktanz und die philosophische Gnosis: Untersuchungen zu Geschichte und Terminologie der gnostischen Erlösungsvorstellung* (AHAWPHK 2; Heidelberg: Carl Winter, 1960)

Wold, B.G., *Women, Men and Angels* (WUNT 201; Tübingen: Mohr Siebeck, 2005)

Wright III, B.G., "Putting the Puzzle Together: Some Suggestions concerning the Social Location of the Wisdom of Ben Sira," in *Conflicted Boundaries in Wisdom and Apocalypticism* (eds. B.G. Wright III and L.M. Wills; SBLSymS 35; Atlanta: Society of Biblical Literature, 2005), 89-112

Wright, J.P. and P. Potter (eds.), *Psyche and Soma: Physicians and Metaphysicians on the Mind-Body Problem from Antiquity to Enlightenment* (Oxford: Clarendon, 2000)

Yadin, Y. and J. Naveh, *The Aramaic and Hebrew Ostraca and Jar Inscriptions* (vol. 1 of *Masada: The Yigael Yadin Excavations 1963-1965, Final Reports*; Jerusalem: Israel Exploration Society/The Hebrew University of Jerusalem, 1989)

Zeller,, D. "The θεῖα φῦσις of Hippocrates and of Other 'Divine Men," in *Neues Testament und hellenistische Umwelt* (D. Zeller; BBB 150; Hamburg: Philo, 2006), 129-40

Zimmer, H., "Zu den 'Keilinschriften aus Assur religiösen Inhalts,'" *ZA* 30 (1915/1916): 184-229

Zimmermann, J., *Messianische Texte aus Qumran: Königliche, priesterliche und prophetische Messiasvorstellungen in den Schriftfunden von Qumran* (WUNT 104; Tübingen: J.C.B. Mohr [Paul Siebeck], 1998)

INDEX OF MODERN AUTHORS

INDEX OF ANCIENT SOURCES

I. JEWISH TEXTS

1. *Hebrew Bible*

2. *Apocrypha and Pseudepigrapha*

3. *Dead Sea Scrolls*

4. *Ancient Jewish Writers*

5. *Rabbinic and Cairo Genizah Texts*

II. MESOPOTAMIAN TEXTS

III. GREEK AND LATIN TEXTS

IV. CHRISTIAN TEXTS

1. *New Testament*

2. *Ancient Christian Writers*